COMPRESSORS

SELECTION AND SIZING

Gulf Publishing Company
Houston, Texas

SECOND EDITION

COMPRESSORS

SELECTION AND SIZING

ROYCE N. BROWN

Dedication

To June,
for her love and encouragement
to keep me moving.

Gulf Publishing Company
Book Division
P.O. Box 2608 □ Houston, Texas 77252-2608

10   9   8   7   6   5   4   3   2

Library of Congress Cataloging-in-Publication Data
Brown, Royce N.
    Compressors : selection & sizing / Royce N.
    Brown.—2nd ed.
        p. cm.
    Includes bibliographical references and index.
    ISBN 0-88415-164-6
    1. Compressors.   I. Title.
TJ990.B76   1997
621.5′1—dc20   96-35816
                                                    CIP

Contents

Preface _____ xiii

Acknowledgments _____ xv

1. Overview _____ 1

Introduction _____ 1

Compression Methods _____ 2

Intermittent Cycle Compressors _____ 4
Reciprocating Compressors. Rotary Compressors.

Continuous Compression Compressors _____ 9
Ejectors. Dynamic Compressors.

2. Basic Relationships _____ 14

Introduction _____ 14

Gas and Vapor _____ 15
Perfect Gas Equation.

Compressibility _____ 17
Generalized Compressibility Charts.

Partial Pressure _____ 18

Gas Mixtures _____ 18
Specific Heat Ratio. Molecular Weight.

Specific Gravity _____ 19

Mixture Compressibility _____ 20

Humidity ———————————————————— 20

Flow ——————————————————————— 21

Acoustic Velocity ———————————————— 26

Equations of State ——————————————— 26
Mollier Charts. First Law of Thermodynamics. Second Law of
Thermodynamics.

Theoretical Work ————————————————— 30
Real Gas Exponent. Power. Velocity Head.

Intercooling ———————————————————— 41
Isothermal Compression.

References ———————————————————— 46

3. Reciprocating Compressors ———————————— **48**

Description ————————————————————— 48
Classification. Arrangement. Drive Methods.

Performance ———————————————————— 54
Compression Cycle. Cylinder Displacement. Volumetric Efficiency. Piston
Speed. Discharge Temperature. Power. Valve Loss. Application Notes.

Mechanics ————————————————————— 67
Cylinders. Pistons and Rods. Valves. Distance Piece. Rod Packing.
Crankshaft and Bearings. Frame Lubrication. Cylinder and Packing
Lubrication. Cooling. Capacity Control. Pulsation Control.

References ———————————————————— 90

4. Rotary Compressors ———————————————— **93**

Common Features ———————————————— 93
Arrangements and Drivers.

Helical Lobe ———————————————————— 95
History. Operating Principles. Displacement. Dry Compressors. Flooded
Compressors. Flooding Fluid. Application Notes—Dry Compressors.
Application Notes—Flooded Compressors. Casings. Rotors. Bearings and
Seals. Timing Gears. Capacity Control.

Straight Lobe _____ 121
Compression Cycle. Sizing. Applications. Mechanical Construction.

Sliding Vane _____ 126
Compression Cycle. Sizing. Application Notes. Mechanical Construction.

Liquid Piston _____ 130
Operation. Performance. Mechanical Construction.

References _____ 131

5. Centrifugal Compressors _____ 132

Introduction _____ 132
Classification. Arrangement. Drive Methods.

Performance _____ 147
Compression Cycle. Vector Triangles. Slip. Reaction. Sizing. Fan Laws.
Curve Shape. Surge. Choke. Application Notes.

Mechanical Design _____ 188
Introduction. Casings. Diaphragms. Casing Connections. Impellers. Shafts.
Radial Bearings. Thrust Bearings. Bearing Housings. Magnetic Bearings.
Balance Piston. Interstage Seals. Shaft End Seals.

Shaft End Seals _____ 211
Restrictive Seals. Liquid Buffered Seals. Dry Gas Seals. Capacity Control.
Maintenance.

References _____ 222

6. Axial Compressors _____ 224

Historical Background _____ 224

Description _____ 225

Performance _____ 226
Blades. Compression Cycle. Reaction. Stagger. Curve Shape. Surge. Sizing.
Application Notes.

Mechanical Design _____ 247
Casings. Stators. Casing Connections. Rotor. Shaft. Blading. Bearings.
Balance Piston. Seals. Capacity Control. Maintenance.

References _____ 255

7. Drivers _____ 256

Introduction _____ 256

Electric Motors _____ 257
Voltage. Enclosures. Totally Enclosed Motors. Division 1 Enclosures. Inert
Gas-Filled. Insulation. Service Factor. Synchronous Motors. Brushless
Excitation. Motor Equations.

Compressor and Motor _____ 268
Selecting Compressor Motors. Starting Characteristics. Starting Time.
Enclosure Selection. Enclosure Applications.

Variable Frequency Drives _____ 277
Motor.

Steam Turbines _____ 282
Steam Temperature. Speed. Operation Principles. Steam Turbine Rating.

Gas Engines _____ 292

Gas Turbines _____ 292
Gas Turbine Types. Gas Turbine Economics. Sizing Application.

Expansion Turbines _____ 296
Types. Operation Limits. Power Recovery. Refrigeration. Condensation.
Expander Applications.

References _____ 300

8. Accessories _____ 302

Introduction _____ 302

Lubrication Systems _____ 303
Reservoir. Pumps and Drivers. Relief Valves. Pressure Control Valves.
Startup Control. Check Valves. Coolers. Filters. Transfer Valves.
Accumulators. Seal Oil Overhead Tank. Lube Oil Overhead Tank. Seal Oil
Drainers. Degassing Drum. Piping. System Review. Testing of Lubrication
Systems. Commissioning of Lube Oil Systems.

Dry Gas Seal Systems _____ 323
System Design Considerations. Dry Gas Seal System Control. Dry Gas Seal
System Filters.

Gears _____ 328
Gear Design and Application. Rotors and Shafts. Bearings and Seals.
Housing. Lubrication.

Couplings _____ 333
Introduction. Ratings. Spacers. Hubs. Gear Couplings. Alignment. Flexible
Element Couplings. Limited End-Float Couplings.

Instrumentation _____ 342
Overview. Pressure. Temperature. Flow. Torque. Speed. Rod Drop Monitor.
Molecular Weight.

Vibration _____ 349
Vibration Sensors. Seismic Sensors. Proximity Sensors. Axial Shaft Motion.
Radial Shaft Vibration.

Control _____ 356
Analysis of the Controlled System. Pressure Control at Variable Speed.
Volume Control at Variable Speed. Weight Flow Control with Variable
Stator Vanes. Pressure Control at Constant Speed. Volume Control at
Constant Speed. Weight Flow Control at Constant Speed. Anti-Surge
Control.

References _____ 366

9. Dynamics _____ 368

Introduction _____ 368

Balance _____ 369
Basics. Unbalance.

Balance Methods _____ 374
Shop Balance Machine. High Speed Balancing. Field Balancing.

Reciprocating Shaking Forces _____ 378

Rotary Shaking Forces _____ 382

Rotor Dynamics _____ 384
Damped Unbalance Response. Torsionals. Torsional Damping and Resilient
Coupling.

References _____ 400

10. Testing _____ 403

Introduction _____ 403
Objectives. Hydrostatic Test. Impeller Overspeed Test.

Operational Tests _____ 407
General. Mechanical Running Test.

Objectives of Centrifugal Compressor Mechanical Tests _____ 408
Rotor Dynamics Verification. String Testing. Stability. Helical-Lobe
Compressor Test. Reciprocating Compressor Test. Spare Rotor Test. Static
Gas Test. Testing of Lubrication Systems. Shop Performance Test. Test
Codes. Loop Testing. Gas Purity. Sidestream Compressors. Instrumentation.
Test Correlation. Reynolds Number. Abnormalities in Testing. Field
Testing. Planning. Flow Meters. Gas Composition. Location. Power
Measurement. Speed Conducting the Test.

References _____ 435

11. Negotiation and Purchasing _____ 438

Introduction _____ 438
Procurement Steps. Supplier Partnerships.

Preliminary Sizing _____ 440

Specifications _____ 441
Basic Data. Operations.

Writing the Specification _____ 443
Specification Outline. General. Basic Design. Materials. Bearings. Shaft
End Seals. Accessories. Lube and Seal System. Drivers. Gear Units.
Couplings. Mounting Plates. Controls and Instrumentation. Inspection and
Testing. Vendor Data. Guarantee and Warranty.

Bid and Quotation _____ 455

Bid Evaluation _____ 455

Pre-Award Meeting _____ 456

Purchase Specification _____ 457

Award Contract _____ 457

Coordination Meeting _____ 457

Engineering Reviews _____ 458

Inspections _____ 459

Tests _____ 459

Shipment _____ 461

Site Arrival _____ 462

Installation and Startup _____ 462
Commissioning the Compressor. Commissioning the Lube Oil System.

Successful Operation _____ 464

References _____ 464

12. Reliability Issues _____ **466**

General _____ 466
Overview. Robust Design.

The Installation _____ 470
Foundations. Suction Drums. Check Valves. Piping.

Compressors _____ 474
Type Comparison. Reciprocating Compressors. Positive Displacement
Rotary Compressors. Centrifugal Compressors. Axial Compressors.

Drivers _____ 478
Turbines. Motors. Gears. Expanders.

Applications _____ 480
Process. Experience.

Operations _____ 483
General Comments. Gas Considerations. Operating Envelope.

System Components _____ 485
Lubrication. Couplings.

Quality _____ 487
Methodology. Manufacturing Tolerances.

Summary _____ 489

References _____ 490

Appendix A—Conversion Factors _____ **491**

**Appendix B—Pressure-Enthalpy and Compressibility
  Charts** _____ **494**

Appendix C—Physical Constants of Hydrocarbons ___ **528**

**Appendix D—Labyrinth and Carbon Ring Seal Leakage
  Calculations** _____ **533**

Index _____ **543**

Preface to
Second Edition

About the time the first edition was written, the process industries, which represent a large part of the compressor market, were at a low ebb. As a result, the activity in the compressor world was almost at a standstill. Development at best was relatively slow. Currently, however, activity level has increased significantly. A look at the credit lines on many of the suppliers will tell of the many changes that have taken place. Even many of the companies whose names have not changed are now under different ownership than they were at the time of the first edition. Large investments have been made in facilities, in terms of new or remodeled factory buildings and the addition of new improved machine tools. Development funds are being expended and improved designs are becoming available. Management styles have changed and the theme of continuous improvement is quite prevalent. With all this activity, it seemed appropriate to offer an updated edition of this book.

Many of the readers of the first edition have commented that the book was easy to read. I have attempted to maintain that tone in this new edition. The major change to the book is the addition of a chapter on reliability. As in the other chapters, this one also leaves the high power statistics for someone else and instead uses a "common sense" approach. It probably has a "do and don't" flavor, which just seemed appropriate as I was writing it. Because the subject of reliability is so important and so much can be written about it, the chapter had to be limited to what I felt was the more pertinent information. I had to remind myself that the subject of the book was compressors, not just their reliability. It is hoped that a proper balance was obtained.

Another area that is addressed in the new edition is the dry gas seal. The subject of dry gas seals, which are now widely used by the industry, was expanded considerably in Chapter 5, and a discussion of dry gas seal systems has been added to Chapter 8. Also in Chapter 5, I added a section on magnetic bearings, which are emerging in the industry although they are not as quick to catch on. Chapter 8 expands the discussion of dry flexible element couplings to reflect current industry practice. The section on gear couplings was left because gear couplings are still used and I felt the information would provide some useful background.

I touched up some of Chapter 3 by reworking the valve section, and I hope it does a better job of describing the currently available valves. I also expanded the area of unloaders to more adequately cover the different styles available to the industry.

Where current practice seemed to dictate I updated curves, and added a table in Chapter 4 to help with the sizing of the oil-free helical lobe compressors. Instrumentation was updated to take rod-drop monitoring of reciprocating compressors into consideration. Improvements in torque monitoring are also included.

In general, wherever I felt the organization of the material could be improved, I did it. The most notable of this are the changes to the testing chapter to aid in clarity.

Royce N. Brown

Acknowledgments
Second Edition

I would like to thank Alex and Linda Atkins of Alta Systems for coming to my assistance when I got overloaded with the chore of scanning my photographs and line illustrations. They helped get the illustrations organized and kept them in the proper order. Linda also helped with debugging the text and keeping the format consistent. Alex put the finishing touches on the figures and then put them on a CD Rom so they could be transported to the publisher. They were very flexible and made themselves available to fit my schedule.

I also want to thank Dan Beard and his son Sean for computer support and some tedious image editing.

Thanks go to Brown and Root for scanning the first edition, and for giving me an electronic form on which to build the revised edition. Thanks also to Buddy Wachel of EDI for giving me an assist at the reciprocating compressor acoustics, and to Susan Dally, Terryl Matthews, Rick Powell, Kelly Fort, Rich Lewis, Carl Fredericks, and Mary Rivers of Dow Chemical for their reviews of the revised chapters.

Finally, a sincere thanks to all the suppliers who provided material for the figures.

1

Overview

Introduction

A compressor is a device used to increase the pressure of a compressible fluid. The inlet pressure level can be any value from a deep vacuum to a high positive pressure. The discharge pressure can range from subatmospheric levels to high values in the tens of thousands of pounds per square inch. The inlet and outlet pressure are related, corresponding with the type of compressor and its configuration. The fluid can be any compressible fluid, either gas or vapor, and can have a wide molecular weight range. Recorded molecular weights of compressed gases range from 2 for hydrogen to 352 for uranium hexafluoride. Applications of compressed gas vary from consumer products, such as the home refrigerator, to large complex petrochemical plant installations.

The compressors to be covered in this book are those using mechanical motion to effect the compression. These types of compressors are commonly used in the process and gas transport/distribution industries. A partial list of these industries includes chemical, petrochemical, refinery, pulp and paper, and utilities. A few typical applications are air separation, vapor extraction, refrigeration, steam recompression, process and plant air.

Compression Methods

Compressors have numerous forms, the exact configuration being based on the application. For comparison, the different types of compressors can be subdivided into two broad groups based on compression mode. There are two basic modes: intermittent and continuous. The *intermittent* mode of compression is cyclic in nature, in that a specific quantity of gas is ingested by the compressor, acted upon, and discharged, before the cycle is repeated. The *continuous* compression mode is one in which the gas is moved into the compressor, is acted upon, moved through the compressor, and discharged without interruption of the flow at any point in the process.

Compressors using the intermittent compression mode are referred to as positive displacement compressors, of which there are two distinct types: reciprocating and rotary. Continuous-mode compressors are also characterized by two fundamental types: dynamic and ejector.

This chapter will give a brief overview of each of the different compressors commonly used in the process industries. Subsequent chapters will then cover each of the mechanical types in depth. (The ejector, which does not use mechanical action, will not be covered in detail.) Figure 1-1

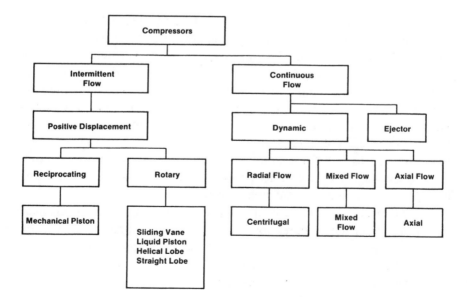

Figure 1-1. Chart of compressor types.

diagrams the relationship of the various compressors by type. Figure 1-2 shows the typical application range of each compressor, and Figure 1-3 compares the characteristic curves of the dynamic compressors, axial and centrifugal, with positive displacement compressors.

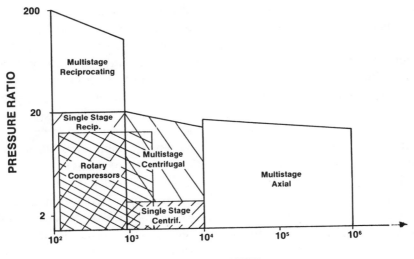

Figure 1-2. Typical application ranges of compressor types.

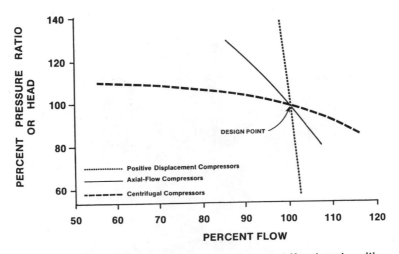

Figure 1-3. General performance curve for axial flow, centrifugal, and positive displacement.

Intermittent Mode Compressors

Reciprocating Compressors

The reciprocating compressor is probably the best known and the most widely used of all compressors. It consists of a mechanical arrangement in which reciprocating motion is transmitted to a piston which is free to move in a cylinder. The displacing action of the piston, together with the inlet valve or valves, causes a quantity of gas to enter the cylinder where it is in turn compressed and discharged. Action of the discharge valve or valves prevents the backflow of gas into the compressor from the discharge line during the next intake cycle. When the compression takes place on one side of the piston only, the compressor is said to be single-acting. The compressor is double-acting when compression takes place on each side of the piston. Configurations consist of a single cylinder or multiple cylinders on a frame. When a single cylinder is used or when multiple cylinders on a common frame are connected in parallel, the arrangement is referred to as a *single-stage compressor.* When multiple cylinders on a common frame are connected in series, usually through a cooler, the arrangement is referred to as a *multistage compressor.* Figures 1-4 and 1-5 are typical reciprocating compressor arrangements, beginning with the single-stage and ending with a more complex multistage.

Figure 1-4. A three-stage single-acting reciprocating compressor. (*Courtesy of Ingersoll Rand*)

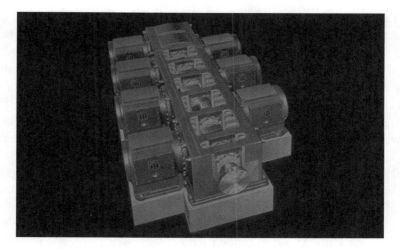

Figure 1-5. Cutaway of the frame end of a large multistage reciprocating compressor. (*Courtesy of Dresser-Rand*)

The reciprocating compressor is generally in the lower flow end of the compressor spectrum. Inlet flows range from less than 100 to approximately 10,000 cfm per cylinder. It is particularly well-suited for high-pressure service. One of the highest pressure applications is at a discharge pressure of 40,000 psi. Above approximately a 1.5-to-1 pressure ratio, the reciprocating compressor is one of the most efficient of all the compressors.

Rotary Compressors

The rotary compressor portion of the positive displacement family is made up of several compressor configurations. The features these compressors have in common are:

1. They impart energy to the gas being compressed by way of an input shaft moving a single or multiple rotating element.
2. They perform the compression in an intermittent mode.
3. They do not use inlet and discharge valves.

The helical and spiral-lobe compressors are generally similar and use two intermeshing helical or spiral lobes to compress gas between the lobes and the rotor chamber of the casing. The compression cycle begins as the open part of the spiral form of the rotors passes over the inlet port and traps a quantity of gas. The gas is moved axially along the rotor to the discharge port where the gas is discharged into the discharge nozzle of the casing. The volume of the trapped gas is decreased as it moves toward the outlet, with the relative port location controlling the pressure ratio. Figure 1-6 shows a cutaway view of a helical-lobe compressor. The spiral-lobe version is the more limited of the two and is used only in the lower pressure applications. Therefore, only the helical-lobe compressor will be covered in depth in this book (see Chapter 4).

The helical-lobe compressor is further divided into a dry and a flooded form. The dry form uses timing gears to hold a prescribed timing to the relative motion of the rotors; the flooded form uses a liquid media to keep the rotors from touching. The helical-lobe compressor is the most sophisticated and versatile of the rotary compressor group and operates at the highest rotor tip Mach number of any of the compressors in the rotary family. This compressor is usually referred to as the "screw compressor" or the "SRM compressor."

The application range of the helical-lobe compressor is unique in that it bridges the application gap between the centrifugal compressor and the reciprocating compressor. The capacity range for the dry configuration is approximately 500 to 35,000 cfm. Discharge pressure is limited to 45 psi in single-stage configuration with atmospheric suction pressure. On

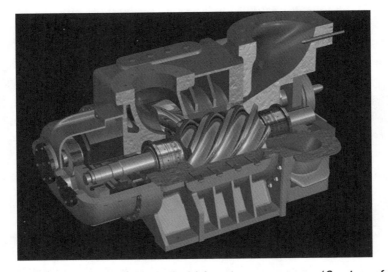

Figure 1-6. Cutaway of an oil-free helical-lobe rotary compressor. (*Courtesy of A-C Compressor Corporation*)

supercharged or multistage applications, pressures of 250 psi are attainable. The spiral-lobe version is limited to 10,000 cfm flow and about 15 psi discharge pressure.

The *straight-lobe compressor* is similar to the helical-lobe machine but is much less sophisticated. As the name implies, it has two untwisted or straight-lobe rotors that intermesh as they rotate. Normally, each rotor pair has a two-lobe rotor configuration, although a three-lobe version is available. All versions of the straight-lobe compressor use timing gears to phase the rotors. Gas is trapped in the open area of the lobes as the lobe pair crosses the inlet port. There is no compression as gas is moved to the discharge port; rather, it is compressed by the backflow from the discharge port. Four cycles of compression take place in the period of one shaft rotation on the two-lobe version. The operating cycle of the straight-lobe rotary compressor is shown in Figure 1-7.

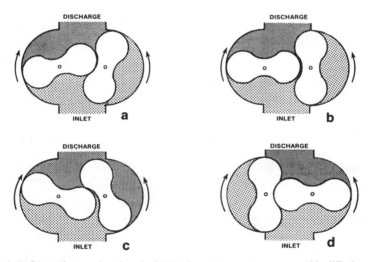

Figure 1-7. Operating cycle of a straight-lobe rotary compressor. (*Modified, courtesy of Ingersoll-Rand*)

Volume range of the straight-lobe compressor is 5 to 30,000 cfm. Pressure ranges are very limited with the maximum single-stage rating at 15 psi. In a few applications, the compressors are used in two-stage form where the discharge pressure is extended to 20 psi.

The *sliding-vane compressor* uses a single rotating element (see Figure 1-8). The rotor is mounted eccentric to the center of the cylinder portion of the casing and is slotted and fitted with vanes. The vanes are free to

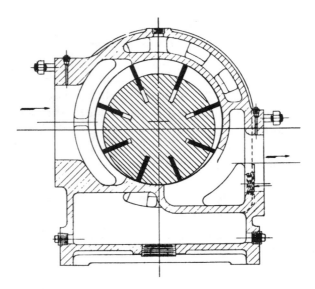

Figure 1-8. Cross section of a sliding vane compressor. (*Courtesy of A-C Compressor Corporation*)

move in and out within the slots as the rotor revolves. Gas is trapped between a pair of vanes as the vanes cross the inlet port. Gas is moved and compressed circumferentially as the vane pair moves toward the discharge port. The port locations control the pressure ratio. (This compressor must have an external source of lubrication for the vanes.)

The sliding-vane compressor is widely used as a vacuum pump as well as a compressor, with the largest volume approximately 6,000 cfm. The lower end of the volume range is 50 cfm. A single-stage compressor with atmospheric inlet pressure is limited to a 50 psi discharge pressure. In booster service, the smaller units can be used to approximately 400 psi.

The *liquid piston compressor,* or liquid ring pump as it is more commonly called, uses a single rotor and can be seen in Figure 1-9. The rotor consists of a set of forward-curved vanes. The inner area of the rotor contains sealed openings, which in turn rotate about a stationary hollow inner core. The inner core contains the inlet and discharge ports. The rotor turns in an eccentric cylinder of either a single- or double-lobe design. Liquid is carried at the tips of the vanes and moves in and out as the rotor turns, forming a liquid piston. The port openings are so located as to allow gas to enter when the liquid piston is moving away from center. The port is then closed as rotation progresses and compression takes place, with the discharge port coming open as the liquid piston approaches the innermost part of the travel. As with some of the other rotary com-

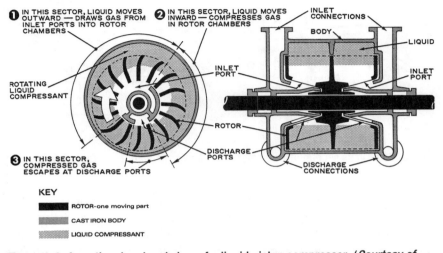

Figure 1-9. A sectional and end view of a liquid piston compressor. (*Courtesy of Nash Engineering Co.*)

pressors, the exact port locations must be tailored to the desired pressure ratio at time of manufacture. In the two-lobe design, two compression cycles take place during the course of one rotor revolution.

The capacity range is relatively large, ranging from 2 to 16,000 cfm. Like the sliding-vane compressors, the liquid piston compressor is widely used in vacuum service. The compressor is also used in pressure service with a normal range of 5 to 80 psi with an occasional application up to 100 psi. Because of the liquid piston, the compressor can ingest liquid in the suction gas without damage. This feature helps offset a somewhat poor efficiency. The compressor is used in multiple units to form a multistage arrangement.

Continuous Compression Compressors

Ejectors

Continuous compression compressors are of two types: ejector and dynamic.

The ejector can first be identified as having no moving parts (see Figure 1-10). It is used primarily for that feature as it is not as efficient as most of the mechanical compressors. Simplicity and the lack of wearing parts contribute to the unit's inherent reliability and low-maintenance expense.

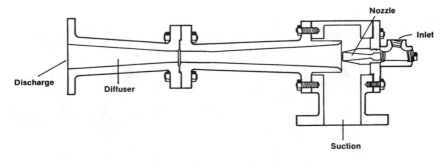

Figure 1-10. Cross section of an ejector. (*Courtesy of Graham Manufacturing Co., Inc.*)

The ejector is operated directly by a motive gas or vapor source. Air and steam are probably the two most common of the motive gases. The ejector uses a nozzle to accelerate the motive gas into the suction chamber where the gas to be compressed is admitted at right angles to the motive gas direction. In the suction chamber, also referred to as the mixing chamber, the suction gas is entrained by the motive fluid. The mixture moves into a diffuser where the high velocity gas is gradually decelerated and increased in pressure.

The ejector is widely used as a vacuum pump, where it is staged when required to achieve deeper vacuum levels. If the motive fluid pressure is sufficiently high, the ejector can compress gas to a slightly positive pressure. Ejectors are used both as subsonic and supersonic devices. The design must incorporate the appropriate nozzle and diffuser compatible with the gas velocity. The ejector is one of the few compressors immune to liquid carryover in the suction gas.

Dynamic Compressors

In dynamic compressors, energy is transferred from a moving set of blades to the gas. The energy takes the form of velocity and pressure in the rotating element, with further pressure conversion taking place in the stationary elements. Because of the dynamic nature of these compressors, the density and molecular weight have an influence on the amount of pressure the compressor can generate. The dynamic compressors are further subdivided into three categories, based primarily on the direction of flow through the machine. These are radial, axial, and mixed flow.

The *radial-flow,* or *centrifugal compressor* is a widely used compressor and is probably second only to the reciprocating compressor in usage in the process industries. A typical multistage centrifugal compressor can be seen in Figure 1-11. The compressor uses an impeller consisting of

Figure 1-11. Radial-flow horizontally split multistage centrifugal compressor.
(*Courtesy of Nuovo Pignone*)

radial or backward-leaning blades and a front and rear shroud. The front shroud is optionally rotating or stationary depending on the specific design. As the impeller rotates, gas is moved between the rotating blades from the area near the shaft and radially outward to discharge into a stationary section, called a diffuser. Energy is transferred to the gas while it is traveling through the impeller. Part of the energy converts to pressure along the blade path while the balance remains as velocity at the impeller tip where it is slowed in the diffuser and converted to pressure. The fraction of the pressure conversion taking place in the impeller is a function of the backward leaning of the blades. The more radial the blade, the less pressure conversion in the impeller and the more conversion taking place in the diffuser. Centrifugal compressors are quite often built in a multistage configuration, where multiple impellers are installed in one frame and operate in series.

Centrifugal compressors range in volumetric size from approximately 1,000 to 150,000 cfm. In single-wheel configuration, pressures vary considerably. A common low pressure compressor may only be capable of 10 to 12 psi discharge pressure. In higher-head models, pressure ratios of 3 are available, which on air is a 30-psi discharge pressure when the inlet is at atmospheric conditions.

Another feature of the centrifugal is its ability to admit or extract flow to or from the main flow stream, at relatively close pressure intervals, by means of strategically located nozzles. These flows are referred to as sidestreams. Pressures of the multistage machine are quite varied, and difficult to generalize because of the many factors that control pressure. Centrifugals are in service at relatively high pressures up to 10,000 psi either as a booster or as the result of multiple compressors operating in series.

Axial compressors are large-volume compressors that are characterized by the axial direction of the flow passing through the machine. The energy from the rotor is transferred to the gas by blading (see Figure 1-12). Typically, the rotor consists of multiple rows of unshrouded blades. Before and after each rotor row is a stationary (stator) row. For example, a gas particle passing through the machine alternately moves through a stationary row, then a rotor row, then another stationary row, until it completes the total gas path. A pair of rotating and stationary blade rows define a stage. One common arrangement has the energy transfer arranged to provide 50% of the pressure rise in the rotating row and the other 50% in the stationary row. This design is referred to as 50% reaction.

Axial compressors are smaller and are significantly more efficient than centrifugal compressors when a comparison is made at an equivalent flow rating. The exacting blade design, while maintaining structural integrity, renders this an expensive piece of equipment when compared to centrifugals. But it is generally justified with an overall evaluation that includes the energy cost.

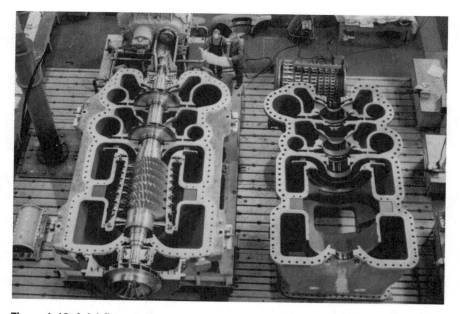

Figure 1-12. Axial-flow compressor. (*Courtesy of Demag Delaval Turbomachinery Corp.*)

The volume range of the axial starts at approximately 70,000 cfm. One of the largest sizes built is 1,000,000 cfm, with the common upper range at 300,000 cfm. The axial compressor, because of a low-pressure rise per stage, is exclusively manufactured as a multistage machine. The pressure for a process air compressor can go as high as 60 psi. Axial compressors are an integral part of large gas turbines where the pressure ratios normally are much higher. In gas turbine service, discharge pressures up to 250 psi are used.

The mixed-flow compressor is a relatively uncommon form, and is being mentioned here in the interest of completeness. At first glance, the mixed-flow compressor very much resembles the radial-flow compressor. A bladed impeller is used, but the flow path is angular in direction to the rotor; that is, it has both radial and axial components (see Figure 1-13). Because the stage spacing is wide, the compressor is used almost exclusively as a single-stage machine. The energy transfer is the same as was described for the radial-flow compressor.

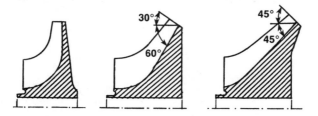
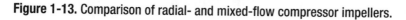

Centrifugal impeller   60° mixed-flow impeller   45° mixed-flow impeller

Figure 1-13. Comparison of radial- and mixed-flow compressor impellers.

The compressor size is flexible and covers the centrifugal compressor flow range, generally favoring the higher flow rates. The head per stage is lower than available in the centrifugal. The compressor finds itself in the marketplace because of the unique head-capacity characteristic, which can be illustrated by its application in pipeline booster service. In this situation the pressure ratio needed is not high, and as a result the head required is low. However, because of the high inlet pressure of the gas, a relatively high pressure rise is taken across the machine. Thus, there is a real need for a more rugged and less expensive alternative to the axial compressor.

2

Basic Relationships

Introduction

This chapter presents some basic thermodynamic relationships that apply to all compressors. Equations that apply to a particular type of compressor will be covered in the chapter addressing that compressor. In most cases, the derivations will not be presented, as these are available in the literature. The references given are one possible source for additional background information.

The equations are presented in their primitive form to keep them more universal. Consistent units must be used, as appropriate, at the time of application. The example problems will include conversion values for the units presented. The symbol g will be used for the universal gravity constant to maintain open form to the units.

Gas and Vapor

A gas is defined as the state of matter distinguished from solid and liquid states by very low density and viscosity, relatively great expansion and contraction with changes in pressure and temperature, and the ability to diffuse readily, distributing itself uniformly throughout any container.

A vapor is defined as a substance that exists below its critical temperature and that may be liquefied by application of sufficient pressure. It may be defined more broadly as the gaseous state of any substance that is liquid or solid under ordinary conditions.

Many of the common "gases" used in compressors for process plant service are actually vapors. In many cases, the material may change states during a portion of the compression cycle. Water is a good example, since a decrease in temperature at high pressure will cause a portion of the water to condense. This is a common occurrence in the first intercooler of a plant air compressor. Conversely, lowering the pressure in a reservoir of liquid refrigerant at a fixed temperature will cause the vapor quantity to increase.

Perfect Gas Equation

Charles and Gay-Lussac, working independently, found that gas pressure varied with the absolute temperature. If the volume was maintained constant, the pressure would vary in proportion to the absolute temperature [1]. Using a proportionality constant R, the relationships can be combined to form the equation of state for a perfect gas, otherwise known as the perfect gas law.

$$Pv = RT \tag{2.1}$$

where

P = absolute pressure
v = specific volume
R = constant of proportionality
T = absolute temperature

If the specific volume v is multiplied by mass m, the volume becomes a total volume V. Therefore, multiplying both sides of Equation 2.1 by m, yields

$$PV = mRT \tag{2.2}$$

In process engineering, moles are used extensively in performing the calculations. A mole is defined as that mass of a substance that is numerically equal to its molecular weight. Avogadro's Law states that identical volumes of gas at the same temperature and pressure contain equal numbers of molecules for each gas. It can be reasoned that these identical volumes will have a weight proportional to the molecular weight of the gas. If the mass is expressed as

$$m = n \times mw \tag{2.3}$$

where

n = number of moles

mw = molecular weight

then,

$$PV = n \; mw \; RT \tag{2.4}$$

If the value $mw \; R$ is the same for all gases, the universal gas constant Ugc is defined and R becomes the specific gas constant.

$$R = \frac{Ugc}{mw} \tag{2.5}$$

Another useful relationship can be written using Equation 2.2.

$$\frac{P_1 V_1}{T_1} = mR = \frac{P_2 V_2}{T_2} \tag{2.6}$$

If in Equation 2.2 both sides are divided by time, the term V becomes Q, volumetric flow per unit time, and the mass flow per unit time becomes w.

$$PQ = wRT \tag{2.7}$$

Compressibility

A term may now be added to Equation 2.1 to correct it for deviations from the ideal gas or perfect gas law.

$$Pv = ZRT \tag{2.8}$$

Solving for Z:

$$Z = \frac{Pv}{RT} \tag{2.9}$$

Equation 2.7 may be modified in a similar manner by the addition of the compressibility term Z as follows:

$$PQ = wZRT \tag{2.10}$$

Generalized Compressibility Charts

The vapor definition introduces another concept, that of critical temperature. Critical temperature is defined as that temperature above which a gas will not liquefy regardless of any increase in pressure. Critical pressure is defined as the pressure required at the critical temperature to cause the gas to change state.

The following two equations are used to define reduced temperature and reduced pressure:

$$T_r = \frac{T}{T_c} \tag{2.11}$$

$$P_r = \frac{P}{P_c} \tag{2.12}$$

The generalized compressibility charts may be used with values obtained in the use of Equations 2.7 and 2.8 to determine the compressibility of a wide range of gases. The charts were derived from experimental data and are a good source of information for use in compressor calculations [1].

Partial Pressure

Avogadro's Law states that equal volumes of gas at identical pressure and temperature contain equal numbers of molecules. Avogadro's Law can be used in a similar manner to develop gas mixture relationships. A mixture of gases occupying a given volume will have the same number of molecules as a single gas. The weight will be a sum of the proportionate parts of the gases in the mixture. If the gas proportion is presented as a mole percent, this value is the same as a volume percent.

When one pure liquid exists in the presence of another pure liquid, where the liquids neither react nor are soluble in each other, the vapor pressure of one liquid will not affect the vapor pressure of the other liquid. The sum of the partial pressures P_n is equal to the total pressure P. This relationship is formalized in Dalton's Law, which is expressed as

$$P = P_1 + P_2 + P_3 + \ldots \tag{2.13}$$

Gas Mixtures

If the total pressure of a mixture is known, the partial pressure of each component can be calculated from the mole fraction. The total number of moles in the mixture M_m is the sum of the individual component moles.

$$M_m = M_1 + M_2 + M_3 + \ldots \tag{2.14}$$

The mole fraction x_n is

$$x_1 = \frac{M_1}{M_m} ; x_2 = \frac{M_3}{M_m} ; x_3 = \frac{M_3}{M_m} \tag{2.15}$$

The partial pressure can be calculated by use of the following:

$$P_1 = x_1 P; P_2 = x_2 P; P_3 = x_3 P \tag{2.16}$$

Specific Heat Ratio

The value k is defined as the ratio of specific heats.

$$k = \frac{c_p}{c_v} \tag{2.18}$$

where

c_p = specific heat at constant pressure
c_v = specific heat at constant volume

$$\text{Also, } k = \frac{Mc_p}{Mc_p - 1.99} \tag{2.19}$$

where

Mc_p = molal specific heat at constant pressure.

$$Mc_{pm} = x_1 Mc_{p1} + x_2 Mc_{p2} + x_3 Mc_{p3} + \ldots \tag{2.20}$$

Substitute into Equation 2.19

$$k_m = \frac{Mc_{pm}}{Mc_{pm} - 1.99} \tag{2.21}$$

Molecular Weight

To calculate the mixture molecular weight (mw_m) use the following equation:

$$mw_m = x_1 mw_1 + x_2 mw_2 + x_3 mw_3 \tag{2.22}$$

The weight fraction y_n of the mixture is

$$y_1 = \frac{x_1 mw_1}{mw_m}; y_2 = \frac{x_2 mw_2}{mw_m}; y_3 = \frac{x_3 mw_3}{mw_m} \tag{2.23}$$

$$y_1 + y_2 + y_3 + \ldots = 1.0 \tag{2.24}$$

Specific Gravity

The specific gravity, SG, is the ratio of the density of a given gas to the density of dry air at the same temperature and pressure. It can be calculated from the ratio of molecular weights if the given gas is a perfect gas.

$$SG = \frac{mw}{28.96} \tag{2.25}$$

Mixture Compressibility

The simplest and most common method of establishing pseudocriticals for a mixture is Kay's Rule.

$$T_{cm} = x_1 T_{c1} + x_2 T_{c2} + x_3 T_{c3} + \ldots \tag{2.26}$$

$$P_{cm} = x_1 P_{c1} + x_2 P_{c2} + x_3 P_{c3} + \ldots \tag{2.27}$$

Substituting Equations 2.26 and 2.27 into Equations 2.11 and 2.12:

$$T_{rm} = \frac{T}{T_{cm}} \tag{2.28}$$

$$P_{rm} = \frac{P}{P_{cm}} \tag{2.29}$$

Humidity

Although air is a mixture of gases, it is generally treated as an individual gas with accounting made only for other components such as moisture when present.

When a mixture is saturated, the proper terminology is that the volume occupied by the mixture is saturated by one or more of the components. For air space, which is partially saturated by water vapor, the actual partial pressure of the water vapor may be determined by multiplying the saturation pressure at the space temperature by the relative humidity.

Relative humidity can be calculated from the following:

$$RH = \frac{P_v}{P_{satv}} \times 100 \tag{2.30}$$

Specific humidity, which is the weight of water vapor to the weight of dry air, is given by the following ratio:

$$SH = \frac{W_v}{W_a} \tag{2.31}$$

Psychrometric charts plot wet bulb and dry bulb data for air-water vapor mixtures at atmospheric pressure. These charts are quite useful for moisture corrections in air compressors with atmospheric inlets (see Figures B-2 and B-3 in Appendix B).

Flow

There are several different flow terminology conventions in common use. The following discussion is presented in order to eliminate any confusion this may cause.

The most important thing to remember in compressor calculations is that compressor flow is a volumetric value based on the flowing conditions of pressure, temperature, relative humidity (if moisture is present), and gas composition at the compressor inlet nozzle. The flow units are inlet cubic feet per minute (icfm).

Process calculations, where material balances are performed, normally produce flow values in terms of a weight flow. The flow is generally stated as pounds per hour. Equation 2.10 can be used either with a single-component gas or with a mixture.

Pipeline engineers use the flow value stated as standard cubic feet per day. This is an artificial weight flow because flowing conditions are referred to a standard pressure and temperature. The balance of the flow specification is then stated in terms of specific gravity.

A common method of stating flow is standard cubic feet per minute where the flowing conditions are referred to an arbitrary set of standard conditions. Unfortunately, standard conditions are anything but standard. Of the many used, two are more common. The ASME standard uses 68°F and 14.7 psia. The relative humidity is given as 36%. The other standard that is used by the gas transmission industry and the API Mechanical Equipment Standards is 60°F at 14.7 psia. As can be seen from this short discussion, a flow value must be carefully evaluated before it can be used in a compressor calculation.

Example 2-1

A pipeline is flowing 3.6 standard million cubic feet per day. The gas is made up of the following components: 85% methane, 10% ethane, 4% butane, 1% nitrogen. The values are given as a mole percent. The flowing temperature is 80°F and the pressure is 300 psig.

The problem is to calculate the suction conditions for a proposed booster compressor. Values to calculate are flow in cfm at the flowing conditions, the mixture molecular weight, mixture specific heat ratio, and the compressibility of the mixture.

Step 1. Convert the flow to standard cfm using 24 hours per day and 60 minutes per hour.

$$Q_{std} = \frac{3.6 \times 10^6}{24 \times 60}$$

$$Q_{std} = 2500$$

Step 2. Convert scfm to flowing conditions using Equation 2.6. Standard conditions:

$P_2 = 14.7$ psia

$T_2 = 60°F + 460°R = 520°R$

Flowing conditions:

$P_1 = 300 + 14.7 = 314.7$ psia

$T_1 = 80°F + 460°R = 540°R$

Step 3. Substituting into Equation 2.6, using Q_1 for V_1 and solving for Q_1.

$$Q_1 = \frac{14.7}{314.7} \times \frac{540}{520} \times 2500$$

$Q_1 = 121.3$ cfm (flow at the compressor inlet)

Step 4. Change the molal percentages to fractions and substitute for x_n, then use Equations 2.20, 2.22, 2.26, and 2.27 to construct Table 2-1.

Step 5. Solve for mixture specific heat ratio k_m, using Equation 2.21.

$$k_m = \frac{9.59}{9.59 - 1.99}$$

$k_m = 1.26$

Table 2-1
Gas Mixture Data

Gas	x_n	m_{cp}	$x_n m_{cp}$	mw	$x_n mw$	T_c	$x_n T_c$	p_c	$x_n p_c$
Methane	.85	8.60	7.31	16.04	13.63	344	292.4	673	572.1
Ethane	.10	12.64	1.26	30.07	3.01	550	55.0	708	70.8
Butane	.04	23.82	.95	58.12	2.33	766	30.6	551	22.0
Nitrogen	.01	6.97	.07	28.02	0.28	227	2.3	492	4.9
Mixture	1.00		9.59		19.25		380.		670.

Step 6. Using $T_{cm} = 380°R$ and $P_{cm} = 670$ psia, substitute into Equations 2.28 and 2.29.

$$T_{rm} = \frac{540}{380}$$

$$T_{rm} = 1.42$$

$$P_{rm} = \frac{314.7}{670}$$

$$P_{rm} = .47$$

Step 7. From the general compressibility charts in the Appendix, $Z = .95$.

Example 2-2

Determine the volumetric flow to use in sizing a compressor to meet the following suction requirements:

Weight flow = 425 lb/min dry air
Inlet pressure = 14.7 psia ambient air
Inlet temperature = 90°F
Inlet relative humidity = 95%

Step 1. Determine the total moist air flow to provide the dry air needed. Because the air is at atmospheric pressure, psychrometric charts may be used to determine the amount of water vapor contained in the dry air (see Figures B-2 and B-3 in Appendix B).

From the psychrometric chart, for a dry bulb temperature of 90°F with a relative humidity of 95%,

Specific humidity = .0294 lbs of water vapor/lb of dry air
For the 425 lb/min of dry air, the water vapor content is

$$w_2 = 425 \times .0294$$

$$w_2 = 12.495 \text{ lb/min water vapor}$$

Therefore,

$$w_m = 425 + 12.495$$

$$w_m = 437.5 \text{ lb/min total weight flow}$$

Step 2. Determine the molecular weight of the moist air mixture using Equation 2.3.

$$M_1 = \frac{425 \text{ lb/min}}{28.95 \text{ lb/lb - mol}}$$

$$M_1 = 14.68 \text{ lb - mols/min dry air}$$

$$M_2 = \frac{12.495 \text{ lb/min}}{18.02 \text{ lb/lb - mol}}$$

$$M_2 = .693 \text{ lb - mols/min water vapor}$$

$$M_m = 14.68 + .693$$

$$M_m = 15.373 \text{ total mols/min mixture}$$

Step 3. Using Equation 2.15, calculate the mol fraction of each component.

$$x_1 = \frac{14.68}{15.373}$$

$$x_1 = .955 \text{ mol fraction dry air}$$

$$x_2 = \frac{.693}{15.373}$$

$$x_2 = .045 \text{ mol fraction dry air}$$

Step 4. Calculate the molecular weight using Equation 2.22.

$$mw_m = .955 \times 28.95 + .045 \times 18.02$$

$$mw_m = 28.46 \text{ mol weight mixture}$$

Step 5. Calculate the compressor inlet volume using Equation 2.10. First use Equation 2.5 to calculate the specific gas constant.

$$R_m = 1545/28.46$$

$$R_m = 54.29$$

Convert to absolute temperature.

$$T_1 = 460 + 90$$

$$T_1 = 550°R$$

Substitute into Equation 2.10 and, using 144 in^2/ft^2,

$$Q_1 = 437.5 \frac{1 \times 54.29 \times 550}{14.7 \times 144}$$

$$Q_1 = 6171 \text{ cfm air mixture}$$

For comparison, assume the moisture had been ignored.

$$R_m = 1545/28.95$$

$$R_m = 53.37$$

$$Q_1 = 425 \frac{1 \times 53.37 \times 550}{14.7 \times 144}$$

$$Q_1 = 5893 \text{ cfm}$$

The calculation would indicate that the volume would have been short by approximately 5% if the moisture in the air was ignored.

Acoustic Velocity

A relationship that is useful in compressor and compressor systems is the speed of sound of the gas at the flowing conditions. The *acoustic velocity,* a, can be calculated using the following equation:

$$a = \sqrt{kRgT} \tag{2.32}$$

where

 k = ratio of specific heats
 R = specific gas constant
 g = gravitational constant
 T = absolute temperature of the fluid

The Mach number is given by

$$M_a = \frac{V}{a} \tag{2.33}$$

The relationship for uniform flow velocity V in a cross-sectional area, A, such as a compressor flow channel or nozzle is

$$V = \frac{Q}{A} \tag{2.34}$$

where

 Q = volumetric flow.

Equations of State

Gases can be treated individually or as mixtures by the methods just outlined for most applications including evaluation of vendor proposals. More sophisticated equations of state can be used for real gas applications when large deviations from the perfect gas law are anticipated. For mixtures, more sophisticated mixing rules can be paired with the equation of state when required. For hydrocarbons, the most widely used equation of state is the Benedict-Webb-Rubin (BWR) equation [2]. For a gas mixture, the pseudocritical constants used in the BWR equation may be developed using Kay's mixing rule. If the application is outside Kay's Rule guidelines, a more complex rule such as Leland-Mueller may be substituted [3]. An alternate approach is the Starling BWR implementation [4]. Starling includes gas mixing in formulation of the equation of state. Another, the Redlich-Kwong equation, is widely used because of its simplicity. Finally, for chlorinated compounds and halocarbon refrigerants, the Martin-Hou equation yields results generally superior to the previously mentioned equations, which were developed primarily for hydrocarbons [5]. The equations of state discussed are by no means a complete list, but they have proved to be especially accurate in direct application.

The equations of state will not be further described or presented in more detail as they are unfortunately somewhat difficult to solve without the use of a computer. Full details are available in the referenced material for those wishing to pursue this subject further. In the past, these equations required the use of a mainframe computer not only to solve the equations them-selves, but to store the great number of constants required. This has been true particularly if the gas mixture contains numerous components. With the power and storage capacity of personal computers increasing, the equations have the potential of becoming more readily available for general use.

Mollier Charts

Another form in which gas properties are presented is found in plots of pressure, specific volume, temperature, entropy, and enthalpy. The most common form, the Mollier chart, plots enthalpy against entropy. A good example of this is the Mollier chart for steam. Gases are generally plotted as pressure against enthalpy (P-h charts). These are also sometimes referred to as Mollier charts. The charts are readily available for a wide range of pure gases, particularly hydrocarbons and refrigerants. Some of the more common charts are included in Appendix B.

First Law of Thermodynamics

The first law of thermodynamics states that energy cannot be created or destroyed, although it may be changed from one form to another. Stated in equation form, it is written as follows:

$$Q_h - W_t = \Delta E \tag{2.35}$$

where

Q_h = heat supplied to a system
W_t = work done by the system
ΔE = change in energy of the system

If the change in energy to the system is expanded, then

$$\Delta E = \Delta U + \Delta PE + \Delta KE \tag{2.36}$$

where

ΔU = change in internal energy
ΔPE = change in potential energy
ΔKE = change in kinetic energy

If the work term W_t is expanded to breakdown shaft work done to or from the system and the work done by the system, then

$$W_t = W + (pv\Delta m)_{out} - (pv\Delta m)_{in} \tag{2.37}$$

where

W = shaft work in or out of the system
p = fluid pressure in the system
v = specific volume of the fluid in the system
Δm = mass of fluid working in the system

If Equation 2.36 is rewritten in a general form using specific energy notation,

$$e = u + \frac{V^2}{2g} + z \tag{2.38}$$

where

u = specific form of internal energy
V = velocity of the gas
z = height above some arbitrary reference

By substituting Equations 2.37 and 2.38 into Equation 2.35, maintaining the specific energy form, and regrouping, the following equation can be written:

$$u_1 + P_1 v_1 + \frac{V_1^2}{2g} + z_1 + Q_h = u_2 + P_2 v_2 \frac{V_2^2}{2g} + z_2 + W \tag{2.39}$$

By defining enthalpy as

$$h = u + Pv \tag{2.40}$$

and substituting into Equation 2.39,

$$h_1 + \frac{V_1^2}{2g} + z_1 + Q_h = h_2 + \frac{V_2^2}{2g} + z_2 + W \tag{2.41}$$

Equation 2.41 is the general energy equation for a steady flow process.

Second Law of Thermodynamics

The second law of thermodynamics was actually postulated by Carnot prior to the development of the first law. The original statements made concerning the second law were negative—they said what would *not* happen. The second law states that heat will not flow, in itself, from cold to hot. While no mathematical relationships come directly from the second law, a set of equations can be developed by adding a few assumptions for use in compressor analysis. For a reversible process, entropy, s, can be defined in differential form as

$$ds = \frac{dQ_h}{T} \tag{2.42}$$

It is recognized that a truly reversible process does not exist in the real world. If it is further recognized that real processes result in an increase in entropy, the second law can be stated.

$$\Delta s \gtreqless 0 \tag{2.43}$$

If work done in a system is distributed over an area, for example, pressure P is acting through volume v, then in specific notation and in differential form the Equation 2.44 results.

$$dW_t = Pdv \tag{2.44}$$

If further $\Delta U = \Delta E$ when the kinetic and potential energies in Equation 2.36 do not change, Equation 2.35 can be rewritten, substituting U for E, changing to the specific notation and putting the equation in differential form.

$$du = dQ_h - W_t \tag{2.45}$$

Combining Equations 2.42, 2.44 and 2.45 yields

$$du = Tds - Pdv \tag{2.46}$$

Theoretical Work

Theoretical work or compressor head is the heart and substance of compressor design. Some basic form of understanding must be developed even if involvement with compressors is less than that of design of the machine itself. Proper applications cannot be made if this understanding is absent. The following theoretical evaluations will be abbreviated as much as possible to reduce the length and still present the philosophy. For the reader with the ambition and desire, the presentation will be an outline to which the reader can fill in the spaces.

In deriving the head equation, the general energy Equation 2.41 will be used. The equation can be modified by regrouping and eliminating the z terms, as elevation differences are not significant with gas.

$$\left(h_2 + \frac{V_2^2}{2g} \right) - \left(h_1 + \frac{V_1^2}{2g} \right) = -W + Q_h \tag{2.47}$$

The velocity term can be considered part of the enthalpy if the enthalpy is defined as the stagnation or total enthalpy. The equation can be simplified to

$$h_2 - h_1 = -W + Q_h \tag{2.48}$$

If the process is assumed to be adiabatic (no heat transfer), then

$$Q_h = 0$$

For the next step the enthalpy equation is written in differential form:

$$dh = du + Pdv + vdP \tag{2.49}$$

Recalling Equation 2.46,

$$du = Tds - Pdv \tag{2.46}$$

and substituting Equation 2.46 into 2.49,

$$dh = Tds + vdP \tag{2.50}$$

The process is assumed reversible. This defines entropy as constant and therefore $ds = 0$, making $Tds = 0$. The enthalpy equation is simplified to

$$dh = vdP \tag{2.51}$$

For an isentropic, adiabatic process,

$$Pv^k = constant = C \tag{2.52}$$

solving for P,

$$P = Cv^{-k} \tag{2.53}$$

Taking the derivative of P with respect to v yields,

$$dP = C(-k)v^{-k-1}dv \tag{2.54}$$

Substituting into the enthalpy Equation 2.51,

$$dh = C(-k)v^{-k}dv \tag{2.55}$$

Integrating from state point 1 to 2 and assuming k is constant over the path yields,

$$h_2 - h_1 = C \frac{v_2^{1-k} - v_1^{1-k}}{(k-1)/k} \tag{2.56}$$

Substitute

$$C = P_1 v_1^k = P_2 v_2^k \tag{2.57}$$

into Equation 2.55, which yields

$$h_2 - h_1 = \frac{P_2 v_2 - P_1 v_1}{(k-1)/k} \tag{2.58}$$

Using the perfect gas Equation 2.1 and substituting into Equation 2.58 yields

$$h_2 - h_1 = \frac{R(T_2 - T_1)}{(k-1)/k} \qquad (2.59)$$

As a check on the assumptions made, a comparison can be made to a different method of checking the derivation of the head. Enthalpy difference, as a function of temperature change, for an adiabatic process is expressed by

$$h_2 - h_1 = c_p(T_2 - T_1) \qquad (2.60)$$

Specific heat c_p can be calculated using specific gas constant R and specific heat ratio k.

$$c_p = \frac{Rk}{k-1} \qquad (2.61)$$

Substitute Equation 2.61 into Equation 2.60 with the result,

$$h_2 - h_1 = \frac{R(T_2 - T_1)}{(k-1)/k} \qquad (2.59)$$

This equation is identical with Equation 2.59 previously derived, giving a check on the method.

By regrouping Equation 2.59, substituting into Equation 2.48, and maintaining the adiabatic assumption $Q_h = 0$, Equation 2.62 is developed.

$$h_2 - h_1 = \frac{RT_1 k}{k-1}\left[\frac{T_2}{T_1} - 1\right] = -W \qquad (2.62)$$

The $-W$ signifies work done to the system, a driven machine, as contrasted to $+W$, which would indicate work done by the system as with a driver.

If the adiabatic head is defined by the following equation:

$$H_a = h_2 - h_1 \qquad (2.63)$$

and the term r_p is introduced as the ratio of discharge pressure to inlet pressure,

$$r_p = \frac{P_2}{P_1} \qquad (2.64)$$

Next, the temperature ratio relationship in Equation 2.65 will be used.

This relationship is the result of combining Equations 2.6 and 2.57 as well as a half dozen algebraic steps:

$$\frac{T_2}{T_1} = r_p^{\frac{k-1}{k}} \tag{2.65}$$

When substituting Equation 2.65 into Equation 2.62, the result is the classical form of the adiabatic head equation.

$$H_a = RT_1 \frac{k}{k-1} (r_p^{\frac{k-1}{k}} - 1) \tag{2.66}$$

An interesting note is that if in Equation 2.58, Equation 2.8 were used in place of 2.1, the result would be

$$h_2 - h_1 = \frac{R(T_2 Z_2 - T_1 Z_1)}{(k-1)/k} \tag{2.67}$$

Since the compressibility does not change the isentropic temperature rise, it should be factored out of the ΔT portion of the equation. To achieve this for moderate changes in compressibility, an assumption can be made as follows:

$$Z_{avg} = (Z_2 + Z_1)/2 \tag{2.68}$$

By replacing the values of Z_2 and Z_1, with Z_{avg} in Equation 2.67 and factoring, Equation 2.67 is rewritten as

$$h_2 - h_1 = \frac{Z_{avg} R (T_2 - T_1)}{(k-1)/k} \tag{2.69}$$

Now with the same process used to obtain Equation 2.66, the final form of the head equation with compressibility is

$$H_a = Z_{avg} RT_1 \frac{k}{k-1} (r_p^{\frac{k-1}{k}} - 1) \tag{2.70}$$

For a polytropic (reversible) process, the following definitions need to be considered:

$$\frac{n-1}{n} = \frac{k-1}{k} \times \frac{1}{\eta_p} \tag{2.71}$$

where

η_p = polytropic efficiency
n = polytropic exponent

By regrouping Equation 2.71, a polytropic expression can be

$$\eta_p = \frac{n/(n-1)}{k/(k-1)} \tag{2.72}$$

By substituting n for k, the head equation becomes

$$H_p = Z_{avg}\, RT_1 \frac{n}{n-1}\, (r_p^{\frac{n-1}{n}} - 1) \tag{2.73}$$

One significant practical difference in use of polytropic head is that the temperature rise in the equation is the actual temperature rise when there is no jacket cooling. The other practical uses of the equation will be covered as they apply to each compressor in the later chapters.

Real Gas Exponent

About the time it appears that there is some order to all the chaos of compressible flow, there comes another complication to worry about. It has been implied that k is constant over the compression path. The sad fact is that it is not really true. The k value has been defined in Equation 2.18 as

$$k = \frac{c_p}{c_v} \tag{2.18}$$

It has played a dual role, one in Equation 2.18 on specific heat ratio and the other as an isentropic exponent in Equation 2.53. In the previous calculation of the speed of sound, Equation 2.32, the k assumes the singular specific heat ratio value, such as at compressor suction conditions. When a non-perfect gas is being compressed from point 1 to point 2, as in the head Equation 2.66, k at 2 will not necessarily be the same as k at 1. Fortunately, in many practical conditions, the k doesn't change very much. But if one were inclined to be a bit more judicious about it and calculate a k at both state points, and if the values differed by a small amount, then one could average the two and never look back. This could not be done, however, with a gas near its critical pressure or one that's somewhat unruly like ethylene, where the k value change from point 1 to 2 is highly nonlinear. For a situation like this, the averaging approach is just not good enough and the following modification will be presented to help make the analysis more accurate.

To calculate a single compression exponent to represent the path from point 1 to point 2, the following equations will be used. Substitute γ for k and Equation 2.64 into Equation 2.65

$$\frac{T_2}{T_1} = \left(\frac{P_2}{P_1}\right)^{\frac{\gamma-1}{\gamma}} \tag{2.74}$$

where γ = compression path exponent.

The expression in Equation 2.52 can be modified to Equation 2.75 to show the basic relationship for the exponent.

$$\left(P\frac{v}{Z}\right)^{\gamma} = C \tag{2.75}$$

To solve for γ use the following equation:

$$\gamma = \frac{\ln(P_2/P_1)}{\ln(P_2/P_1) - \ln(T_2/T_1)} \tag{2.76}$$

To solve for the compression exponent, use a Mollier diagram to establish the T_2 temperature value. By establishing a starting point at P_1, and T_1, and taking a path of constant entropy to P_2, the T_2 value can be read from the diagram. For a gas mixture or gas with no convenient Mollier diagram available, the problem becomes more acute. There are two alternatives: one is to use an equation of state and the other is to use a method suggested by Edmister and McGarry [6]. The latter is somewhat tedious, making the equation of state the preferred method.

Power

Input shaft power is the head of the compressor multiplied by the weight flow and divided by an appropriate efficiency with the result being added to the mechanical losses. The head portion covers the fluid or thermodynamic portion of the cycle, whereas the mechanical losses cover items such as bearings and liquid seals that are not directly linked to the fluid process. The form shown here is generalized. Each compressor type has its own unique considerations and will be covered in the appropriate chapter. The adiabatic shaft work can be expressed as

$$W_a = \frac{wH_a}{\eta_a} + \text{mech losses} \tag{2.77}$$

For polytropic shaft work,

$$W_p = \frac{wH_p}{\eta_p} + \text{mech losses} \tag{2.78}$$

Velocity Head

The determination of pressure losses at compressor nozzles and other peripheral points must be made when performing an analysis of the system. It is common in the compressor industry to state the losses as a function of velocity head. An expression for velocity head may be derived from Equation 2.39 and the following: (1) Assume flow is incompressible, which is reasonable since the change in density is negligible; therefore, $v_1 = v_2$, (2) because there is no heat added or work done, u, W, Q, $= 0$. When these assumptions are factored into Equation 2.39,

$$vP_2 = \frac{V_2^2}{2g} = vP_1 + \frac{V_1^2}{2g} \tag{2.79}$$

Equation 2.79 contains two pairs of head terms, the Pv head terms and the $V^2/2g$ or velocity head terms. When a flow stream passes through a nozzle, the flow is accelerated. This flow phenomenon can be further examined by regrouping Equation 2.79.

$$v(P_1 - P_2) = \frac{V_2^2}{2g} - \frac{V_1^2}{2g} \tag{2.80}$$

The left term of Equation 2.80 represents a head drop required to accelerate the flow from an initial velocity to the final velocity V_2. If the initial velocity is low it can be assumed negligible and if density $\rho = 1/v$ is substituted into Equation 2.80, it can be written as

$$P_1 - P_2 = \rho \frac{V_2^2}{2g} \tag{2.81}$$

When gas flows through pipe, casing openings, valves, or fittings, a pressure drop is experienced. This pressure drop can be defined in terms of an equivalent velocity head. The velocity head is, therefore, the pressure drop necessary to produce a velocity equal to the flowing stream velocity. The term K will be used to describe the pressure dropping potential of various restrictive elements, regardless of density or velocity. The term K is a multiplier equal to one at a value of one velocity head and can be greater than or less than one. Typical values of K are presented in Table 2-2. By substituting $\Delta P = P_1 - P_2$ and dropping the subscript on the velocity term, the working equation to use in the calculation of pressure drop for K velocity heads is

$$\Delta P = K\rho \frac{V^2}{2g} \tag{2.82}$$

Example 2-3

An example will help illustrate one use of velocity head. A compressor is being considered for reuse in another application, and the question was raised as to the size of the inlet nozzle. The original conditions are stated as follows:

Inlet nozzle size: 18 inches
Inlet flow: 10,000 CFM
Inlet pressure: 25 psia
Inlet temperature: 80°F
Molecular weight: 29
Specific heat ratio: 1.35

The new conditions:

Inlet flow: 11,000 CFM
Inlet pressure: 31 psia
Inlet temperature: 40°F
Molecular weight: 31
Specific heat ratio: 1.30

Table 2-2
Velocity Head Multipliers

Description	K-Factor
Reducer contraction	
0.75	0.2
0.50	0.3
Reducer enlargement	
0.75	0.5
0.50	0.6
0.25	0.9
Gate valve	
Fully open	0.15
0.25 open	25.0
Elbow	
Long radius	0.15
Short radius	0.25
Miter	1.10
Close return bend	0.5
Swing check or ball valve	2.2
Tee flow through bull-head	1.8
Angle valve, open	3.0
Globe valve, open	5.0
Filters	
Clean	4.0
Foul	20.0
Intercoolers	17.0
Gas separators	7.0
Surge bottles	
No choke tube	4.0
With choke tube	12.0
Casing inlet nozzle	0.5
Sidestream inlet nozzle (diaphragm)	1.0
Sidestream inlet nozzle (stage space)	0.8
Casing discharge nozzle	0.5
Extraction nozzle	0.8

Source: Modified from [16].

By consultation with the original equipment maker, it has been determined that the vendor used a value of .2 velocity heads in the original design. From this information, K = .2. The effect of the rerate conditions on the inlet will be

$$V = \frac{10,000 \times 144}{233.7 \times 60}$$

$V = 102.7 \text{ fps}$

Calculate the sonic velocity using Equation 2.32

where

$R = 1545/29 = 53.3$

$T = 80 + 460 = 540°F$

$a = \sqrt{1.35T \times 53.3 \times 32.2 \times 540}$

$a = 1118.3 \text{ fps}$

Using Equation 2.33, calculate the Mach number

$$M_a = \frac{102.7}{1118.3}$$

$M_a = .09$

This is a low value, therefore, the possibility exists of an up-rate relative to any nozzle flow limits. At this point, a comment or two is in order. There is a rule of thumb that sets inlet nozzle velocity limit at approximately 100 fps. But because the gases used in the examples have relatively high acoustic velocities, they will help illustrate how this limit may be extended. Regardless of the method being used to extend the velocity, a value of 150 fps should be considered maximum. When the sonic velocity of a gas is relatively low, the method used in this example may dictate a velocity for the inlet nozzle of less than 100 fps. The pressure drop due to velocity head loss of the original design is calculated as follows:

$$v = \frac{53.3 \times 540}{25 \times 144}$$

$v = 7.99 \text{ ft}^3/\text{lb}$

$$\rho = \frac{1}{7.99}$$

$$\rho = .13 \text{ lb/ft}^3$$

$$\Delta P = .2 \times .13 \frac{(102.7)^2}{64.4 \times 144}$$

$$\Delta P = .03 \text{ psi}$$

This is also a low value, so the proposed rerate conditions will be:

$$V = \frac{11,000 \times 144}{233.7 \times 60}$$

$$V = 113 \text{ fps}$$

$$R = 1545 / 31 = 49.8$$

$$T = 40 + 460 = 500°F$$

$$a = \sqrt{1.30 \times 49.8 \times 32.2 \times 500}$$

$$a = 1020 \text{ fps}$$

$$M_a = \frac{113}{1020}$$

$$M_a = .11$$

$$v = \frac{49.8 \times 500}{31 \times 144}$$

$$v = 5.58 \text{ ft}^3 / \text{lb}$$

$$\rho = \frac{1}{5.58}$$

$$\rho = .18 \text{ lb/ft}^3$$

$$\Delta P = .2 \times .18 \frac{(113)^2}{64.4 \times 144}$$

$\Delta P = .05$ psi

The up-rate looks feasible considering that none of the inlet nozzle guidelines have been exceeded, the Mach number is still a low value, and the pressure drop is not significant. If the pressure drop had been significant, the effect of the drop could have been evaluated with respect to the compressor head and possibly a usable compromise worked out.

Intercooling

Cooling between compressor stages limits the value of the discharge temperature and reduces energy demands. Normally there would be no argument against intercooling because of the obvious operating cost saving. However, in some process applications, the higher temperature of the gas leaving the compressor can have additional uses such as driving a reboiler. Since heat would have to be added to the gas anyway, it is more economical to use the heat in the gas at the compressor discharge and forego the benefit of the intercooling. However, each application must be evaluated if there is a temperature limit for the gas, or the power savings from cooling overshadows the alternate heat sources available to drive the reboiler.

The capital cost of the coolers, the piping, and the installation must become a part of any evaluation. Figure 2-1 shows a two-intercooler compressor. Cooling water must be added as an operating expense. Air cooling is an alternative, subject, however, to higher outlet temperature and higher capital cost. The extremes in ambient temperature and their effect on operation should not be ignored. An additional consideration is to observe the gas stream for possible condensation of components during cooling. If there is an objection to these components coming out of the stream, then some form of temperature control must be provided. If the condensation is acceptable, provision must still be made to remove this liquid fraction from the coolers before it enters the compressor. Most compressors are quite sensitive to liquid with some more so than others. In the sizing procedure, the loss of the fraction must also be considered in the resulting gas properties for the succeeding stages.

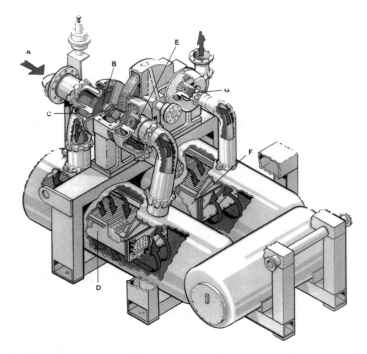

Figure 2-1. Two-water cooled intercoolers on a three-stage air compressor. (*Courtesy of Elliott Company*)

Isothermal Compression

Isothermal compression is presented here to represent the upper limits of cooling and horsepower savings. It is the equivalent of an infinite number of intercoolers and is not achievable in the practical types of compressors described in this book. For an isothermal process,

$$Pv = C \tag{2.83}$$

From this, a theoretical value for the power used by the compressor can be found.

$$W_{it} = wRT \ \text{In}(P_2/P_1) \tag{2.84}$$

This equation is useful when evaluating the benefit of multiple intercoolers, because it establishes the theoretical power limit that can be achieved by cooling. Figure 2-2 is a plot comparing the effect of different numbers of intercoolers in terms of uncooled horsepower. Note the diminishing effect as the number of coolers is increased.

An example illustrates the benefits of intercooling.

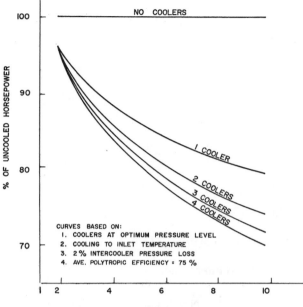

Figure 2-2. Percent of uncooled horsepower required with intercoolers. (*Courtesy of Elliott Company*)

Example 2-4

To keep from complicating the example with real world considerations, a few simplifying assumptions will be made. In all cases, the compressor will be considered to be 100% efficient. Intercooling will be perfect, that is, no pressure drop will be considered and the cooler return gas will be the same temperature to the first stage of the compressor.

Gas: nitrogen
Molecular weight: 29
Ratio of specific heats: 1.4
Inlet pressure: 20 psia
Outlet pressure: 180 psia
Inlet temperature: 80°F
Weight flow: 100 lb/min

Calculate the theoretical power for each case: (1) no intercooling, (2) one intercooler, (3) two intercoolers, (4) isothermal compression.

Because of the assumption that efficiency is 100%, Equations 2.70 and 2.73 yield the same results.

$R = 1545/29$

$R = 53.3$ ft lb/lb °R

$T_1 = 80 + 460$

$T_1 = 540°R$

$k/(k - 1) = 1.4/.4$

$k/(k - 1) = 3.5$

$(k - 1)/k = .4/1.4$

$(k - 1)/k = .286$

$H_a = 1.0 \times 53.3 \times 540 \times 3.5(9.^{286} - 1)$

$H_a = 88,107$ ft lb/lb (no intercooler)

Before proceeding with the power calculation, the head for each cooled case will be calculated. In the idealized case, the most efficient division of work for minimum power is achieved by taking the nth root of the pressure ratio, where n is the number of uncooled sections or compression stages in the parlance of the process engineer. For one cooler, $n = 2$.

$r_p = 9^{1/2}$

$r_p = 3$

For two coolers $n = 3$.

$r_p = 9^{1/3}$

$r_p = 2.08$

$H_a = 100737(3^{.286} - 1)$

$H_a = 37,189$ ft lb/lb (one cooler)

$H_a = 100,737(2.08^{.286} - 1)$

$H_a = 23,473$ ft lb/lb (two coolers)

Calculate the power using Equation 2.77 and setting $\eta_a = 1.0$ and mechanical losses $= 0$.

$$W_a = \frac{100 \times 87,983}{33,000 \times 1.0}$$

$W_a = 267.0\text{hp}$ (no cooler)

Multiply the head by n for the cooled cases.

$$W_a = \frac{100 \times 37,189}{33,000 \times 1.0} \times 2$$

$W_a = 225.4 \text{ hp}$ (one cooler)

$$W_a = \frac{100 \times 23,473}{33,000 \times 1.0} \times 3$$

$W_a = 213.3 \text{ hp}$ (two coolers)

Using Equation 2.84 to establish the theoretical limit of isothermal compression,

$$W_{it} = \frac{100 \times 53.3 \times 540 \times \ln(9)}{33,000}$$

$W_{it} = 191.6 \text{ hp}$

Taking the horsepower values due to cooling, comparing them to the uncooled case, and converting to percentage,

$$= \frac{225.4}{267.0} \times 100$$

$= 84.4\%$ of the uncooled hp, one cooler

$$= \frac{213.3}{267.0} \times 100$$

$= 79.9\%$ of the uncooled hp, two coolers

$$= \frac{191.6}{267.0} \times 100$$

$= 71.8\%$ of the uncooled hp, isothermal case

It can be seen by comparing the percentages that the benefit of cooling diminishes as each cooler is added. This is particularly noticeable in light of the comparatively small horsepower reduction brought about by isothermal compression since this represents the effect of an infinite number of coolers. The first step was a 15.6% decrease in power, while the second cooler only reduced it another 4.5 percentage points. Even the addition of an infinite number of coolers (isothermal case) added just 12.6 percentage points, a decrease less than the percentage achieved with the first cooler. While the economic impact must be evaluated in each case, this illustration does demonstrate that intercooling does save horsepower. In a practical evaluation, some of the idealized values used in the illustration must be replaced with anticipated actual values, such as real efficiency for the compressor, pressure drops in the coolers and piping, and the true outlet temperature expected from the coolers, based on cooling medium temperature. Because the point was made about the decrease in outlet temperature, these values will be calculated to make the example complete. Note that with temperature as with horsepower, the first increment of cooling yields the largest return. The Equation 2.65 is used to make the calculation.

$$t_2 = 540(9)^{.286} - 460$$

$$t_2 = 552°F \text{ (no cooler)}$$

$$t_2 = 540(3)^{.286} - 460$$

$$t_2 = 279°F \text{ (one cooler)}$$

$$t_2 = 540(2.08)^{.286} - 460$$

$$t_2 = 205°F \text{ (two coolers)}$$

The equations presented in this chapter should have general application to most compressors, particularly the ones to be discussed in the following chapters. As each compressor is covered, additional equations will be introduced.

References

1. Nelson, L. C. and Obert, E. F., "How to Use the New Generalized Compressibility Charts," *Chemical Engineering,* July 1954, pp. 203–208.

2. Benedict, Manson, Webb, George B., and Rubin, Louis C., "An Empirical Equation for Thermodynamic Properties of Light Hydrocarbons and Their Mixtures," *Chemical Engineering Progress,* Vol. 47, No. 8, August, 1951, pp. 419–422.

3. Reid, R. C. and Sherwood, T. K., *The Properties of Gases and Liquids,* Second Edition, New York: McGraw-Hill Book Company, 1966, p. 314.

4. Starling, Kenneth E., *Fluid Thermodynamic Properties for Light Petroleum Systems,* Houston, TX: Gulf Publishing Company, 1973.

5. Martin, Joseph J. and Hou, Yu-Chun, "Development of an Equation of State for Gases," *A.I.Ch.E. Journal,* June 1955, pp. 142–151.

6. Edmister, Wayne C. and McGarry, R. J., "Gas Compressor Design, Isentropic Temperature and Enthalpy Changes," *Chemical Engineering Progress,* Vol. 45, No. 7, July, 1949, pp. 421–434.

7. Edmister, Wayne C. and Lee, Bying Ik, *Applied Hydrocarbon Thermodynamics,* Vol. 1, Second Edition, Houston, TX: Gulf Publishing Company, 1984.

8. Boyce, Meherwan P., *Gas Turbine Engineering Handbook,* Houston, TX: Gulf Publishing Company, 1982.

9. *Compressed Air and Gas Handbook,* Third Edition, New York, NY: Compressed Air and Gas Institute, 1961.

10. Dodge, Russell A. and Thompson, Milton, *Fluid Mechanics,* McGraw-Hill, 1937.

11. Evans, Frank L. Jr., *Equipment Design Handbook for Refineries and Chemical Plants,* Vol. 1, Second Edition, Houston, TX: Gulf Publishing Company, 1979.

12. Gibbs, C. W., Editor, *Compressed Air and Gas Data,* Woodcliff Lake, NJ: Ingersoll-Rand, 1969.

13. *Compressor Handbook for Hydrocarbon Processing Industries,* Houston, TX: Gulf Publishing Company, 1979.

14. Perry, R. H., Editor-in-Chief, *Engineering Manual,* McGraw-Hill Book Co., 1959, pp. C-44, 8–51.

15. Scheel, Lyman F., *Gas Machinery,* Houston, TX: Gulf Publishing Company, 1972.

16. Shepherd, D. G., *Principles of Turbomachinery,* The Macmillan Company, 1969, pp. 100–148.

3

Reciprocating Compressors

Description

The reciprocating compressor is the patriarch of the compressor family. In the process industry, the reciprocating compressor is probably the oldest of the compressors with wide application ranging from consumer to industrial usage. This compressor is manufactured in a broad range of configurations and its pressure range is the broadest in the compressor family extending from vacuum to 40,000 psig. The reciprocating compressor declined in popularity from the late 1950s through the mid 1970s. Higher maintenance cost and lower capacity, when compared to the centrifugal compressor, contributed to this decline. However, recent rises in energy cost and the advent of new specialty process plants have given the more flexible, higher efficiency, though lower capacity, reciprocating compressor a more prominent role in new plant design.

The reciprocating compressor is a positive displacement, intermittent flow machine and operates at a fixed volume in its basic configuration.

48

One method of volume variations is by speed modulation. Another, more common method, is the use of clearance pockets, with or without valve unloading. With clearance pockets, the cylinder performance is modified. With valve unloading, one or more inlet valves are physically open. Capacity may be regulated in a single- or double-acting cylinder with single or multiple cylinder configuration.

A unique feature of the reciprocating compressor is the possibility of multiple services on one compressor frame. On a multistage frame, each cylinder can be used for a separate gas service. For example, one cylinder may be dedicated to propane refrigeration, while the balance of the cylinders may be devoted to product gas.

Lubrication of compressor cylinders can be tailored to the application. The cylinders may be designed for normal hydrocarbon lubricants or can be modified for synthetic lubricants. The cylinder may also be designed for self lubrication, generally referred to as nonlubed. A compromise lubrication method that uses the nonlubed design but requires a small amount of lubricant is referred to as the mini-lube system.

An unusual nonlubed compressor is a labyrinth piston compressor, shown in Figure 3-1. The piston does not touch the sides of the cylinder

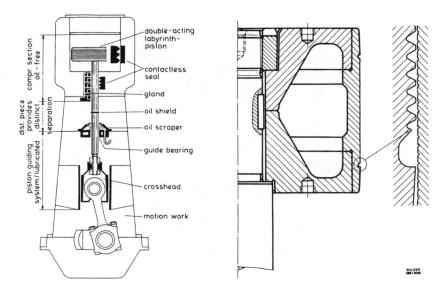

Figure 3-1. Labyrinth piston compressor. This non-lubed piston's circumferential labyrinths operate with a close clearance to the cylinder wall instead of rubbing. (*Courtesy of Sulzer*)

that is equipped with a series of circumferential labyrinths operating with a close clearance to the cylinder wall. Efficiency is sacrificed (due to gas bypass) in order to obtain a low maintenance cylinder. This design is mentioned primarily because it is unique and not widely manufactured.

Another feature necessary to the reciprocating compressor is cylinder cooling. Most process compressors are furnished with water jackets as an integral part of the cylinder. Alternatively, particularly in the smaller size compressors, the cylinder can be designed for air cooling.

Classification

Reciprocating compressors can be classified into several types. One type is the trunk or automotive piston type (see Figure 3-2). The piston is connected to a connecting rod, which is in turn connected directly to the crankshaft. This type of compressor has a single-acting cylinder and is limited to refrigeration service and to smaller air compressors. Most of the smaller packaged refrigeration system compressors are of this type. The compressors may be single or multistage. Approximate capacity is 50 tons in water-chilled refrigeration service and 75 scfm in air service.

Figure 3-2. Trunk-piston type two-stage compressor with fins for air cooling. (*Courtesy of Ingersoll-Rand*)

The more common type of compressor used in process service is the crosshead type, as shown in Figure 3-3. The piston is driven by a fixed piston rod that passes through a stuffing or packing box and is connected to a crosshead. The crosshead, in turn, is connected to the crankshaft by a connecting rod. In this design, the cylinder is isolated from the crankcase by a distance piece. A variable length or double distance piece is used to keep crankcase lubrication from being exposed to the process gas. This design has obvious advantages for hazardous material. The cylinder can be either single- or double-acting. The double-acting construction uses both sides of the piston and compresses on both strokes of the piston during one revolution. Except for very small compressors, most reciprocating compressors furnished to the process industry use the double-acting configuration.

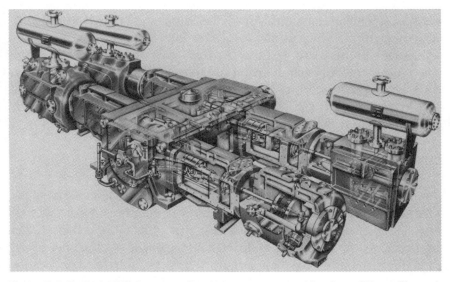

Figure 3-3. Typical multistage crosshead type compressor. (*Courtesy of Nuovo Pignone*)

Arrangement

The trunk type compressor is generally arranged with the cylinder vertical in the basic single-stage arrangement. In the vertical, "in line," multistage configuration, the number of cylinders is normally limited to two. Most multi-cylinder arrangements are in pairs in the form of a V, usually at 45° from the vertical. These compressors usually have up to eight cylinders and are normally used in compressing organic refrigerants.

The few single-acting crosshead compressors are normally single-stage machines with vertical cylinders. The more common double-acting type, when used as a single-stage, commonly has a horizontal cylinder. The double-acting cylinder compressor is built in both the horizontal and the vertical arrangement. There is generally a design trade-off to be made in this group of compressors regarding cylinder orientation. From a ring wear consideration, the more logical orientation is vertical; however, taking into account size and the ensuing physical location as well as maintenance problems, most installations normally favor the horizontal arrangement.

There is wide variation in multistage configuration. The most common is the horizontally opposed. Probably the next most common is the vertical arrangement. Other variations include V, Y, angle or L type. These later arrangements are not too common and are mentioned only to complete possible configurations. Another modification is the tandem-cylinder arrangement, which is almost always horizontal. In this configuration, the cylinders are oriented in line with one another with the innermost cylinder having a piston rod protruding from both ends. This outboard rod in turn drives the next cylinder. While somewhat compact and more competitive in price than the side-by-side arrangements, it is not too popular with maintenance people.

Drive Methods

Another feature of reciprocating compressors that is somewhat unique when compared to the rest of the compressor family is the number of available drive arrangements, which is almost as complex as the cylinder arrangements. In single and multistage arrangement small compressors, particularly the trunk type, are usually V-belt driven by electric motors. The single-acting crosshead type and the small, double-acting, single-stage compressor are also driven in a similar manner. Larger, multistage, trunk type compressors can be sized to operate at common motor speeds and therefore are direct coupled. The larger, crosshead, double-acting, multistage compressors present the most variations in drive arrangements. If it has an integral electric motor sharing a common shaft with the compressor, it is called an engine type. These compressors can also be directly coupled to a separate electric motor in a more conventional manner. Gear units may be involved in the drive train where speed matching is required. Multiple frames are sometimes used with a common crankshaft in a compound arrangement to use a common driver.

Variable frequency motor drives are becoming more popular because of the ability to provide capacity control.

Reciprocating compressors are available with a large variety of other drivers, which include the piston engine, steam turbine, or, in rare cases, a gas turbine. Next in popularity to the electric motor is the piston engine. The arrangement lends itself to skid mounting, particularly with the semi-portable units found in the oilfield. The unit is also popular as a "lease" unit, which may be lifted onto a flat bed trailer and moved from one location to another as needed. The engine is either direct-coupled or, as with smaller compressors, it may be belt-connected.

A variation of the smaller, skid-mounted, engine-driven compressor is a larger, engine-driven version in the form of the integral engine compressor (see Figure 3-4). The compressor and the engine share a common frame and crankshaft. When the engine cylinders are vertical or in a V configuration and the compressor cylinders are horizontal, the machine is called an angle engine compressor.

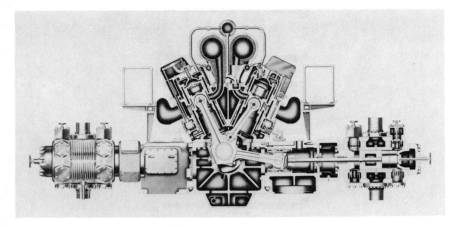

Figure 3-4. Cutaway of a two-stage piston engine driven compressor. (*Courtesy of Dresser-Rand*)

A more rare form of driver is the steam cylinder. Most arrangements combine the steam driver and compressor on the same frame with the steam cylinder opposite the compressor cylinder. Each cylinder's connecting rod is connected to a common throw on the crankshaft. A flywheel is used to provide inertia. For air service, the units are built as single- and two-stage units, with other combinations available for process service.

Performance

Compression Cycle

For the following discussion, refer to Figure 3-5, which shows an ideal indicator diagram followed by a series of cylinder illustrations depicting piston movement and valve position. The figure shows in diagram form one

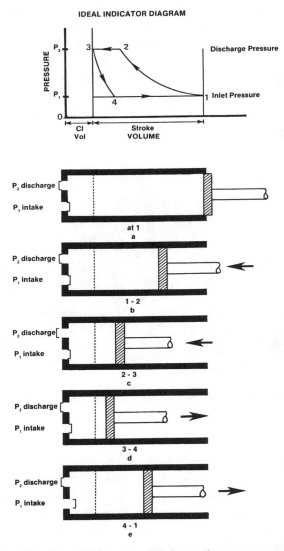

Figure 3-5. Steps in the cycle of reciprocating compressor.

complete crankshaft revolution and encompasses a complete compression cycle. To begin the cycle, refer to the figure at (a) the location where the piston is at the lower end of the stroke (bottom dead center) and is at path point 1 on the indicator diagram. At this point, the cylinder has filled with gas at intake pressure P_1. Note that the valves are both closed. At (b), the piston has started to move to the left. This is the compression portion of the cycle and is illustrated by Path 1-2. When the piston reaches point 2 on the indicator diagram, the exhaust valve starts to open. The discharge portion of the cycle is shown at (c). This is shown on the indicator diagram Path 2-3. Note that the discharge valve is open during this period while the intake valve is closed. The gas is discharged at the discharge line pressure P_2. When the piston reaches point 3, it has traveled to the upper end of its stroke (top dead center). Physically, at this point in the stroke, there is a space between the piston face and the head. This space results in a trapped volume and is called the clearance volume. Next in the cycle, the piston reverses direction and starts the expansion portion of the cycle, as illustrated at (d) in the figure. Path 3-4 shows this portion of the cycle. Here the gas trapped in the clearance volume is re-expanded to the intake pressure. Note that the discharge valve has closed, and the intake valve is still closed. At point 4, the expansion is complete and the intake valve opens. The intake portion of the cycle is shown at (e). This is indicated by Path 4-1 on the indicator diagram. The cylinder fills with gas at intake line pressure P_1. When the piston reaches point 1, the cycle is complete and starts to repeat.

Cylinder Displacement

The calculation of the cylinder displacement is a straightforward geometric procedure. It is the product of three factors, namely, the piston area minus rod area (when appropriate), the stroke, and the number of strokes in a given time. There are four options, which can be covered by three equations.

For a single-acting cylinder compressing at the outer end of the cylinder,

$$Pd = S_t \times N \times \frac{\pi D^2}{4} \tag{3.1}$$

where

Pd = piston displacement
S_t = stroke
N = speed of the compressor
D = cylinder diameter

For a double-acting cylinder without a tail rod,

$$Pd = S_t \times N \times \frac{\pi(2D^2 - d^2)}{4} \qquad (3.2)$$

where

d = piston rod diameter

For a double-acting cylinder with a tail rod,

$$Pd = S_t \times N \times \frac{2\pi(D^2 - d^2)}{4} \qquad (3.3)$$

For the application requiring a single-acting cylinder compressing on the frame end only, use Equation 3.3 deleting the 2 in the expression.

Volumetric Efficiency

To determine the actual inlet capacity of a cylinder, the calculated displacement must be modified. There are two reasons why modification is needed. The first is because of the clearance at the end of the piston travel.

Earlier in the chapter, when the compression cycle was described, a portion of the indicator, Path 3-4, was referred to as the expansion portion of the cycle. The gas trapped in the clearance area expands and partly refills the cylinder taking away some of the capacity. The following equation reflects the expansion effect on capacity and is referred to as the theoretical volumetric efficiency E_{vt}.

$$E_{vt} = 1.00 - [(1/f)r_p^{1/k} - 1]c \qquad (3.4)$$

where

f = ratio of discharge compressibility to inlet compressibility as calculated by Equation 3.6
r_p = pressure ratio
c = percent clearance
k = isentropic exponent

The limit of the theoretical value can be demonstrated by substituting zero for the clearance c, which results in a volumetric efficiency multiplier of 1.0.

The second reason for modification of the displaced volume is that in real world application, the cylinder will not achieve the volumetric performance predicted by Equation 3.4. It is modified, therefore, to include empirical data. The equation used here is the one recommended by the Compressed Air and Gas Institute [1], but it is somewhat arbitrary as there is no universal equation. Practically speaking, however, there is enough flexibility in guidelines for the equation to produce reasonable results. The 1.00 in the theoretical equation is replaced with .97 to reflect that even with zero clearance the cylinder will not fill perfectly. Term L is added at the end to allow for gas slippage past the piston rings in the various types of construction. If, in the course of making an estimate, a specific value is desired, use .03 for lubricated compressors and .07 for nonlubricated machines. These are approximations, and the exact value may vary by as much as an additional .02 to .03.

$$E_v = .97 - [(1/f)r_p^{1/k} - 1]c - L \tag{3.5}$$

$$f = Z_2/Z_1 \tag{3.6}$$

The inlet capacity of the cylinder is calculated by

$$Q_1 = E_v \times Pd \tag{3.7}$$

Piston Speed

Another value to be determined is piston speed, PS. The average piston speed may be calculated by

$$PS = 2 \times S_t \times N \tag{3.8}$$

The basis for evaluation of piston speed varies throughout industry. This indicates that the subject is spiced with as much emotion as technical basics. An attempt to sort out the fundamentals will be made. First, because there are so many configurations and forms of the reciprocating compressor, it would appear logical that there is no one piston speed limit that will apply across the board to all machines. The manufacturer is at odds with the user because he would like to keep the speed up to keep the size of the compressor down, while the user would like to keep the speed down for reliability purposes. As is true for so many other cases, the referee is the economics. An obvious reason to limit the speed is maintenance expense. The lower the piston speed, the lower the maintenance and the higher the reliability. The relationship given by Equation 3.1 defines the size of the cylinder. Therefore, if the speed is reduced to lower the piston speed, then the diameter of the cylinder must increase to compensate for the lost displacement to maintain the desired capacity. As cylinder size goes up, so does the cost of the cylinder. It is not difficult to see why the user and manufacturer are at somewhat of a cross purpose. If the user's service requires a high degree of reliability and he wants to keep cylinder and ring wear down, he must be aware of the increase in cost.

To complicate the subject of piston speed, look at Equations 3.1 and 3.8. Note the term S_t (stroke). The piston speed can be controlled by a shorter stroke, but because of loss of displacement, the diameter and/or the speed must be increased. If only speed is increased, the whole exercise is academic as the piston speed will be back up to the original value. If, however, diameter alone or both diameter and speed are increased, the net result can be a lower piston speed. Another factor comes to bear at this point concerning valve life, that decreases with the increase in the number of strokes and can negate the apparent gain in maintenance cost if not adequate. It would appear that the engineer trying to evaluate a compressor bid just can't win. The various points are not tendered just to frustrate the user but rather are given to help show that this is another area that must have a complete evaluation. All facets of a problem must be considered before an intelligent evaluation can be made.

After all the previous statements, it would seem very difficult to select a piston speed. For someone without direct experience, the following guidelines can be used as a starting point. Actual gas compressing experience should be solicited when a new compressor for the same gas is being considered. These values will apply to the industrial process type of compressor with a double-acting cylinder construction. For horizontal compressors with lubricated cylinders, use 700 feet per minute (fpm) and for nonlubricated cylinders use 600 fpm. For vertical compressors with lubricated cylinders, use 800 fpm and for nonlubricated cylinders use 700 fpm.

Another factor to consider is the compressor rotative speed relative to valve wear. The lower the speed, the fewer the valve cycles, which contribute to longer valve life. A desirable speed range is 300 to 600 rpm.

Discharge Temperature

While head is normally not a particularly significant value in the selection of the reciprocating compressor, it is used for comparison with other types of compressors. Equation 2.66, the equation for adiabatic head, is recalled as

$$H_a = RT_1 \frac{k}{k-1} (r_p^{\frac{k-1}{k}} - 1) \tag{2.66}$$

The *discharge temperature* can be calculated by rewriting Equation 2.65.

$$T_2 = T_1 \left(r_p^{\frac{k-1}{k}} \right) \tag{3.9}$$

where

T_1 = absolute inlet temperature
T_2 = absolute discharge temperature

Why use an adiabatic relationship with a compressor whose cylinder is almost always cooled? An assumption made in Chapter 2 on adiabatic isentropic relationships was that heat transfer was zero. In practical applications, however, the cooling generally offsets the effect of efficiency. As a side note, cylinder cooling is as much cylinder stabilization for the various load points as it is heat removal.

Power

The work-per-stage can be calculated by multiplying the adiabatic head by the weight flow per stage.

$$\text{Work} = H_a \times w \tag{3.10}$$

then,

$$W_{cyl} = wRT_1 \frac{k}{k-1} \left(r_p^{\frac{k-1}{k}} - 1 \right) \tag{3.11}$$

Substituting $P_1 Q_1$ for wRT_1 from Equation 2.7,

$$\text{Work} = P_1 Q_1 \frac{k}{k-1} \left(r_p^{\frac{k-1}{k}} - 1 \right) \tag{3.12}$$

For two stages, the above equation can be expanded to add the interstage conditions for the second stage. Note the subscript i is added to the second set of terms to reflect the second-stage inlet.

$$\text{Work} = P_1 Q_1 \frac{k}{k-1}\left(r_p^{\frac{k-1}{k}} - 1\right) + P_i Q_i \frac{k}{k-1}\left(r_{pi}^{\frac{k-1}{k}} - 1\right) \tag{3.13}$$

For a first trial at sizing or for estimates, the Equation 3.13 can be differentiated and solved for P_i, with the result,

$$P_i = \sqrt{P_1 \times P_2} \tag{3.14}$$

This expression can be changed to

$$\frac{P_i}{P_1} = \frac{P_2}{P_i} \tag{3.15}$$

Substituting the term r for the pressure ratio, the following results

$$r_{pl} = r_{pi} \tag{3.16}$$

Equation 3.16 can be generalized for optimum work division by dividing the pressure ratio into a set of balanced values,

$$r_{p\text{-stage}} = (r_{p\text{-overall}})^{1/n\text{-stage}} \tag{3.17}$$

The values for pressure ratio in a practical case must include allowance for pressure drop in the interstage piping. In the sizing procedure used by manufacturers, certain adjustments must be made to the ideal for incremental cylinder sizes and allowable rod loading. Efficiency is represented by η_{cyl}.

$$W_{cyl} = \frac{P_1 Q_1}{\eta_{cyl}} \frac{k}{k-1} (r_p^{\frac{k-1}{k}} - 1) \tag{3.18}$$

To assist the engineer in making estimates, the curve in Figure 3-6 gives values of efficiency plotted against pressure ratios. The values on the curve include a 95% mechanical efficiency and a valve velocity of 3,000 feet per minute. Table 3-1 and Table 3-2 are included to permit a correction to be made to the compressor horsepower for specific gravity and low inlet pressure. They are included to help illustrate the influence of these factors to the power required. The application of these factors to

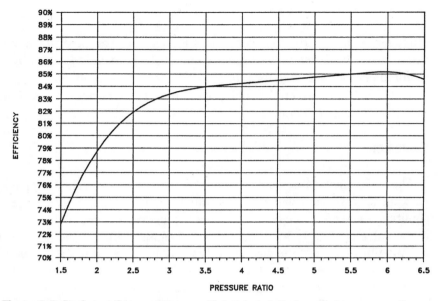

Figure 3-6. Reciprocating compressor efficiencies plotted against pressure ratio with a valve velocity of 3,000 fpm and a mechanical efficiency of 95%.

Table 3-1
Efficiency Multiplier for Specific Gravity

	SG				
r_p	1.5	1.3	1.0	0.8	0.6
2.0	0.99	1.0	1.0	1.0	1.01
1.75	0.97	0.99	1.0	1.01	1.02
1.5	0.94	0.97	1.0	1.02	1.04

Source: Modified courtesy of the Gas Processors Suppliers Association.

Table 3-2
Efficiency Multiplier for Low Pressure

	Pressure Psia							
r_p	10	14.7	20	40	60	80	100	150
3.0	.990	1.00	1.00	1.00	1.00	1.00	1.00	1.00
2.5	.980	.985	.990	.995	1.00	1.00	1.00	1.00
2.0	.960	.965	.970	.980	.990	1.00	1.00	1.00
1.5	.890	.900	.920	.940	.960	.980	.990	1.00

Source: Modified courtesy of the Gas Processors Association and Ingersoll-Rand.

efficiency value is arbitrary. While it is recognized that the efficiency is not necessarily the element affected, the desire is to modify the power required per the criteria in the tables.

The efficiency correction accomplishes this. These corrections become more significant at the lower pressure ratios.

Valve Loss

The efficiency values are affected by several losses: ring slippage, packing leakage, and valve losses. Valve losses are generally the most significant and are made of several components such as channel loss, loss in the valve opening, and leakage. Also, because of inertia and imperfect damping properties of the gas, the valve may have transient losses due to *bounce.* The manufacturer, therefore, modifies the valve lift to suit the gas specified. For example, an air compressor might be furnished with a lift of .100 inch. The same compressor being furnished for a low molecular service such as a hydrogen-rich gas, might use a lift of .032 inches. The problem with the higher lift is that hydrogen lacks the damping properties of air and, as a result, the valve would experience excessive bounce. The effect on the compressor would be loss in efficiency and higher valve maintenance.

The valve porting influences volumetric efficiency by contributing to the minimum clearance volume. If the porting must be enlarged to reduce the flow loss, it is done at the expense of minimum clearance volume.

This is just one example of the many compromises the engineer is faced with while designing the compressor. The subject of valve design is involved and complex. For individuals wishing to obtain more information on the subject, references discussing additional aspects of valves are included at the end of the chapter [2, 3, 4, 5, 6].

To calculate the valve velocity for evaluation purposes, use the following equation. This equation is based on the equation given in API 618.

$$v = 144 \, \frac{Pd}{A} \tag{3.19}$$

where

   v = average gas velocity, fpm
   Pd = piston displacement per cylinder, ft^3/min
   A = total inlet or discharge valve area per cylinder, in.^{2}

To calculate Pd, use Equation 3.1 for single-acting cylinders, Equation 3.2 for double-acting cylinders without a tail rod, and Equation 3.3 for double-acting cylinders with a tail rod.

The area, A, is the product of actual lift and the valve opening periphery and is the total for all inlet or discharge valves in a cylinder. The lift is a compressor vendor-furnished number.

Example 3-1

Calculate the suction capacity, horsepower, discharge temperature, and piston speed for the following single-stage double acting compressor.

Bore: 6 inches
Stroke: 12 inches
Speed: 300 rpm
Rod diameter: 2½ inches
Clearance: 12%
Gas: CO_2
Inlet pPressure: 1,720 psia
Discharge pressure: 3,440 psia
Inlet temperature: 115°F

Calculate the piston displacement using Equation 3.2 and dividing by 1,728 in.^{3} per ft^3 to convert the output to cfm.

$$Pd = \frac{12 \times 300 \times \pi \, [2(6)^2 - (2.5)]}{1,728 \times 4}$$

$$Pd = 107.6 \text{ cfm}$$

Step 1. Calculate volumetric efficiency using Equations 3.5 and 3.6. To complete the calculation for volumetric efficiency, the compressibilities are needed to evaluate the f term of Equation 3.6. Using Equations 2.11 and 2.12 for the inlet conditions,

$$T_1 = 460 + 115$$

$$T_1 = 575°R$$

$$T_r = \frac{575}{548}$$

$$T_r = 1.05$$

$$P_r = \frac{1,720}{1,073}$$

$$P_r = 1.6$$

From the generalized compressibility charts (see Appendix B),

$$Z_1 = .312$$

Step 2. At this point the discharge temperature must be calculated to arrive at a value for the discharge compressibility.

$$r_p = \frac{3,440}{1,720}$$

$$r_p = 2.0$$

$$T_2 = 575 \ [2^{(1.3-1)/1.3}]$$

$$T_2 = 674 \ 7°R$$

$$t_2 = 674.7 - 460$$

$$t_2 = 214.7°F \text{ discharge temperature}$$

Calculate the discharge compressibility:

$$T_r = \frac{674.7}{548}$$

$$T_r = 1.23$$

$$P_r = \frac{3,440}{1,073}$$

$$P_r = 3.21$$

From the generalized compressibility charts,

$$Z_2 = 0.575$$

From Equation 3.6, calculate f:

$$f = \frac{.575}{.312}$$

$$f = 1.842$$

Calculate the volumetric efficiency using Equation 3.5. Use .05 for L because of the high differential pressure:

$$E_v = .97 - [(1/1.842)(2)^{1/1.3} - 1].12 - 0.5$$

$$E_v = .93 \text{ volumetric efficiency}$$

Now calculate suction capacity using Equation 3.7:

$$Q_1 = .93 \times 107.6$$

$$Q_1 = 100.1 \text{ cfm suction capacity}$$

Step 3. Piston speed is calculated using Equation 3.8 converting the stroke to feet by dividing the equation by 12 inches per foot:

$$PS = \frac{2 \times 12 \times 300}{12}$$

$$PS = 600 \text{ fpm piston speed}$$

Step 4. Calculate the power required. Refer to Figure 3-6 and select the efficiency at a pressure ratio of 2.0. The value from the curve is 79%. Equation 3.18 is used to calculate power. The constants 144 in^2ft^2 and 33,000 ft-lbs/min/hp have been used to correct the equation for the unit from the example.

$$W_{cyl} = \frac{144 \times 1,720 \times 100.1}{33,000 \times .79} \times \frac{1.3}{.3} (2^{.3/1.3} - 1)$$

$$W_{cyl} = 714.8 \text{ hp cylinder horsepower}$$

Application Notes

There are several items regarding the application of reciprocating compressors that must be considered. These items are minor, but if neglected may cause a great deal of concern when the inevitable problem occurs.

Reciprocating compressors are not fond of liquids of any sort, particularly when delivered with the inlet gas stream. For any application, a good-sized suction drum with a drain provision is in order. It may be a part of the pulsation control if properly done. The pulsation control will be covered in more detail later in the chapter. If the stream is near saturation or has a component near saturation, consideration should be given to using a horizontally oriented cylinder configuration, with the discharge nozzle on the bottom side of the cylinder. While on the subject of condensation, for the same gas near saturation, cylinder cooling must be monitored and controlled. It would not do to let the gas condense inside the cylinder after all the care has been taken not to let it condense outside the cylinder. A rule of thumb is to keep the cooling water temperature 10°F above the gas inlet temperature.

It would appear obvious for startup, and in some cases full-time operation, that a suction strainer or filter is mandatory. The reason for the strainer is to keep junk and pipe scale out of the compressor. Fines from pipe scale and rust will make short work of the internal bore of a cylinder and are not all that good for the balance of the components. In some severe cases, cylinders have been badly damaged in a matter of a few weeks. The strainer should be removable in service for cleaning, particularly when it is intended for permanent installation. Under all circumstances, provision must be made to monitor the condition of the strainer. Much frustration has been expended because a compressor overheated or lost capacity and no one knew if the strainer had fouled or blinded.

The discharge temperature should be limited to 300°F as recommended by API 618. Higher temperatures cause problems with lubricant coking and valve deterioration. In nonlube service, the ring material is also a factor in setting the temperature limit. While 300°F doesn't seem all that hot, it should be remembered that this is an average outlet temperature, whereas the cylinder will have "hot" spots exceeding this temperature.

Finally, planning may save money and time if process changes are foreseeable. For instance, capacity increase, or an increase in molecular weight due to a catalyst change, results in decreased volumetric flow. Although the cylinders must be sized for economical operation at the present rate, the frame can be sized for future applications. When the future conditions become a reality, the cylinders can be changed while keeping the same frame. This saves the investment cost and delivery time of a complete new compressor without the penalty of oversizing and its inherent inefficient operation.

Mechanics

Cylinders

Cylinders for compressors used in the process industries are separable from the frame. They are attached to the frame by way of an intermediate part known as the distance piece and can be seen in Figure 3-3. Piloting is provided to maintain alignment of all moving elements. A requirement of API 618 is for the cylinders to be equipped with replaceable liners. The purpose of the liner is to provide a renewable surface to the wearing portion of the cylinder. This saves the cost of replacing a complete cylinder once the bore has been worn or scored. In the larger, more complex compressors, this feature is standard or readily available as an option. On the smaller frames, particularly the single-stage models, the smaller cylinder size is such that the replaceable liner is not economical and may not be available.

All cylinders are equipped for cooling, usually by means of a water jacket. Those not having a water jacket are finned to provide air cooling. The latter method is limited to either small or special purpose machines.

The most common material used in cylinder construction is cast iron for the larger, low-pressure cylinders and steel for the smaller, high pressure cylinders. In some cases, nodular or ductile iron can be used in lieu of cast iron. For hydrocarbon service, steel is most desirable, although not universally available.

Larger cylinders normally have enough space for clearance pockets. An additional location is the head casting on the outboard end of the cylinder. Figure 3-7 is an illustration of a cylinder with an unloading pocket in the head. On smaller cylinders, this feature must be provided external to the cylinder.

Pistons and Rods

The lowly piston, one of the more simple items, has one of the most important functions of the entire compressor. The piston must translate the energy from the crankshaft to the gas in the cylinder. The piston is

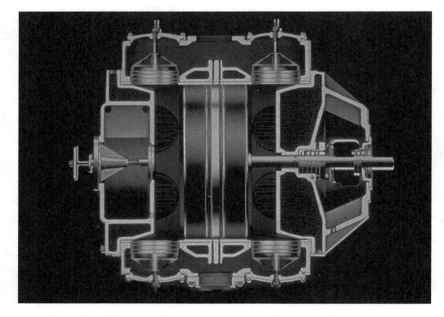

Figure 3-7. Cylinder with clearance pocket. (*Courtesy of Dresser-Rand*)

equipped with a set of sliding seals referred to as piston rings. Rings are made of a material that must be reasonably compliant for sealing, yet must slide along the cylinder wall with minimum wear. Different rings are used for lubricated or nonlubricated service, with the rings in the nonlubed cylinders needing good dry lubricating qualities. For lubricated service, metallic rings such as cast iron or bronze as well as nonmetallic materials such as filled nylon are used. The nonmetallic materials are becoming more common. For nonlubricated service, the ring material is nonmetallic, ranging from carbon to an assortment of fluorocarbon compounds. Horizontal cylinder pistons feature the addition of a wear band, sometimes referred to as a rider ring (see Figure 3-8).

Pistons may be of segmented construction to permit the use of one-piece wear bands. One-piece wear bands are a requirement in API 618. Pistons have a problem in common with humans—a weight problem. Weight in a piston contributes directly to the compressor shaking forces and must be controlled. For this reason, aluminum pistons are often found in larger low pressure cylinders. Hollow pistons are used but can pose a hazard to maintenance personnel if not properly vented. If trapped, the gas will be released in an unpredictable and dangerous manner when the piston is dismantled.

Figure 3-8. Piston rings and wear band. (*Courtesy of Nuovo Pignone*)

The piston rod is threaded to the piston and transmits the reciprocating motion from the crosshead to the piston. The piston rod is normally constructed of alloy steel and must have a hardened and polished surface, particularly where it passes through the cylinder packing (double-acting cylinders). Rod loading must be kept within the limits set by the compressor vendor because overloading can cause excess runout of the rod resulting in premature packing wear. This in turn leads to leakage, reduced efficiency, and increased maintenance expense.

In unloaded or part-load operation, rod reversals must be of sufficient magnitude to provide lubrication to the crosshead bearings. The bearings are lubricated by the pumping action of the opening and closing of the bearing clearance area.

Tail rods are dummy rods that protrude from the head end of the cylinder (see Figure 3-9). The purpose of the rod is to pressure-balance a piston or to stabilize a particular piston design. Because of the personnel hazard, a guard must be specified and provided. In a tandem cylinder arrangement, the outboard cylinders are driven with a rod similar to the tail rod.

Valves

The compressor cylinder valves are of the spring-loaded, gas-actuated type in all but a limited number of portable compressors. This kind of

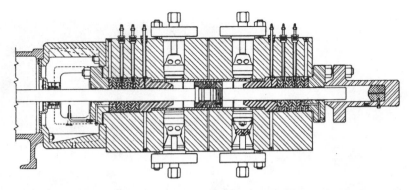

Figure 3-9. Diagram of cylinder with piston tail rod. (*Courtesy of Dresser-Rand*)

valve is used in contrast to the cam-actuated poppet type normally found in piston engines. Reciprocating compressors generally use one of four basic valve configurations:

- rectangular element
- concentric ring
- ported plate
- poppet

The rectangular element valve, as the name implies, uses rectangular-shaped sealing elements. These valves are the feather valve, channel valve, and the reed valve. These valves are applied to the industrial air machines for the most part. A channel valve is shown in Figure 3-10.

The concentric ring valve uses one or more relatively narrow rings arranged concentrically about the centerline of the valve (see Figure 3-11). These valves have the advantage of a low stress level due to the lack of stress concentration points. The disadvantage is that it is difficult to maintain uniform flow control with the independent rings. These valves work well with plug type unloaders. Space for the unloader is obtained by eliminating one or more of the innermost rings.

The ported plate valves, as shown in Figure 3-12, are similar to the concentric ring valve except that the rings are joined into a single element. The advantage is that the valve has a single element making flow control somewhat easier. Because of the single element, the number of edges available for impact is reduced. The valve may be mechanically damped, as this design permits the use of damping plates. It has the disadvantage that because of the geometry used, the stress is higher due to the potential of higher stress concentrations. This valve element is probably one of the most commonly used in process reciprocating compressors.

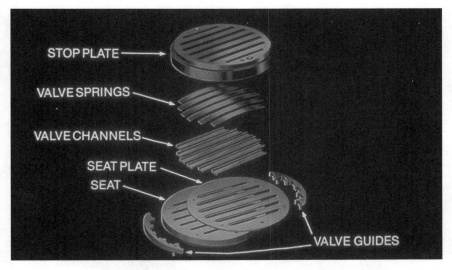

Figure 3-10. An exploded view of a cushioned channel valve. (*Courtesy of Dresser-Rand*)

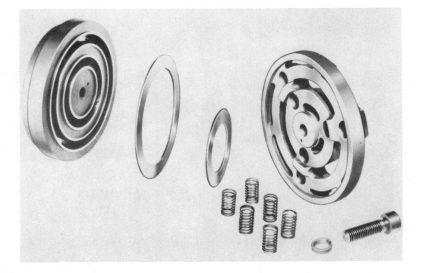

Figure 3-11. Exploded view of a concentric plate valve. (*Courtesy of Dresser-Rand*)

The poppet valve (see Figure 3-13) consists of multiple, same-size ports and sealing elements. The advantage of the valve is that has a high flow efficiency due to the high lift used and the streamlined shape of the sealing element. The disadvantage is that the valve is not tolerant of

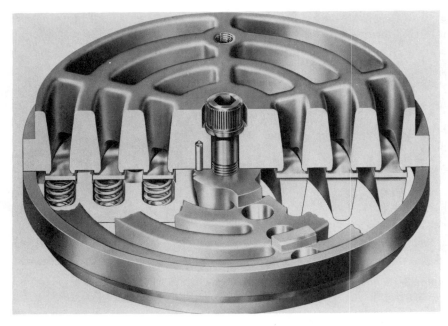

Figure 3-12. Cutaway of a ported plate valve. (*Courtesy of Dresser-Rand*)

Figure 3-13. Cutaway of a poppet valve. (*Courtesy of Dresser-Rand*)

uneven flow distribution. The valve is most commonly used in gas transmission service and in low speed, low-to-medium compression ratio compressors. There appears to be an increase in the use of poppet valves in hydrocarbon process service because of the ease of maintenance.

Valve materials must be selected for durable, long-term operation and must also be compatible with the gas being handled. The use of polymer nonmetallic sealing elements is quite common. The valves are symmetrically placed around the outer circumference of the cylinder and can normally be removed and serviced from outside the cylinder without dismantling any other portion. A good design will have the valve and associated parts so arranged that an assembly cannot be installed backwards. The inlet and discharge valves should not be physically interchangeable and should be so constructed as to keep the valve assembly or its parts from entering the cylinder should they become unbolted or break.

Distance Piece

The distance piece is a separable housing that connects the cylinder to the frame. The distance piece may be open or closed and may have multiple compartments. It may be furnished as single, double, or extra long. The purpose of a longer distance piece is to isolate that part of the rod entering the crankcase and receiving lubrication from the part entering the cylinder and contacting the gas. This prevents lubricant from entering the cylinder and contaminating the gas, particularly necessary in nonlubricated cylinders. It can also keep a synthetic lubricant in a cylinder from being corrupted by the crankcase lubricant.

Compartments in the distance piece collect and control packing leakage when the gas is toxic or flammable. Today, the toxic category covers many of the gases that were allowed to freely escape into the atmosphere not many years ago. With the pollution laws becoming more stringent, leakage control takes on a much greater significance. The leakage can be directed to a flare or other disposal point and, as with multiple compartments, a buffer of inert gas can be used together with the collection compartment to further prevent gas leakage.

Rod Packing

A packing is required on double-acting cylinders to provide a barrier to leakage past the rod where it passes through the crank end cylinder closure. The same arrangement is needed at the head end if a tail rod or tandem cylinder is used. The packing may consist of a number of rings of packing material and may include a lantern ring (see Figure 3-14). The lantern ring provides a space into which a gas or liquid may be injected to aid in the sealing process. If cooling of the packing is required, the packing box may be jacketed for liquid coolant.

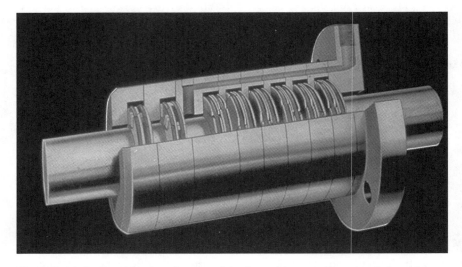

Figure 3-14. Rod packing box. Lantern rings in packing provide space into which a buffer may be injected to aid in sealing. (*Courtesy of Dresser-Rand*)

Crankshaft and Bearings

Larger compressors, normally above 150 to 200 horsepower, have forged steel crankshafts. Cast crankshafts are used in medium-size machines. Crankshafts should have removable balance weights to compensate for rotary unbalance as well as reciprocating unbalance. The crankshaft should be dynamically balanced when above 800 rpm.

When pressure lubrication is used, the crankshaft oil passages should be drilled rather than cored in the cast construction. Figure 3-15 shows a drilled crankshaft. On machines above 150 horsepower, the main and connecting rod bearings should be split-sleeve, steel-backed, babbitted-insert type. Figure 3-16 shows a connecting rod. The main bearings of smaller compressors are the rolling element type. Crosshead pins should

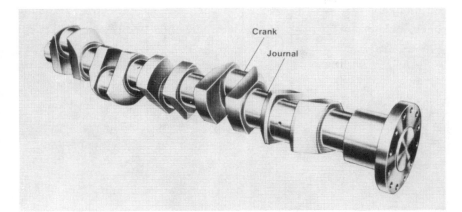

Figure 3-15. A five throw crankshaft with drilled oil passages. (*Courtesy of Dresser-Rand*)

Figure 3-16. A large connecting rod.

have replaceable bushings if available. See Figure 3-17 for some typical crossheads. Figure 3-18 shows split sleeve main bearing caps. Replaceable bushings are standard on larger, multistage compressors and option-

Figure 3-17. Crossheads.

Figure 3-18. Split sleeve bearing caps.

al as the size decreases. On the smaller, standardized single-stage machines, they are not available at all. On large multistage compressors, flywheels are sometimes used to dampen torque pulsations, minimize transient torque absorbed by the driver, and to tune torsional natural frequencies. In most applications however, flywheels are not used and the driver inertia must absorb torque pulses.

Frame Lubrication

Frame lubrication is integral on most reciprocating compressors. The small, horizontal, single-stage compressors, particularly 100 horsepower and smaller, use the splash lubrication system. This system distributes lubricating oil by the splashing of the crankthrow moving through the lubricant surface in the sump. Dippers may be attached to the crankshaft to increase this effect.

The pressurized lubrication system is a more elaborate lubrication method (see Figure 3-19). The system has a main oil pump, either crankshaft or separately driven, a pump suction strainer, a cooler when needed, a

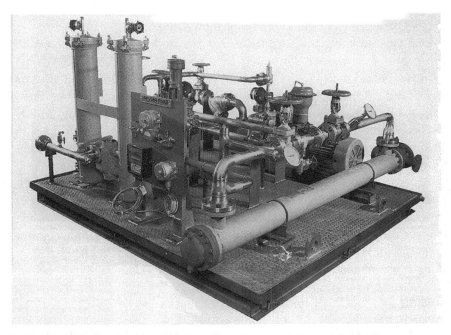

Figure 3-19. Pressurized lubrication system for a multistage reciprocating compressor. (*Courtesy of Dresser-Rand*)

full-flow oil filter and safety instrumentation. Options that should be considered when purchasing a new compressor are an auxiliary oil pump, which can also be used for startup, and dual oil filters with a non-shutoff type of transfer valve. Safety instrumentation should include a crankcase low-oil-level switch, a low-oil-pressure switch, and a high-oil-temperature switch. The switches can be duplicated and set for different operating points to provide an alarm, or early warning signal, and a shutdown signal. In this system, as in the splash system, the crank case acts as the oil sump.

Cylinder and Packing Lubrication

Lubricated cylinders use a separate mechanical lubricator to force feed, in metered droplet form, a very precise amount of lubricant to specified points. This minimizes the amount of lubricant in the cylinder and allows a lubricant most compatible with the gas to be selected without compromising the frame lubrication system. Lubricant is fed to a point or points on the cylinder to service the piston rings and the packing when required. In a few cases, as in air compressors, the packing is lubricated from the crankcase. On some applications involving wet CO_2 or H_2S in the gas stream, special materials may be avoided if one of the lubrication points is connected to the suction pulsation dampener.

One type of mechanical lubricator is the multiplunger pump, which has a plunger dedicated to each feed point (see Figure 3-20). This arrangement normally includes a sight glass per feed point. Another type of mechanical lubricator is the single metering pump sized for total flow, with a divider block arrangement to separate the lubricants going to different feed points (see Figure 3-21). While there are pros and cons to each system, the compressor vendor will normally recommend a system for any given application.

Because of the small amount of lubricant dispensed, the divider block system must be employed with the mini-lube method of lubrication previously discussed. The special divider block is usually connected to one plunger on a multiplunger pump, taking advantage of the smaller output to do the initial flow reduction. The balance of the pump's plungers may be used where a more conventional quantity of lubricant can be used.

Cooling

Three methods of cooling are in common use, the pressurized cooling fluid system, the thermosyphon, and the static system. The *static* system is used on smaller compressors and is probably the least common. Cool-

Figure 3-20. A multiplunger lubrication pump.

Figure 3-21. A single lubrication pump with a divider block.

ing fluid is used as a static heat sink and can be thought of more as a heat stabilizer than a cooling system. There is some heat transferred from the system by normal conduction to the atmosphere.

The *thermosyphon* is a good system for remote areas where utilities are limited, but requires some careful design to ensure proper operation. This is a circulating system with the motive force derived from the change in density of the cooling fluid from the hot to the cold sections of the system. API 618 permits this system for discharge gas temperatures below 210°F or a temperature rise across the compressor of 150°F or less.

The most common system is the *pressurized cooling fluid* system. In a plant or refinery environment where cooling tower water is available, this system has the highest heat removal capability. In locations where cooling water is not available, a self-contained, closed-cooling fluid may be used. The system consists of a circulating pump, a surge tank, and a fan-cooled radiator or air-to-liquid heat exchanger. The radiator may have multiple sections, one for frame oil cooling and another for inter- or after-cooling. The cooling fluid is either water or an ethylene glycol and water mixture. Allowance must be made in the design to accommodate the inherently higher temperature coming from the air-cooled radiator and also ambient temperature variations.

On all systems using water, the obvious is overlooked all too often. A method of draining the equipment during periods when the equipment is idle and freezing temperatures are a possibility should be provided. The consequences of failing to provide this feature are obvious.

Capacity Control

The reciprocating compressor is a fixed displacement compressor in its basic configuration; however, several methods are used to overcome this limitation to permit running at multiple operating points. In the discussion on cylinders, mention was made of clearance pockets. By use of the clearance pockets, the cylinder capacity can be lowered (see Equation 3.5). If the pocket is connected directly to the clearance area, the clearance term c can be increased. Increasing the clearance reduces the capacity by lowering the volumetric efficiency. Control of the pocket addition is by either a manual valve or by a remotely operated valve. If multiple pockets are used, a step unloading system can be designed (see Figure 3-22). The variable volume clearance pocket can provide an alternate unloading method. This device is normally attached to the outboard head. It consists of a piston-cylinder arrangement where the piston rod is threaded and

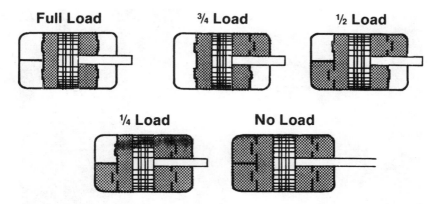

Figure 3-22. A 5-step clearance pocket unloading scheme. (*Courtesy of Dresser-Rand*)

attached to a handwheel. Turning the handwheel changes the clearance volume in an infinite number of steps up to the total pocket volume.

On cylinders lacking the physical space for pockets, the same effect can be achieved by using external bottles and some piping. Care must be taken to keep the piping close-coupled and physically strong enough to prevent accidental breakage. Remotely operated valves permit the capacity reduction to be integrated into an automatic control system.

An additional capacity control method is the *unloader.* This method can be used in conjunction with clearance pockets to extend the range of control to zero capacity. On double-acting cylinders, unloading the individual sides one at a time will provide a two-step unloading of the cylinder. On multicylinder arrangements, the cylinders can be unloaded one at a time providing as many steps as cylinders operating in parallel. The unloaders can also be used to totally unload the compressor, as is necessary for electric motor driver startup.

Three types of unloaders will be described, the plug type, the port type, and the plunger type. The plug type, shown in Figure 3-23, is normally used on all inlet valves for the unloaded end. The center of the valve is used for the unloader plug and port. The port type, shown in Figure 3-24, is used to replace one of the inlet valves on multiple inlet valve cylinders. It is normally used with low molecular weight applications. This unloader consists of a plug and port using the entire space of the valve it replaces. The plunger type, shown in Figure 3-25, is used on heavier gas applications where the maximum unloaded flow area is needed. The unloader operates by using the plunger fingers to hold the valve

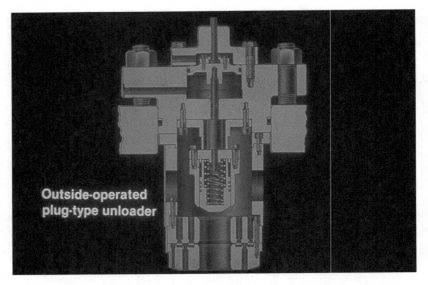

Figure 3-23. Plug type unloader. (*Courtesy of Dresser-Rand*)

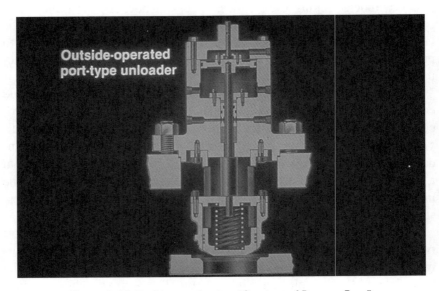

Figure 3-24. Port type unloader. (*Courtesy of Dresser-Rand*)

plates open. Control of all the described unloaders is the same, in that a piston operator is used. Additional control may be obtained by using a cooled bypass line from the discharge to the compressor suction. The bypass is normally used with discrete unloading steps.

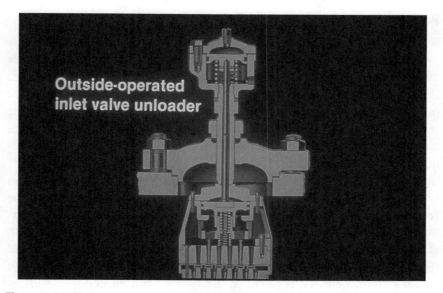

Figure 3-25. Plunger type unloader. Note the plunger finger used to hold the valve open when energized. (*Courtesy of Dresser-Rand*)

A few words of caution when using the valve unloading method: A problem arises with the possible loss of rod load reversals. Rod reversals are needed to provide lubrication to some of the bearings, as discussed earlier in this chapter. While the reversal problem is generally associated with unloading a double-acting cylinder from one side, it should be checked for all unloaded cases, including pocket unloading. If operation without rod reversals is absolutely mandatory, auxiliary lubrication must be brought to the bearings affected. The second caution is the anticipated duration of a totally unloaded condition. While the capacity has been reduced to zero, the gas in the outer end of the cylinder is being moved about in a reciprocating manner following piston movement. The movement of uncompressed gas will generate heat, and prolonged unloaded operation without proper cooling may cause severe overheating. In any case, investigation of potential problems should be undertaken with the equipment manufacturer.

From the foregoing discussion, it should be clear that cylinder capacity can be controlled. While the automatic control is normally limited to certain finite steps, the steps can be selected in size or number to minimize any adverse effect especially in conjunction with prudent use of the variable volume pocket.

Pulsation Control

The intermittent personality of the reciprocating compressor becomes evident when the subject of pulsations is broached. Because discharge flow is interrupted while the piston is on the suction stroke, pressure pulses are superimposed on the discharge system's mean pressure. At the suction side of the system, the same type of interruption is going on, causing the suction pressure to take on a non-steady component. The frequency of the pulses is constant when the speed is constant, which is the most normal condition. The pulses are literally that, not sinusoidal in characteristic; therefore, if the frequency spectrum is analyzed, it will be found to contain the fundamental frequency and a rich content of harmonics. When a forcing phenomenon is superimposed on a system with elastic and inertial properties (a second order system), a resonant response is likely to occur. This is particularly true when the band of exciting frequencies is as broad as the type of system under consideration. The gas system meets the criteria of the second order system, as gas is compressible (elastic) and has inertia (mass). If left unchecked, and a resonant response were to occur, the pressure peaks could easily reach a dangerous level. Because the oscillations are waves, standing waves will form, and interference with valve action may occur, adversely affecting the cylinder performance.

While a single, low pressure compressor may require little or no treatment for pulsation control, the same machine with an increased gas density, pressure, or operational changes may develop a problem with pressure pulses or standing wave performance deterioration. As an installation becomes more complex, such as with an increase in the number of cylinders connected to one header and the use of multiple stages, the possibility of a problem can increase.

When an installation is being planned, it is recommended that the API Standard 618 be reviewed in detail. The pulsation level for API 618 at Design Approach 1, the outlet side of any pulsation control device regardless of type, should be no larger than 2% peak-to-peak of the line pressure, or the value given by the following equation, whichever is less.

$$P\% = \frac{10}{P_{line}^{1/3}} \tag{3.20}$$

where

P% = maximum allowable peak-to-peak pulsation level at any discrete frequency, as a percentage of average absolute pressure.

P_{line} = average absolute line pressure.

The objective of this approach is to improve the reliability of the system without having to design acoustical filters. For many systems, this is all that is needed. API 618 contains a chart that recommends the type of analysis that should be performed, based on horsepower and pressure.

The pulsation control elements can have several forms, such as plain volume bottles, volume bottles with baffles, bottles and orifices, and proprietary acoustical filters. See Figure 3-26 for an example of a compressor with a set of attached volume bottles. Regardless of which device or element is selected, a pressure loss evaluation must be made before the selection is finalized because each of these devices causes a pressure drop.

For those installations where a detailed pulsation analysis, API 618 Design Approach 2 or 3, is required, several consulting companies offer these services. Until the 1980s, the most common method was to perform the pulsation analysis on the analog simulator of the Pipeline and Compressor Research Council of the Southern Gas Association. The

Figure 3-26. Manifold-type volume bottles are used where cylinders are operated in parallel, as on this two-stage, motor driven compressor. (*Courtesy of Dresser-Rand*)

procedures used in planning a new installation were to include the pulsation study in the contract with the compressor vendor. During the analog pulsation study, the isometric piping drawings were used to create a lumped model of the piping connected to the compressor cylinders. The purchaser's representative was required to be present for the analog study and had to be familiar with the piping requirements for the compressor area. This was necessary so that decisions relative to the space available and location for the bottles, as well as feasibility of piping modifications could be made during the study. The representative helped to expedite the completion of a final configuration for the piping system and bottle location since the analog components were disassembled after the study was completed. The analog method is still used, although much less frequently.

With the advent of modern workstations and faster PC computers, the solution of the differential equations of motion for acoustical waves in piping system on a digital computer has become feasible. In current practice, pulsation design studies using digital computer technology can produce the same results as obtained with a dynamic simulation on the analog system. The results from digital simulation satisfy the requirements of API 618. The digital computer has the advantage of data file storage. With storage capability and the ability to readily manipulate the data, it is not as necessary to have immediate decisions made. Piping changes that are recommended for acoustical control can be evaluated in a more comprehensive manner taking into account safety, cost, maintenance, and operational considerations. An additional benefit is realized if system changes are anticipated at a later time. The data files can be retrieved and the system rerun with the changes to the thermophysical properties or in the piping system itself without the need to remodel the entire system.

The interpretation of the results and the quality of the design from the pulsation study, whether performed on the analog simulator or with digital computer simulation, depends quite heavily on the experience and skill of the analyst performing the study. A purchaser of a compressor system who may be a novice at this type of analysis should give serious consideration to using the services of a competent consultant.

For the purpose of quick estimates or field evaluation of existing systems, consider the curve in Figure 3-27. This curve is not meant to supersede a comprehensive analysis as previously discussed. It should be used in checking vendor proposals or in revising existing installations where a single cylinder is connected to a header without the interaction of multiple cylinders. While not a hard rule, the curve should be conservative for

VOLUME BOTTLE SIZING

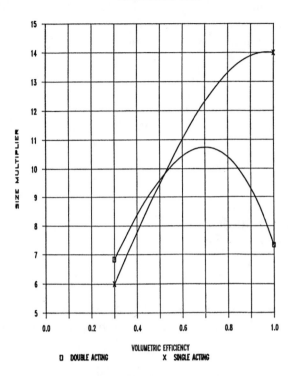

Figure 3-27. Volume bottle sizing graph.

compressors under 1,000 psi and 500 hp. The volume bottle defined is of the simple, unbaffled type.

To calculate the discharge volumetric efficiency necessary to use the curve for discharge volume bottles, use the following equation. Use Equation 3.5 to obtain the inlet volumetric efficiency and Equation 3.6 to calculate the factor.

$$E_{vd} = \frac{E_v}{(r_p)^{1/k}} \times f \qquad (3.21)$$

Once the inlet and discharge volumetric efficiencies are determined, bottles for the inlet and discharge may be sized. Begin the process of sizing the bottle of interest using the appropriate volumetric efficiency (inlet or discharge) and determine a multiplier from Figure 3-27. Use Equation 3.1, 3.2, or 3.3 to determine the piston displacement. In the calculation, use a 1 for the speed and N to determine the piston displacement for a single revolution. Apply the multiplier to the piston displacement per revolution. The product is the bottle volume, Vol, for use in Equation 3.22. This equation will yield the bottle diameter.

$$d_b = .86(\text{Vol})^{1/3} \qquad (3.22)$$

To complete the solution for the volume bottle dimensions, assume 2:1 elliptical heads and use the following relationship:

$$L_b = 2d_b \qquad (3.23)$$

where:

L_b = volume bottle length
d_b = volume bottle diameter

Example 3-2

Approximate the size of a suction and a discharge volume bottle for a single-stage, single-acting, lubricated, reciprocating compressor. The gas being compressed is natural gas at the following conditions:

Cylinder bore:	9 in.
Cylinder stroke:	5 in.
Rod diameter:	2.25 in.
Suction temperature:	80°F
Discharge temperature:	141°F
Suction pressure:	514 psia
Discharge pressure:	831 psia
Isentropic exponent:	1.28
Specific gravity:	.60
Percent clearance:	25.7%

Step 1. Find the suction and discharge volumetric efficiencies using Equations 3.5 and 3.21 with r_p = 831/514 = 1.617. The natural gas compressibility values can be obtained by using the gravity/compressibility charts (see Appendix B-29 through B-35) for a specific gravity of .60. Both Z_1 and Z_2 values are .93. Applying Equation 3.6, the value of f may be obtained as follows:

f = .93/.93

f = 1.0

Using the equation for the suction volumetric efficiency,

$E_v = .97 - [(1/1)1.617^{1/1.28} \ 1] \ .257 - .03$

$E_v = .823$ suction volumetric efficiency

For the discharge volumetric efficiency, use Equation 3.21.

$$E_{vd} = \frac{.823 \times 1}{(1.617)^{1/1.28}}$$

$E_{vd} = .565$ discharge volumetric efficiency

Step 2. Find the total volume displaced per revolution using Equation 3.1 for a single-acting compressor.

$$Pd/Rev = 5 \times \frac{\pi \times 9^2}{4}$$

$Pd/Rev = 318.1$ in.^{3}

Step 3. Using the volumetric efficiencies found in Step 1, find the size multiplier from the volume bottle sizing chart, Figure 3-27.

Suction multiplier = 13.5

Discharge multiplier = 10.4

Step 4. Find the required bottle volume from the displacement and the multiplier.

$Vol = 13.5 \times 318.1$

$Vol = 4,294.2$ in.^{3} suction bottle volume

$Vol_d = 10.4 \times 318.1$

$\text{Vol}_d = 3{,}308.1$ in.^{3} discharge bottle volume

Step 5. Find the bottle dimensions from Equations 3.22 and 3.23 for vessels with 2:1 elliptical heads. Use Equation 3.22 to calculate the diameter.

$d_{bs} = .86 \, (4294.2)^{1/3}$

$d_{bs} = 16.3$ in. diameter of suction bottle

$d_{bd} = .86 \, (3308.1)^{1/3}$

$d_{bd} = 14.9$ in. diameter of discharge bottle

Use Equation 3.23 to calculate the length.

$L_{bs} = 2 \times 16.3$

$L_{bs} = 32.6$ in. length of suction bottle

$L_{bd} = 2 \times 14.9$

$L_{bd} = 29.8$ in. length of the discharge bottle

References

1. *Compressed Air and Gas Handbook,* Third Edition, New York, NY: Compressed Air and Gas Institute, 1961.
2. Joergensen, S. H., *Transient Value Plate Vibration,* Proceedings of the 1980 Purdue Compressor Technology Conference, Purdue University, West Lafayette, IN, 1978, pp. 73–79.
3. Woollatt, D., *Increased Life for Feather Valves of Failure Caused by Impact,* Proceedings of the 1980 Purdue Compressor Technology Conference, Purdue University, West Lafayette, IN, 1980, pp. 293–299.
4. Davis, H., *Effects of Reciprocating Compressor Valve Design on Performance and Reliability,* Presented at Mechanical Engineers, London, England, October 13, 1970 (Reprint, Worthington Corp., Buffalo, NY).
5. Tuymer, W. J., "Maintaining Compressor Valves," *Power,* April 1978, pp. 41–43.

6. White, K. H., "Prediction and Measurement of Compressor Valve Loss," *ASME 72-PET-4,* New York, NY: American Society of Mechanical Engineers, 1972.

7. Szenasi, F. R. and Wachel, J. C., "Analytical Techniques of Evaluation of Compressor-Manifold Response," *ASME 69-PET-31,* New York, NY: American Society of Mechanical Engineers, 1969.

8. Damewood, Glen and Nimitz, Walter, "Compressor Installation Design Utilizing an Electro-Acoustical System Analog," *ASME 61-WA-290,* New York, NY: American Society of Mechanical Engineers, 1961.

9. Nimitz, Walter, "Pulsation Effects on Reciprocating Compressors," *ASME 69-PET-2,* New York, NY: American Society of Mechanical Engineers, 1969, 1.

10. Wachel, J. C., "Consideration of Mechanical System Dynamics in Plant Design," *ASME 67-DGP-5,* New York, NY: American Society of Mechanical Engineers, 1967.

11. Nimitz, Walter W., *Pulsation and Vibration,* Part I. Causes and Effects, Part II. Analysis and Control. *Pipe Line Industry,* Part I, August 1968, pp. 36–39. Part II, September 1968, pp. 39–42.

12. Mowery, J. D., *Rod Loading of Reciprocating Compressors,* Proceedings of the 1978 Purdue Compressor Technology Conference, Purdue University, West Lafayette, IN, 1978, pp. 73–89.

13. Von Nimitz, Walter W., *Reliability and Performance Assurances in the Design of Reciprocating Compressor Installation—Part I Design Criteria, Part II Design Technology,* Proceedings of the 1974 Purdue Compressor Technology Conference, Purdue University, West Lafayette, IN, 1974, pp. 329–346.

14. Safriet, B. E., "Analysis of Pressure Pulsation in Reciprocating Piping Systems by Analog and Digital Simulation," *ASME 76-WA/DGP-3,* New York, NY: American Society of Mechanical Engineers, 1976.

15. API Standard 618, *Reciprocating Compressors for Petroleum, Chemical, and Gas Industry Services,* Fourth Edition, Washington, DC: American Petroleum Institute, 1995.

16. Scheel, Lyman F., *Gas Machinery,* Houston, TX: Gulf Publishing Company, 1972.

17. Evans, Frank L. Jr., *Equipment Design Handbook for Refineries and Chemical Plants,* Vol. 1, Second Edition, Houston, TX: Gulf Publishing Company, 1979.

18. Loomis, A. W., Editor, *Compressed Air and Gas Data,* Third Edition, Woodcliff Lake, NJ: Ingersoll-Rand, 1980.

19. Cohen, R., "Valve Stress Analysis for Fatigue Problems," *ASHRAE Journal,* January 1973, pp. 57–61.

20. Hartwick, W., "Power Requirement and Associated Effects of Reciprocating Compressor Cylinder Ends, Deactivated by Internal By-Passing," *ASME 75-DGP-9,* New York, NY: American Society of Mechanical Engineers, 1975.

21. Hartwick, W., "Efficiency Characteristics of Reciprocating Compressors," *ASME 68-WA /DGP-3,* New York, NY: American Society of Mechanical Engineers, 1968.

22. Carpenter, A. B., "Pulsation Problems in Plant Spotted by Analog Simulator," *The Oil and Gas Journal,* October 30, 1967, pp. 151–152.

23. *Engineering Data Book,* Ninth Edition, 1972, 4th Revision 1979, Tulsa, OK: Gas Processors Suppliers Association, 1972, 1979, pp. 4–12, 4–13.

24. Wachel, J. C. and Tison, J. D., *Vibrations in Reciprocating Machinery and Piping Systems,* Proceedings of the 23rd Turbomachinery Symposium, Texas A&M University, College Station, TX, 1994, pp. 243–272.

4

Rotary Compressors

Common Features

Rotary compressors as a group make up the balance of the positive displacement machines. This group of compressors has several features in common despite differences in construction. Probably the most important feature is the lack of valves as used on the reciprocating compressor. The rotary is lighter in weight than the reciprocator and does not exhibit the shaking forces of the reciprocating compressor, making the foundation requirements less rigorous. Even though rotary compressors are relatively simple in construction, the physical design can vary widely. Both multiple- and single-rotor construction is found. Rotor design is one of the main items that distinguishes the different types. Size and operating range is another area unique to each type of rotary. The following sections cover some of the more common rotary compressors in detail.

Arrangements and Drivers

Rotary compressors are frequently arranged as single units with a driver. Occasionally the compressors are also used in series arrangements, with or without an intercooler. The series configurations may use a form of tandem drive or multiple pinion gears to permit the use of a common driver.

For most of the rotary compressors in process service, the driver is an electric motor. Compressors in portable service, however, particularly the helical-lobe compressor, use internal combustion engines. Many of the rotary compressors require the high speed that can be obtained from a direct-connected motor. The dry type helical-lobe compressor is probably the main exception as the smaller units operate above motor speed and require a speed increasing gear which may be either internal or external (see Figure 4-1). The helical-lobe compressor is the most likely candidate for a driver other than the electric motor. Aside from the portables already mentioned, engines are used extensively as drivers for rotaries located in the field in gas-gathering service. Steam turbines, while not common, probably comprise most of process service alternate drive applications.

Figure 4-1. A skid-mounted oil-free helical-lobe compressor. (*Courtesy of A-C Compressor Corporation*)

Helical Lobe

History

While a form of helical-rotor compressor was invented in Germany in 1878 [1], the helical-lobe compressor as used today is credited to Alf Lysholm, the chief engineer of Svenska Rotor Maskiner AB (SRM). Mr. Lysholm conceived the idea in 1934 as part of gas turbine development at SRM. The original compressor was an oil-free design using timing gears to synchronize the rotors. Three male and four female rotor lobes were used, and a steeper helix angle permitted higher built-in compression ratios and improved operation at higher pressures. The pressures were in the 20 to 30 psig range. Unfortunately, the profile created a trapped pocket where the gas was overcompressed prior to being released. This led to lower efficiency and high noise levels. Despite the disadvantages, the compressor was licensed and used in varying applications. Because of Mr. Lysholm's contributions, the helical-lobe compressor is sometimes referred to as the Lysholm compressor. It also goes by the name of SRM, the company controlling the development and licensing.

Hans Nilson became chief engineer of SRM in the late 1940s and later became president of the company. He made numerous contributions to the technical and commercial growth of the compressor, such as the circular profile invented in 1952. The profile used the four male lobe and six female lobe rotors. The design eliminated the trapped pocket, permitting a steeper helix angle. The resulting higher, built-in pressure ratios also improved efficiency.

The next significant event in the evolution of the SRM compressor was the application of a Holroyd rotor cutting machine to the production of the rotors. Prior to this event, producing the rotors was both slow and costly. In 1952, the first special Holroyd machine was delivered to Howden Company, a licensee in the U.K., who later contributed to the development of the oil-flooded compressor.

The slide valve was invented in the early 1950s, giving the SRM compressor a new dimension by providing a means of flow control. Capacity control had been a limiting factor for applications needing a range of flows. The slide valve provided infinite capacity control while still retaining built-in compression during flow reduction. The slide valve became widely used with the advent of the oil-flooded compressor.

The development and subsequent patent of the oil-flooded compressor was the result of a joint effort on the part of Howden and SRM. The oil-injected prototype was run at SRM on July 4, 1954 and proved to be 8 to 10% better in performance than the dry compressor with timing gears. Performance at low speeds was improved, permitting the use of direct drive motors. The flooding provided both cooling, which permitted higher pressure ratios and lubrication allowing the elimination of the timing gears. The male rotor drives the female rotor through the oil film. The first commercial application was introduced by Atlas Copco in 1957 for air compression. The slide valve was incorporated into the flooded design in the 1960s and was originally used in refrigerant service. More recently, it has also been incorporated into gas compressor service.

Lars Schibbye became chief engineer of SRM in 1950 and contributed to the technical advancement of the compressor. Most significant was the invention of the asymmetric rotor profile, which was introduced commercially by Sullair in the U.S. in 1969. The asymmetric rotor profile reduces the leakage path area and sealing line length resulting in increased efficiency.

The more recent developments have been in the area of manufacture. Rune Nilson of SRM worked with the machine tool manufacturers to develop precision carbide form cutting tools. These permit the rotors to be cut in two to three passes with high accuracy. The machining time has been reduced significantly.

The development of the helical-lobe compressor is quite unique in that it has been controlled primarily by one company, with the cooperation of licensees who, in turn, provide application expertise [2].

Operating Principles

Another name for the helical-lobe compressor is the *screw compressor.* This is probably the most common, even though all the names are used quite interchangeably. While the screw compressor originally fit the area between the centrifugal and the reciprocating compressor, the application areas have expanded. The larger, screw type machines now range to 40,000 cfm, and definitely cross into the centrifugal area. The smaller ones, particularly the oil-flooded type, are being considered for automotive air conditioning service, therefore, completely overlapping the reciprocating compressor in volume. The dry variety generally stops in the 50 cfm area.

Compression is achieved by the intermeshing of the male and female rotor. Power is applied to the male rotor and as a lobe of the male rotor starts to move out of mesh with the female rotor a void is created and gas is taken in at the inlet port (see Figure 4-2). As the rotor continues to turn, the intermesh space is increased and gas continues to flow into the

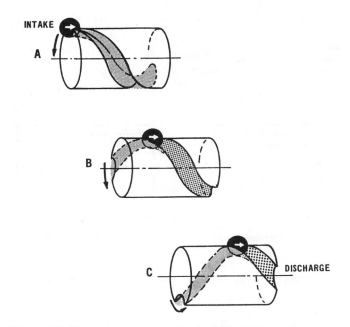

Figure 4-2. The compression cycle of a helical-lobe compressor.

compressor until the entire interlobe space is filled. Continued rotation brings a male lobe into the interlobe space compressing and moving the gas in the direction of the discharge port. The volume of the gas is progressively reduced, increasing the pressure. Further rotation uncovers the discharge port, and the compressed gas starts to flow out of the compressor. Rotation then moves the balance of the trapped gas out while a new charge is drawn into the suction of the unmeshing of a new pair of lobes as the compression cycle begins.

The compressor porting is physically arranged to match the application pressure ratio. To maintain the best efficiency, it is important that the matching be as close as possible.

Figure 4-3 includes four diagrams to show two cases, (A) a low-ratio compressor and (B) a high-ratio compressor. The lower diagram (C) demonstrates a low-volume ratio compressor in a higher-than-design application. Because the gas arriving at the discharge port has not been sufficiently compressed, the resulting negative ratio across the discharge port causes a backflow and resulting loss. Diagram (D) shows a compressor with too high a volume ratio for the process. Here the gas is compressed higher than needed to match the pressure of the gas on the outlet side of the discharge port, resulting in energy waste.

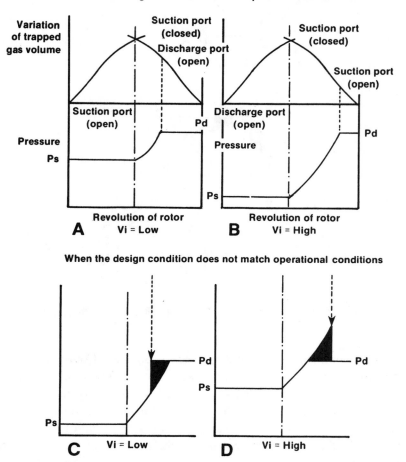

Figure 4-3. Effects of low- and high-volume ratios on the cycle of a screw compressor. (*Courtesy of Mayekawa Manufacturing Company, Ltd.*)

The terms *pressure ratio* and *volume ratio* are used interchangeably in the literature on these machines. To prevent confusion, *volume ratio* r_v, is defined as the volume of the trapped gas at the start of the compression cycle divided by the volume of the gas just prior to the opening of the discharge port. *Pressure ratio* is defined, in Equation 2.64, as the discharge pressure divided by the suction pressure. Their relationship is given in the following equation.

$$r_p = r_v^k \qquad (4.1)$$

where

r_p = pressure ratio
k = isentropic exponent
r_v = volume ratio

Displacement

The displacement of the screw compressor is a function of the inter-lobe volume and speed. The interlobe volume is a function of rotor profile, diameter, and length. Table 4-1 provides some typical rotor diameters and corresponding L/d ratios. The interlobe volume can be expressed by the following equation.

$$Q_r = \frac{d^3 (L/d)}{C}$$ (4.2)

where

Q_r = displacement per revolution
d = rotor diameter
L = rotor length
C = typical profile constant, for 4 + 6 rotor arrangement
C = 2.231 circular profile
   = 2.055 asymmetric profile [3]

Table 4-1
Rotor Diameters
with Available L/d Ratios

Rotor Diameter Inches	L/d 1.0	L/d 1.5
6.75	X	—
8.50	X	X
10.50	X	X
13.25	X	X
16.50	X	X
20.00	—	X
24.80	—	X

Available sizes (X)
Data for table courtesy of A-C Compressor.

$$Q_d = Q_r \times N \qquad (4.3)$$

where

Q_d = displacement
N = compressor speed

$$Q_i = Q_d \times E_v \qquad (4.4)$$

where

Q_i = actual inlet volume
E_v = volumetric efficiency

Because there is no clearance volume expansion, as in the reciprocating compressor, the volumetric efficiency is a function of the rotor slip. This is the internal leakage from the higher pressure to the lower pressure side, reducing potential volume capacity of the compressor.

Dry Compressors

The nonflooded compressor rotor leakage can be related to the rotor tip Mach number. The rotor tip velocity can be calculated by

$$u = \pi \times d \times N \qquad (4.5)$$

The optimum tip speed is .25 Mach at a pressure ratio of 3. The value shifts slightly for other built-in pressure ratios, as shown in Figure 4-4.

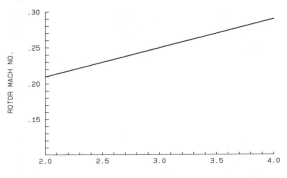

Figure 4-4. Optimum tip speed vs. pressure ratio.

Besides affecting the volumetric efficiency, the leakage also has an effect on the adiabatic efficiency. Figure 4-5 is a plot of the tip speed ratio, u/u_o, (operating to optimum) against the efficiency ratio, off-peak-to-peak effi-

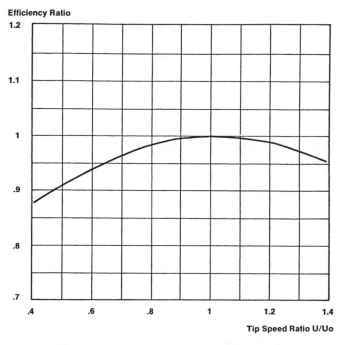

Figure 4-5. Tip speed ratio vs. efficiency ratio.

ciency. Figures 4-6 and 4-7 show a set of typical volumetric and adiabatic efficiency curves for three built-in ratios.

The adiabatic efficiency should be corrected for molecular weight. Generally the efficiency decreases with lower molecular weight and increases with increased molecular weight. As an arbitrary rule of thumb, a straight line relationship can be assumed. The correction is a −3 percentage points at a molecular weight of 2, 0 at 29, and +3 at the molecular weight of 56. For example, a compressor with an air efficiency of 78% would have an adiabatic optimum tip speed efficiency of 75% when operating on hydrogen.

The screw compressor can be evaluated using the adiabatic work equation. Discharge temperature can be calculated by taking the adiabatic temperature rise and dividing by the adiabatic efficiency then multiplying by the

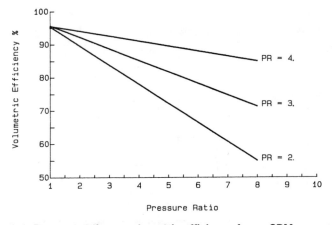

Figure 4-6. Pressure ratio vs. volumetric efficiency for an SRM compressor.

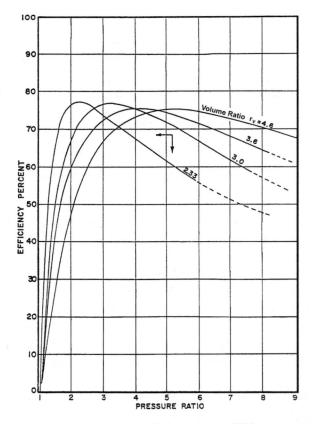

Figure 4-7. Pressure ratio vs. adiabatic efficiency for an SRM compressor. (Modified from [4].)

temperature rise efficiency to account for cooling. To obtain the discharge temperature, add the inlet temperature to the temperature rise. The work equation was developed in Chapter 3 and repeated here for convenience.

$$W_a = P_1 Q_1 \frac{k}{\eta_a (k-1)} (r_p^{\frac{k-1}{k}} - 1) \tag{3.18}$$

where

W_a = adiabatic work input
P_1 = inlet pressure
Q_1 = inlet volume
η_a = adiabatic efficiency

For the discharge temperature, t_2,

$$t_2 = t_1 + \frac{T_1 (r_p^{\frac{k-1}{k}} - 1)}{\eta_a} \times \eta_t \tag{4.6}$$

where

t_1 = inlet temperature
T_1 = absolute inlet temperature
η_a = adiabatic efficiency
η_t = temperature rise efficiency

A typical value for the temperature rise efficiency is .9. For shaft power, W_s,

$$W_s = W_a + \text{mech loss} \tag{4.7}$$

where mech loss = .07 × W_a for estimating purposes.

Example 4-1

Calculate the performance of a compressor using air for the following conditions:

d = 10.5 in. rotor diameter

L/d = 1.5 length-diameter ratio

mw = 23

Q_1 = 2,500 acfm inlet volume

$t_1 = 100°F$ inlet temperature

$P_1, = 14.5$ psia inlet pressure

$P_2 = 43.5$ psia discharge pressure

$r_p = 3.0$ pressure ratio

$k = 1.23$

$w = 138.8$ lbs/min weight flow

Step 1. Using Equation 4.2, solve for the displaced volume per revolution. Convert the units using 1728 in.^{3}/ft^3.

$$Q_r = \frac{10.5^3 \times 1.5}{1728 \times 2.231}$$

$Q_r = .450$ ft^3/ rev

From Figure 4-6, read a volumetric efficiency for pressure ratio, 3, where $E_v = 89\%$. Use Equation 4.4 and solve for the total displaced volume.

$Q_d = 2,500/.89$

$Q_d = 2,809$ cfm total displacement volume

Now calculate the required speed by substituting into Equation 4.3.

$N = 2,809/.450$

$N = 6,242$ rpm rotor speed

Step 2. Find the velocity of sound for air at the inlet conditions given, using Equation 2.32 from Chapter 2.

$R = 1,545/23$

$R = 67.17$ specific gas constant

$a = (1.23 \times 67.17 \times 32.2 \times 560)^{1/2}$

$a = 1,220.6$ fps velocity of sound

Compute the rotor tip velocity using Equation 4.5 and the unit conversions of 12 in./ft and 60 sec/min.

$$u = \frac{\pi \times 10.5 \times 6242}{60 \times 12}$$

$u = 286.0$ fps rotor tip velocity

Refer to Figure 4-4 with pressure ratio $= 3.0$ and read the rotor Mach number $u_o/a = .25$. Calculate u_o, the optimum tip velocity.

$u_o = .25 \times 1220.6$

$u_o = 305$ fps optimum tip velocity

Then calculate the tip speed ratio.

$u/u_o = 286.0/305.0$

$u/u_o = .937$ tip speed ratio

Step 3. Refer to Figure 4-7 and select an efficiency at a pressure ratio of 3 and a volume ratio, r_v of 2.44. The adiabatic efficiency is 74%. Now, from Figure 4-5, select a value of efficiency ratio using the tip speed ratio just calculated. Because the value is .99+, round off to an even 1.0. With a multiplier of 1.0, the final adiabatic efficiency is the value read directly off the curve or $\eta_a = 74$. The molecular weight correction for efficiency, per rule of thumb, is 0.6 for a final efficiency of 73.4.

Step 4. The adiabatic power can be solved by substituting into Equation 3.18.

$k/(k - 1) = 1.23/.23$

$k/(k - 1) = 5.34$

$(k - 1)/k = .187$

Calculate the power using the conversions of 144 in.^{2}/ft^2 and 33,000 ft lbs/min/hp for a net value of 229.

$$W_a = \frac{14.5 \times 2500 \times 5.34 \, (3^{.187} - 1)}{.734 \times 229}$$

$W_a = 262.6$ hp

Substitute into Equation 4.6 for the discharge temperature.

$$t_2 = 100 + \frac{560 \, (3^{.187} - 1)}{.734} \times .9$$

$t_2 = 256.6°F$ discharge temperature

Step 5. Solve for the shaft power substituting into Equation 4.7.

$P_S = 262.6 + 18.4$

$P_S = 281.0$ hp shaft horsepower

Example 4-2

Rerate the compressor considered in Example 4-1 for an alternate set of conditions given below. Use all other conditions from the previous example.

$P_2 = 50.75$ psia new discharge pressure

$r_p = 3.5$ new pressure ratio

$r_v = 2.77$ new volume ratio

Step 1. Calculate a new inlet volume using the value of displaced volume from the previous example.

$Q_d = 2,809$ cfm displaced volume

Refer to Figure 4-6 and, at a pressure ratio = 3.5, read the volumetric efficiency = 87%. Use Equation 4.4 to develop the inlet volume.

$Q_i = 2,809 \times .87$

$Q_i = 2,444$ cfm inlet volume

By proportion, obtain a new inlet weight flow.

$w = 2,444/2,500 \times 138.8$

$w = 135.7$ lb/min new weight flow

Step 2. Reuse the rotor tip speed and sonic velocity from Example 4-1 as the conditions used in their development that have not changed.

a = 1220.6 fps sonic velocity

u = 286.0 fps rotor tip speed

Refer to Figure 4-4 and, at a pressure ratio = 3.5, read the Mach number, u_o/a = 0.27. Calculate the optimum tip speed u_o.

u_o = .27 × 1220.6

u_o = 329.6 fps optimum tip speed

Now calculate the tip speed ratio.

u/u_o = 286.0/329.6

u/u_o = .868 tip speed ratio

Step 3. Use Figure 4-7 to obtain the adiabatic efficiency for pressure ratio = 3.5 and volume ratio of 2.77. From the curve, adiabatic efficiency = 73%. Next, look up the efficiency ratio on Figure 4-5 for the tip speed ratio just developed and obtain a value of .98. Use this value as a multiplier to derate the adiabatic efficiency for operation at other than the optimum tip speed.

η_a = .98 × 73.0

η_a = 71.5%

Step 4. Solve for the adiabatic power, making the same conversions used in Step 4 of Example 4-1.

$$W_a = \frac{14.5 \times 2444 \times 5.34(3.5^{.187} - 1)}{.715 \times 229}$$

W_a = 326.2 hp

Use Equation 4.6 to solve for the discharge temperature.

$$t_2 = 100 + \frac{560\,(3.5^{.187} - 1)}{.715} \times .9$$

t_2 = 399°F discharge temperature

Step 5. For the final step, compute the new shaft power value using Equation 4.7.

$$W_S = 325.0 + 21.4$$

$$W_s = 346.4 \text{ hp new shaft horsepower}$$

The example demonstrates that operating the compressor off the built-in pressure ratio means operating at a lower efficiency. This could be anticipated from Figure 4-3. The optimum port configuration for the various types of screw compressors was determined from a series of prototype tests. The change in volumetric efficiency is not a result of the built-in volume ratio, but is due to the increased slip (internal leakage) from the higher operating pressure ratio. In the last example, a slight loss of efficiency was shown for operation at other than the optimum tip speed. While, in the example, the penalty was not too severe, it does give a directional indication of the potential problems with off-design operation.

Figure 4-8 shows a comparison of the two currently used rotor profiles. Figure 4-8a shows the circular profile used in the past for both the dry and flooded compressor. The newer asymmetric profile shown in Figure 4-8b is being adopted for use in both dry and flooded service by various vendors because of the improved efficiency due to a lower leakage in the discharge area of the compressor. Because size is a factor, the improvement in efficiency is more dramatic in the smaller compressors.

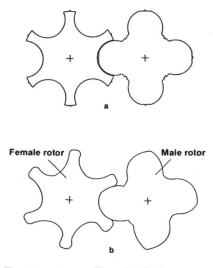

Figure 4-8. The two rotor profiles of helical screw compressors.

Flooded Compressors

The oil-flooded version is an increasingly popular variation of the screw compressor and is seeing a variety of applications. This type of compressor is less complex than the dry version because of the elimination of the timing gears. It also has the advantage of the oil acting as a seal to the internal clearances, which means a higher volumetric and overall efficiency. The sealing improvement also results in higher efficiency at lower speeds. This means quiet operation and the possibility of direct connection to motor drivers, eliminating the need for speed increasing gears. (When gears *are* needed they are available as internal on some models, see Figure 4-9.) Higher pressure ratios can also be realized because of the direct cooling from the injected oil. Pressure ratios as high as 21 to 1 in one casing are possible [3]. Besides the inherently quiet operation from lower speed, the oil dampens some of the internal pulses aiding the suppression of noise. The timing gears can be eliminated because the female rotor is driven by the male through the oil film. To take advantage of the 3-to-2 speed increase, development work is in progress to drive the female rotor. Alterations to the 90-to-10 power division for male and female rotor must be made to shift more of the power

Figure 4-9. An oil-flooded, integrally geared screw compressor package. (*Courtesy of Sullair*)

to the female rotor. The contact surfaces should also be improved to better transfer the additional power.

The injected oil is sheared and pumped in the course of moving through the compressor. These losses can be minimized by taking advantage of the slower speed performance. Figure 4-10 shows the operating speed plotted against the shaft input power. There is an optimum operating speed where the improvement in operation from the oil offsets the potential losses. The points of injection are quite important for efficient operation. The oil is injected in the casing wall at or near the intersection of the rotor bores on the discharge side of the machine. The orifices are lined up axially in the region where compression is taking place. Also, oil enters from each bearing. Good drainage control will keep oil from recycling back to contact the inlet charge and transferring unwanted heat to the uncompressed gas. The inlet port, as well, must be designed to prevent slip oil (oil traveling in the rotor clearance area) from heating the inlet gas.

Test data indicate that for the *pumpless oil system* compressor, the discharge temperature remains constant over a wide range of operation, at varying pressure ratios, staying close to 176°F (80°C)[1]. In contrast, on a *pumped system,* the outlet temperature can be maintained at a desired level. The amount of oil injected must be carefully controlled, admitting enough for operation and not too much to cause high pumping losses. For an application of 100 psig air compression, rates of 6 to 7 gpm are used

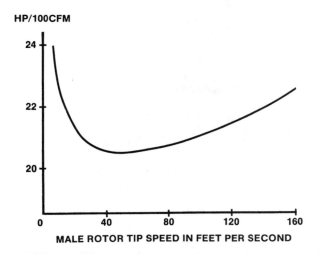

Figure 4-10. Tip speed vs. shaft input power for an oil-flooded screw compressor discharging at 100 psig [3].

per 100 cfm. This results in about 450 BTU/min heat rejection per 10 hp of input energy [3]. Because heat is rejected to the oil and the oil is recirculated in the flooded compressor, a larger lubrication system is required, along with more cooling water than would be needed for a dry compressor. There may be a potential trade-off in some applications such as refrigeration or air compression where dry compressors use inter and aftercoolers. In the case of refrigeration, the heat load is reduced to the refrigerant condenser and, in the air compression, the load is less for an aftercooler.

For areas where cooling water is either scarce or not available, direct liquid injection may be a possibility. The liquid coolant should be injected near the discharge end of the compressor to minimize lubricant dilution. Alternatively, the liquid can be flashed in a separate exchanger and used to cool the lubricant. While the cooling may appear to decrease the power to the compressor, the net effect is an increase in the power due to the additional weight flow of the extra refrigerant needed to perform the cooling.

The following equation provides a way of estimating the discharge temperature if the shaft power is known, or it can be used to estimate the shaft power if the temperature rise and quantity of lubricant is known. The equation assumes 85% of the heat of the compressor is absorbed by the lubricant.

$$.85 W_s = q_L \times \rho_L \times c_{pL} \times \Delta t \tag{4.8}$$

where

W_s = work input to the shaft
q_L = volume of lubricant
ρ_L = specific weight of the lubricant
c_{pL} = specific heat of the lubricant
Δt = lubricant temperature rise

Flooded compressors use the asymmetric profile rotor extensively because the rotor's efficiency is most apparent in this size range. Flooded compressor size has, over the more recent times, been increased. The upper range is in the 7000 cfm range. While most applications are in air and refrigeration, certain modifications can make it applicable for process gas service. One of the considerations is the liquid used for the flooding.

Flooding Fluid

The fluid used in the compressor is normally a petroleum based lubricating oil, but this is not universal. Factors to consider when selecting the lubricant include the following:

1. Oxidation
2. Condensation
3. Viscosity
4. Outgassing in the inlet
5. Foaming
6. Separation performance
7. Chemical reaction

Some of the problems can be solved with specially selected oil grades. Another solution is synthetic oils, but cost is a problem particularly with silicone oils. Alternatives must be reviewed to match service life of the lubricant with lubrication requirements in the compressor.

For chemical service, some lubrication qualities may be sacrificed in order to obtain a fluid compatible with the process gas. In these applications, alternate bearing materials such as graphite or silver have been required. While the requirements may make the operation somewhat special and require considerable care, the life of the compressor and service can be greatly improved.

Application Notes—Dry Compressors

Screw compressors of the dry type generate high frequency pulsations that move into the system piping and can cause acoustic vibration problems. These would be similar to the type of problems experienced in reciprocating compressor applications, except that the frequency is higher. While volume bottles will work with the reciprocator, the dry type screw compressor would require a manufacturer-supplied proprietary silencer that should take care of the problem rather nicely.

There is one problem the dry compressor can handle quite well. Unlike most other compressors, this one will tolerate a moderate amount of liquid. Injection of liquids for auxiliary cooling can be used, normally at a lower level than would be used in the flooded compressor. The compressor also takes reasonably well to fouling service, if the material is not abrasive. The foulant tends to help seal the compressor and, in time, may improve performance. One other application for which the dry machine is particularly well-suited is for hydrogen-rich service, where the molecular weight is low, with a resulting high adiabatic head. For larger flow streams, within the centrifugal compressor's flow range, the screw compressor is a good alternative. While the high adiabatic head requires expensive, multiple centrifugal casings, the positive displacement characteristic of the screw compressor is not compromised by the low molecular weight. For very low molecular weight gas, such as pure hydrogen or helium, a good seal is important to keep the slip in control. This can be tedious and, in extreme cases, a liquid injection is used for leakage control to maintain performance.

Application Notes—Flooded Compressor

One consideration for the flooded compressor is the recovery of the liquid. In the conventional arrangement, the lubricating oil is separated at the compressor outlet, cooled, filtered, and returned to the compressor. Figure 4-11 is a diagram of a typical oil-flooded system. This is fine for air service where oil in the stream is not a major problem, but when oil-free air is needed, the separation problem becomes more complex. Because the machine is flooded and the discharge temperature is not high, separation is much easier than with compressors that send small amounts of fluid at high temperature downstream. Usually part of the lubricant is in a vaporized form and is difficult to condense except where it isn't wanted. To achieve quality oil free air, such as that suitable for a desiccant type dryer, separators to the tertiary level should be considered (see Figure 4-12). Here, the operator must be dedicated to separator maintenance, because these units require more than casual attention. Separation in refrigeration is not as critical if direct expansion chillers are used. In these applications, the oil moves through the tubes with the

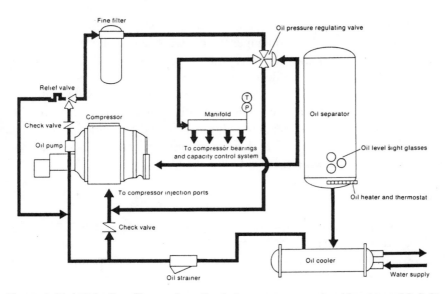

Figure 4-11. Lubrication diagram for a flooded screw compressor. (*Courtesy of Sullair*)

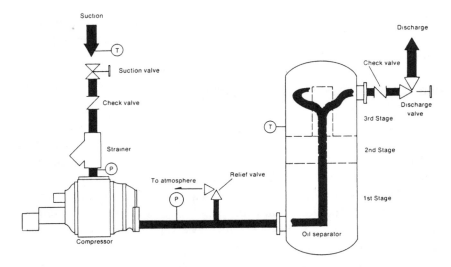

Figure 4-12. Oil and gas separation system for a flooded screw compressor. (*Courtesy of Sullair*)

refrigerant and comes back to the compressor with no problem, if the temperature is not too low for the lubricant. In the case of kettle type chillers, things tend to be more complex. Oil can be removed from a kettle with a *skimmer* arrangement, but getting the skimmer working on a wildly boiling and foaming fluid tends to make the refrigerant maintenance people somewhat testy. Again, it is necessary to match the lubricant to the temperature to get it back at all. No lubricant will return if it's frozen to the walls of an exchanger. If the evaporator is not field-located and can be made a part of the refrigeration package, some of these problems of oil handling can be deferred to the compressor vendor.

A nice feature of the flooded or dry screw compressor is its ability to achieve the required outlet pressure regardless of the molecular weight. The compressor process that starts on nitrogen and then gradually brings in a hydrogen-rich gas mixture does not change in performance as a centrifugal compressor would.

Casings

Most casings on both flooded and dry compressors are cast, normally of grey cast iron. API 619 [5] limits the use of cast iron by specifying steel for services in excess of 400 psig, discharge temperatures in excess of 500°F, and for flammable or toxic gases. While rare, austenitic and high nickel casings have been furnished. On dry compressors, the casing normally includes a water jacket. While referred to as a cooling jacket, the cooling water or alternative fluid is used as a heat sink or casing stabilizer to help control distortions and clearance changes. While castings are used for the iron casings, steel casings may be fabricated or cast and fabricated. Nozzle connections and allowable forces and moments are specified by the API standard, using the NEMA steam turbine equations for the force and moment basis. Most casings are vertically split, using end closures and withdrawing the rotors axially for maintenance. On the larger dry machines, the casing is horizontally split, to facilitate the removal of the heavier rotors.

Rotors

The rotor is the working portion of the compressor and, as was pointed out, is machined to both generate the helix and form the profile (see Figure 4-13). Some dry compressors are furnished with hollow rotors

Figure 4-13. A rotor set for an oil-free helical-lobe compressor. (*Courtesy of A-C Compressor Corporation*)

through which cooling fluid is circulated. As with the cooling jacket, the cooling is somewhat misnamed because the more important aspect of the fluid is to help stabilize the rotor. Materials of construction are steel in most applications. The material may be either a forging or bar stock, based on size availability of the bar stock in the quality needed. Other materials are used whenever carbon steel is not compatible with the gas being compressed. These range from stainless, either of the austenitic or 12 chrome type, to more exotic nickel alloys.

Some vendors furnish coatings for the rotors in order to keep the rotor from wearing and losing seal clearance. One such coating is TFE. There is wide diversity of opinion on the value of coatings in light of varying performance. For the purpose of renewing clearances at the time of maintenance, some vendors use a renewable seal strip. This is a very good feature on dry compressors where seal strip clearance is .0005 to .001 inch per inch of rotor diameter at 350°F. If the compressor is designed for 450°F and run in at 500°F, the loss in area would result in an efficiency loss of ½ to 1%.

Rotor speeds are such that dynamic balancing is required for proper vibration control. Also, while the critical speeds are generally above the operating speed, review of the rotor dynamics should not be ignored, particularly for the dry type.

Bearings and Seals

For a general discussion of bearings and seals refer to Chapter 5. The coverage at this point will be limited to the identification of the various types used on the screw compressor.

In the larger, dry process compressors, the radial bearings are of the *sleeve* or *tilting pad* type. Bearing surfaces use a high tin babbitt on a steel backing. API 619 requires the bearings to be removable without removing the rotors or the upper half on the horizontally split machine. Thrust bearings are generally tilt pad type, though not necessarily symmetric. On standardized compressors for air or refrigeration, the bearings are normally the *rolling element* type. Some standardized dry compressors use a *tapered land* thrust bearing. Most of the flooded compressors and some of the standardized dry compressors use rolling element thrust bearings. In all cases, the bearings are pressure-lubricated with some compressors using the gas differential pressure to circulate the lubricant and thus pressurize the bearings. For difficult services mentioned previously, unusual bearings may be used, such as graphite with a sulfuric acid flooding medium.

In dry compressors, shaft end seals are generally one of five types. These are labyrinth, restrictive ring, mechanical contact, liquid film, and dry gas seal. The *labyrinth* type is the most simple but has the highest leakage. The labyrinth seal is generally ported at an axial point between the seals in order to use an eductor or ejector to control leakage and direct it to the suction or a suitable disposal area. Alternatively, a buffer gas is used to prevent the loss of process gas. Appendix D presents a calculation method for use with labyrinth seals.

Probably the most common seal is the *restrictive ring type,* normally used in the form of *carbon rings.* This seal controls leakage better than the non-floating labyrinth type, although it wears faster. The carbon ring seal does not tolerate dirt as well as the labyrinth seal. The carbon ring seal and the labyrinth seal may be ported for gas injection, ejection, or a combination of both. Any injection gas should be clean.

The *mechanical contact seal* is a very positive seal. The seal is normally oil-buffered. The mechanical seal, which is the most complex and expensive, is used where gas leakage to the atmosphere cannot be tolerated. This may be due to the cost of the gas, as in closed-loop refrigeration, or where the process gas is toxic or flammable. The mechanical contact seal requires more power than the other seals, which is a deterrent to its use on lower power compressors.

The *liquid film seal* uses metallic sealing rings and is liquid buffered to maintain a fluid film in the clearance area and thereby preclude gas leakage. It is not unusual in the screw compressor to find the radial bearing and seal combined.

The *dry gas seal* is a variation of the mechanical contact seal. It differs in that it uses a microscopically thin layer of gas to separate and lubricate the faces. The seal is configured in a tandem or double-opposed seal arrangement. More complete details are covered in Chapter 5 under Dry Gas Seals.

Timing Gears

In screw compressors of the dry type, the rotors are synchronized by timing gears. Because the male rotor, with a conventional profile, absorbs about 90% of the power transmitted to the compressor, only 10% of the power is transmitted through the gears. The gears have to be of good quality both to maintain the timing of the rotors and to minimize noise. Because the compressor will turn in reverse on gas backflow, keeping gear backlash to a minimum is important. A check valve should be included in the compressor installation to prevent gas backflow. To control the backlash in the gears, a split-driven gear is used to provide adjustment to the gear lash and maintain timing on reverse rotation. To provide timing adjustment, the female rotor's timing gear is made to be movable relative to its hub. A close-up of a timing gear set is shown in Figure 4-14.

Timing gears are machined from low alloy steel, normally consisting of a chrome, nickel, and molybdenum chemistry. API 619 mandates an AGMA quality 12 gear, which is commonly used. The gears are of the helical type, which also help control noise. The pitch line runout must be minimized to control torsional excitation. The gears are housed in a chamber outboard from the drive end and are isolated from the gas being compressed.

Figure 4-14. Timing gears for an oil-free helical-lobe compressor. Note the timing adjustment capability on the right side gear. (*Courtesy of A-C Compressor Corporation*)

Capacity Control

Until the slide valve came on the scene, suction throttling was about the only capacity control available to the fixed speed flooded screw compressor. Suction throttling is generally not appropriate on positive displacement compressors because of potential high pressure ratios and resulting high temperatures. The previous examples showed that off-design operation is not energy efficient. Control by use of a bypass is not satisfactory due to the lack of power turndown with the net capacity reduction. If a variable speed driver can be justified, speed control can be used to provide good control in an energy efficient manner. The slide valve offers a more economical alternative to speed control, particularly for small compressors.

The slide valve moves parallel to the rotor axis and changes the area of the opening in the bottom of the rotor casing. This lengthens or shortens the region of compression of the rotor and returns gas to the suction side before compression has taken place. Figure 4-15 shows this in diagram

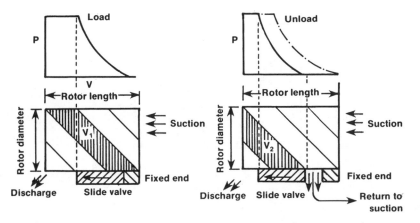

Figure 4-15. Operation of a slide valve. (*Courtesy of Mayekawa Manufacturing Company, Ltd.*)

form and makes it somewhat easier to follow the logic. Figure 4-16 is a cross section of a flooded screw compressor with a slide valve. The slide valve is readily adaptable to the flooded compressor because of ease of lubrication. It can be used in the dry compressor, as well, if provision is made for lubrication of the slide. Figure 4-17 compares the slide valve, labeled as capacity-controlled compressor, with the ideal and a fixed-capacity machine using suction throttling. The energy savings are signifi-

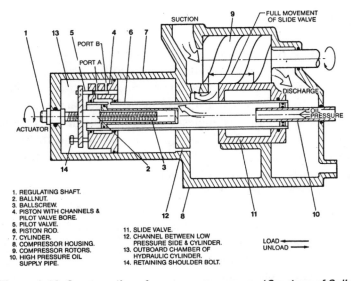

1. REGULATING SHAFT.
2. BALLNUT.
3. BALLSCREW.
4. PISTON WITH CHANNELS &
   PILOT VALVE BORE.
5. PILOT VALVE.
6. PISTON ROD.
7. CYLINDER.
8. COMPRESSOR HOUSING.
9. COMPRESSOR ROTORS.
10. HIGH PRESSURE OIL
    SUPPLY PIPE.

11. SLIDE VALVE.
12. CHANNEL BETWEEN LOW
    PRESSURE SIDE & CYLINDER.
13. OUTBOARD CHAMBER OF
    HYDRAULIC CYLINDER.
14. RETAINING SHOULDER BOLT.

LOAD ◄──────
UNLOAD ──────►

Figure 4-16. Cross section of a screw compressor. (*Courtesy of Sullair*)

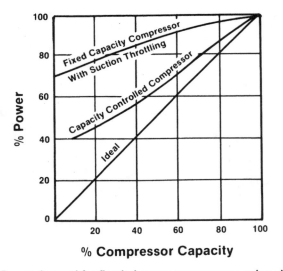

Figure 4-17. Power demand for flooded screw compressors using slide valve and suction throttling [1]. (*Reprinted by permission of the Council of the Institution of Mechanical Engineers from "The Place of the Screw Compressor in Refrigeration"*)

cant. Figure 4-18 is a schematic of the oil system and the slide valve with a piston operator.

Variation to the slide valve is the *turn valve*. The valve functions by turning rather than sliding but has the same effect as the slide valve. The plug is

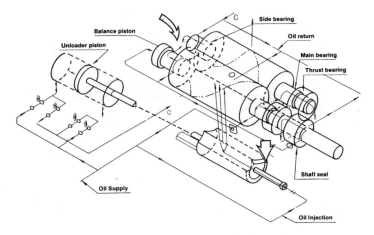

Figure 4-18. Schematic of oil system and slide valve with piston operator. (*Courtesy of Mayekawa Manufacturing Company, Ltd.*)

grooved and the compressor casing is slotted. As the valve turns, the grooves move away from the slots, passing a quantity of gas. The leakage in this valve is greater because of the imperfect matching of grooves and slots.

A control innovation that has potential to broaden the screw compressor's application range is the variable-volume-ratio compressor.[4] A diagram of a low- and a high-volume-ratio compressor is shown together with a *movable slide stop,* which is used with the slide valve to form a variable volume ratio compressor in Figure 4-19. While the "volume ratio" terminology is more descriptive, it is more commonly known as "built-in pressure ratio." By controlling inlet volume with the slide valve and the discharge volume by use of the slide stop, an infinite number of volume ratios can be achieved (see Figures 4-20 and 4-21).

Straight Lobe

Compression Cycle

Straight-lobe compressors, or blowers, as they are commonly called are low-pressure machines. The feature unique to these compressors is that the machines do not compress the gas internally as do most of the other rotaries. The straight-lobe compressor uses two rotors that intermesh as they rotate (see Figure 1-7). The rotors are timed by a set of gears. The lobe shape is either an involute or cyclodial form. A rotor may have either two or three lobes. As the rotors turn and pass the inlet port, a volume of gas is trapped and carried between the lobes and the outer

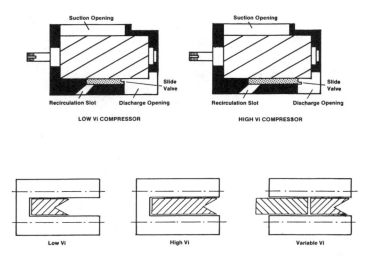

Figure 4-19. Side and top view of slide valve arrangements for low- high- and variable-volume ratio compressors [4].

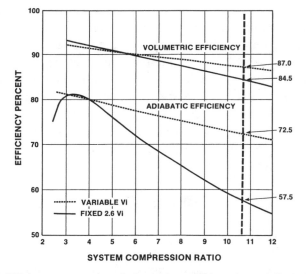

Figure 4-20. Efficiency comparison between a variable-volume ratio and a fixed-volume ratio compressor. These compressors have an asymmetric rotor profile [4].

cylinder wall. When the lobe pushes the gas toward the exit, the gas is compressed by the back pressure of the gas in the discharge line.

Volumetric efficiency is determined by the tip leakage past the rotors, not unlike the rotary screw compressor. The leakage is referred to as slip.

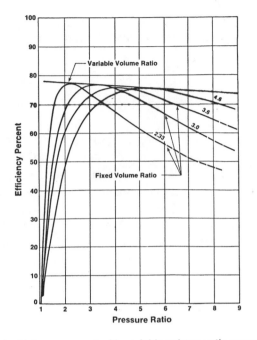

Figure 4-21. Efficiency improvement with variable volume ratio screw compressors [4].

Slippage is a function of the rotor diameter and differential pressure for a given gas. Slippage is determined by test. For the test, the differential pressure is imposed on the blower and the speed gradually increased until the point is reached where the output just matches the slip leakage. This point is detected by watching for the machine to just begin to give a positive output. The speed at which this occurs is called the slip speed. A slip speed is determined for each of several pressure differentials.

Output volume is calculated from displaced volume multiplied by the difference between a desired speed and the slip speed, rather than using the concept of volumetric efficiency as in the other positive displacement compressors. Output volume is expressed by

$$Q_1 = Q_r (N - N_s) \tag{4.9}$$

where

Q_1 = delivered volume
Q_r = displaced volume per revolution
N = rated speed
N_s = slip speed

While volumetric efficiency is not commonly used, the following relationship is an approximation for comparison purposes.

$$E_v = 1.0 - N_s/N \qquad (4.10)$$

Sizing

Sizing for the straight-lobe compressor is normally done using catalog data. Rotor lengths range from approximately one to two times the rotor diameter. Individual frame sizes within a given vendor's line may exceed these limits. Maximum tip speeds are in the 125 fps range with some units approaching 140 fps. The following relationship will permit size approximation if no catalog is available. It should be remembered that this is a rough size and that it may not be a standard with any of the vendors. It is helpful, however, to know at least one dimension from the vendor's line to make an estimate more meaningful.

$$Q_r = d^2 \times l_r/c \qquad (4.11)$$

where

d = rotor diameter
l_r = rotor length
c = sizing constant

A sizing constant of 1.2 can be used to make a reasonable approximation of many commercial sizes. The constant, c, varied from 1.11 to 1.27 for a number of the frames investigated. With the displaced size approximated, the delivered volume can be calculated. Use Equation 4.10 and an assumed volumetric efficiency of .90. This is arbitrary, as the actual volumetric efficiency varies from .95 to .75 or lower for the higher differential pressure applications. Once a slip speed has been determined, Equation 4.9 can be used to complete the calculation. The tip speed should stay near 125 fps.

Discharge temperature and horsepower can be calculated using the expressions given in the rotary screw compressor section, Equations 4.6 and 4.7. Efficiencies are a function of the differential pressure. For differential pressures of 5 psi and less, the efficiency is .80 or higher. It falls off fast with increased pressure rise, dropping to .60 in the 10 to 12 psi differential range. Mechanical losses are near .07 of the gas horsepower in the smaller sizes and .05 to .06 in the larger sizes. Larger is considered as over 5000 cfm.

Applications

The straight-lobe blowers are used both in pressure and vacuum service. Larger units are directly connected to their drivers and the smaller units are belt-driven. The drivers are normally electric motors. Some of the larger models offer an internal gear arrangement to permit the direct connection of a two- or four-pole electric motor. These blowers would normally operate at less than the nominal four-pole speed on a 60 Hz system of 1800 rpm. The main limitation to this rotary compressor is the differential pressure with the longer rotors where deflection is large. For a two-lobe machine, caution should be used when the rotor is more than 1.5 times the rotor diameter at pressures in excess of 8 psi differential. The three-lobe compressors inherently have a stiffer rotor and can sustain a higher differential with less difficulty. The practical upper limit should be 10 psi differential for units above 3000 cfm and 12 psi differential for the smaller units.

Mechanical Construction

Straight-lobe compressor casings, also called housings or cylinders by different manufacturers, are furnished in cast iron by all vendors. There is an optional aluminum construction available for special applications. Inlet and outlet are suitable for a 125 pound standard ANSI flanged connection.

Rotors are cast from ductile iron. Again, the exception is the aluminum construction. Shafts are steel and are cast into the rotors or are pinned to the rotor in a stub shaft construction arrangement. An alternate design has the rotors drilled for through shafts. Rotors are supported by a set of rolling element bearings on the outboard end of each rotor. The timing gears are forged steel on the more competitively priced models and are low alloy steel on the more rugged models. Also on the competitive models, the gears are straight spur, while helical gear teeth are furnished on the heavy duty models.

The most common seals are the lip type and the labyrinth type. Mechanical seals are available where seal leakage must be controlled.

Lubrication is splash type and grease is used on the more competitive models. There are variations available with internal pressure lubrication systems. Some models can be equipped with an external lube system, and for rare cases, API 614 lubrication systems have been proposed.

Sliding Vane

Compression Cycle

The *sliding vane* compressor consists of a single rotor mounted eccentrically in a cylinder slightly larger than the rotor. The rotor has a series of radial slots that hold a set of vanes. The vanes are free to move radially within the rotor slots. They maintain contact with the cylinder wall by centrifugal force generated as the rotor turns.

The space between a pair of vanes and the rotor and the cylinder wall form crescent shaped cells (see Figure 1-8). As the rotor turns and a pair of vanes approach the inlet, gas begins to fill the cell. The rotation and subsequent filling continue until the suction port edge has been passed by both vanes. Simultaneously, the vanes have passed their maximum extension and begin to be pushed back into the rotor by the eccentricity of the cylinder wall. As the space becomes smaller, the gas is compressed. The compression continues until the leading vane crosses the edge of the discharge port, at which time the compressed gas is released into the discharge line.

The port location must be matched to the pressure ratio dictated by the application for efficient compression to take place. Figure 4-22 is an indicator diagram of a compression cycle. If the port has been optimized for a ratio of P_2/P_1, the compression line is a smooth curve, as can be seen as compression proceeds from point 1 to 2. Note point 2 is at the same pressure level as P_2. If the external pressure is higher than the pressure for which the port was cut, so that $P_o/P_1 > P_2/P_1$, then, when the port opens at point 2, discharge gas will return to the compressor from the line and must again be expelled from the compressor. This energy waste is depicted by the shaded area to the left of the line o-2. Conversely, if the external pressure ratio is lower than the pressure ratio for which the port was cut, where $P_U/P_1 < P_2/P_1$, then the gas will be overcompressed to point 2 and when the port opens, it will expand to point u. The lost energy is represented by the shaded area to the right of the line u-2.

Sizing

The displaced volume of the sliding vane compressor can be calculated if certain geometric data are available. Unfortunately the vendor catalog data generally will not be very useful in establishing the geometry because frame designations are not directly related to any convenient measurement. The vendor's literature does give the expected capacity for air at various pressures, which include the displaced volume and the vol-

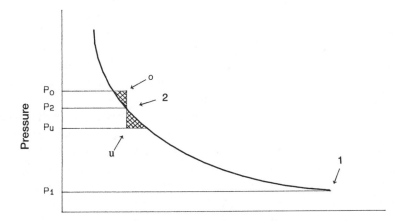

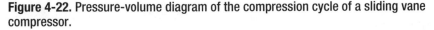

Figure 4-22. Pressure-volume diagram of the compression cycle of a sliding vane compressor.

umetric efficiency. Geometric information might be available, such as may occur with a unit being considered for reuse where measurements can be made. The name plate, of course, would provide the rated conditions. The design ratio would be available and could be used to make a decision regarding the suitability for alternate service. With the bore, the rotor diameter, the cylinder length, and vane number and thickness, a calculation to determine the displaced capacity per revolution may be made. By applying an estimate of volumetric efficiency to the displacement value, the speed needed for a given output can be calculated. If the speed is within the allowable limits, and the pressure ratio is in the range of the original value, the compressor can be reapplied.

The following provides an estimate of Q_r, the displaced volume per revolution.

$$Q_r = 2 \, e \, L \, (\pi D - m \, s) \tag{4.12}$$

where

    e = eccentricity, $R - r$
    R = radius of cylinder bore
    r = radius of rotor
    D = diameter of cylinder bore
    m = number of vanes
    s = vane thickness
    L = cylinder length

Some typical values of geometry are r/R = .88 and e = .12 R. The L/R ranges from 4.5 to 5.8, increasing with the size of the compressor. Volumetric efficiency ranges from approximately .90 at 10 psig to .85 at 30 psig for air service. Volumetric efficiency is slightly better for heavier gases and lower for the lighter gases. Typical maximum vane speed, calculated using the cylinder bore as the diameter, is 50 fps.

Power requirements and discharge temperatures are calculated using the same relationships as used with the other rotary compressors already discussed. The efficiency is .80 for air service and pressure in the 30 psig range. The mechanical losses are higher than the other rotaries. The mechanical loss is variable and dependent on gas, lubrication, and other factors. For an estimate, use .15 of the gas horsepower. This approximation should be close enough for an estimate.

Application Notes

The sliding vane compressor can be used to 50 psig in single-stage form and when staged can be used to 125 psig. An often overlooked application for the sliding vane machine is that of vacuum service where, in single-stage form, it can be used to 28 in hg. Volumes in vacuum service are in the 5000 cfm range. For pressure service, at the lower pressures, volumes are just under 4000 cfm and decrease to around 2000 cfm as the discharge pressure exceeds 30 psig.

The sliding vane compressor is used in gas gathering and gas boosting applications in direct competition with the reciprocating compressor. Efficiency is not as good, but the machine is rugged and light and doesn't have the foundation or skid weight requirement of the reciprocator. The compressor is also widely used as a vapor extraction machine in a wide variety of applications, which include steam turbine condenser service for air extraction.

Vane wear must be monitored in order to schedule replacement before the vanes become too short and wear the rotor slots. If the vanes are permitted to become too worn on the sides or too short, the vane may break and wedge between the rotor and the cylinder wall at the point of eccentricity, possibly breaking the cylinder. Shear pin couplings or equivalent torque limiting couplings are sometimes used to prevent damage from a broken vane under sudden stall conditions.

As in most jacket-cooled compressors, the cooling acts as a heat sink to stabilize the cylinder dimensionally. The jacket outlet temperature should be around 115°F and be controlled by an automatic temperature regulator if the load or the water inlet temperature are prone to change.

Most of the drivers used with the sliding vane compressor are electric motors. Variable speed operation is possible within the limits of vane speed requirements. The vanes must travel fast enough to seal against the cylinder wall but not so fast that they cause excessive wear. For the smaller units, under 100 hp, V-belts are widely used. Direct connection to a motor, however, is possible for most compressors and is used throughout the size range.

Mechanical Construction

The cylinder is generally constructed of cast iron and includes the water jacket. The bore is machined and brought to a good finish to reduce the vane sliding friction. The inlet and outlet connections are flanged. The heads, which also house the bearings and stuffing box, are also made of cast iron.

The rotor and shaft extension are machined from a single piece of bar stock or from a forging in all but the largest sizes, where the rotor and shaft may be made as two separate parts. The material is carbon steel for the single-piece models. The larger compressors, using the two-piece rotor arrangement, use carbon steel for the shaft and cast iron for the rotor body. The rotor body is attached to the shaft using a press fit. Keys are used to lock the rotor body to the shaft. Vanes attach to the rotor body by means of milled slots.

For the lubricated machines, vanes are made of a laminated material impregnated with phenolic resin. For a non-lubricated design, carbon is used. The vane number influences the differential pressure between adjacent vane cells. This influence becomes less as the number of vanes increases.

Rolling element bearings are widely used, generally the roller type. Seals are either a packing or mechanical contact type. Packing and bearings are lubricated by a pressurized system. For the non-flooded, lubricated compressor a multiplunger pump, similar to the one used with reciprocating compressors, is used. Lubrication is directed from the lubricator to drilled passages in the compressor cylinder and heads. One feed is directed to each of the bearings. Other feeds meter lubrication onto the cylinder wall. As the vanes pass the oil injection openings, lubricant is spread around the cylinder walls to lubricate the vane tips and eventually the vanes themselves. The oil entering the gas stream is separated in the discharge line. Because of the high local heat, the lubricant may have broken down and, therefore, is not suitable for recycling.

Flooded compressors pressure feed a large amount of lubricant into the compressor where it both cools the gas and lubricates the compressor. It is separated from the gas at the discharge line and recycled.

Liquid Piston

Operation

The liquid piston compressor is a unique type of rotary compressor in that it performs its compression by use of a liquid ring acting as a piston. Refer to Figure 1-9 for a cross section of this compressor. As with the sliding vane compressor, the single rotor is located eccentrically inside a cylinder or stator. Extending from the rotor is a series of vanes in a purely radial or radial with forward curved tip orientation. Gas inlet and outlet passages are located on the rotor. A liquid compressant partially fills the rotor and cylinder and orients itself in a ring-like manner as the rotor turns. Because of the eccentricity, the ring moves in an oscillatory manner. The center of the ring communicates with the inlet and outlet ports and forms the gas pocket. As the rotor turns, and the pocket is moving away from the rotor, the gas enters through the inlet and fills the pocket. As the rotor turns, it carries the gas pocket with it. Further turning takes the liquid ring from the maximum clearance area toward the minimum side. The ring seals off the inlet port and traps the pocket of gas. As the liquid ring is taken into the minimum clearance area, the pocket is compressed. When the ring uncovers the discharge port, the compressed pocket of gas is discharged.

Performance

Efficiency of the liquid piston is about 50%, which is not very good compared to the other rotary compressors. But because liquid is integral to the liquid piston compressor, taking in liquid with the gas stream does not affect its operation as it would in other types of compressors. Because of significant differences in the construction of the various competitive models of this compressor, no universal sizing data are available. The process engineer will therefore have to rely on catalog data for sizing estimates. The liquid ring compressor is most often used in vacuum service; although, it can also act as a positive pressure compressor. The liquid piston machine can be staged when the application requires more differential pressure than can be generated by a single stage. The liquid piston compressor can be used to compress air to 100 psig. Vacuums of 26 in hg are possible. Flow capacity ranges from 2 cfm to 16,000 cfm.

Mechanical Construction

Standard materials for the compressor are cast iron for the cylinder and carbon steel for the shaft. The rotor parts are steel. The liquid piston compressor has another feature that compensates for low efficiency. By using special materials of construction and compatible liquid compressant, unusual or difficult gases may be compressed. By using titanium internal materials and water as a compressant, gases containing wet chlorine can be compressed. This is a very difficult application for most of the other compressor types.

Both rolling element and split sleeve bearings are used. Normally, packing is used for shaft sealing or for special services. Mechanical contact seals can be used.

References

1. Laing, P. O., *The Place of the Screw Compressor in Refrigeration,* a paper presented to the Institution of Mechanical Engineers at Grimsby, March 1968.

2. Ingram, Walter B., *Notes on SRM Compressor History,* unpublished.

3. Ingram, Walter B., *Screw Compressor Performance,* Frick Company Bulletin, Waynesboro, PA., Form 300-13A.

4. Pillis, J. W., *Development of a Variable Volume Ratio Screw Compressor,* IIAR Annual Meeting, April 17–20, 1983.

5. API Standard 619, *Rotary-Type Positive Displacement Compressors for General Refinery Services,* Second Edition, 1985 Reaffirmed 1991, Washington, DC: American Petroleum Institute, 1975.

6. Barber, A. D., "Computer Techniques in the Design of Rotary Screw Compressors," a paper presented at Conference held at the University of Strathclyde, Glasgow, March 21–22, 1978, in *Design and Operation of Industrial Compressors,* Institution of Mechanical Engineers Conference Publications, 1978.

5

Centrifugal Compressors

Introduction

Centrifugal compressors are second only to reciprocating compressors in numbers of machines in service. In the process plant arena, the leader in numbers is too close to call with any degree of certainty. Where capacity or horsepower rather than numbers is considered as a measure, the centrifugal, without a doubt, heads the compressor field. During the past 30 years, the centrifugal compressor, because of its simplicity and larger capacity/size ratio, compared to the reciprocating machine, became much more popular for use in process plants that were growing in size. The centrifugal compressor does not exhibit the inertially induced shaking forces of the reciprocator and, therefore, does not need the same massive foundation. Initially, the efficiency of the centrifugal was not as high as that of a well-maintained reciprocating compressor. However, the centrifugal established its hold on the market in an era of cheap energy, when power cost was rarely, if ever, evaluated.

The centrifugal compressor has been around for quite a long time. Originally, it was used in process applications at relatively low-pressure, high-volume service. In the early 1930s, the main application was in the steel industry, where it was used chiefly as an oxidation air compressor for blast furnaces. The centrifugal displaced the reciprocating blowing engines that were being used at the time. The centrifugal was employed in the coal-to-coke conversion process, where it was used to draw off the gas from the coke ovens. In the late 1930s, the beginning of air conditioning for movie theaters, department stores, and later office buildings, gave birth to a generation of small centrifugals, which gained the advantage because of smaller size and absence of shaking forces. These forces were difficult to contain when a comparable capacity reciprocating compressor was used in a populated environment. It was the smaller compressor design that was able to penetrate the general process plant market, which had historically belonged to the reciprocating compressor. As stated previously, the growth of plant size and low-cost energy helped bring the centrifugal compressor into prominence in the 1950s. As the compressor grew in popularity, developments were begun to improve reliability, performance, and efficiency. With the increase in energy cost in the mid 1970s, efficiency improvements moved from last to first priority in the allocation of development funds. Prior to this turn of events, most development had concentrated on making the machine reliable, a goal which was reasonably well achieved. Run time between overhauls currently is three years or more with six-year run times not unusual. As plant size increased, the pressure to maintain or improve reliability was very high because of the large economic impact of a nonscheduled shutdown. This being the case, even with an increase in the efficiency emphasis, there is no sympathy for an energy versus reliability trade-off. The operating groups tend to evaluate reliability first, with the energy cost as secondary.

The centrifugal compressor has been applied in an approximate range of 1,000 cfm to 150,000 cfm. Plant air package centrifugals are available somewhat lower in capacity but have problems competing because of other more efficient compressors that are available in the lower ranges. Pressure ratios and pressure levels are difficult to describe in general terms because of the wide range of applications. Pressure ratio is probably the best parameter for comparing the centrifugal compressor to other types of compressors. Polytropic head, as defined in Chapter 2, is much more definitive to the dynamic machine but does not mean much numerically to a user. Pressure ratios of up to 3 and higher are available for single-stage compressors, operating on air or nitrogen. Multistage machines, of the process type, generally operate at a pressure ratio of less than 2 per impeller.

Classification

A better definition of a compressor stage can be made here to prevent confusion later on. Up to this point, in the positive displacement compressors, a compressing entity and a stage were one and the same; for example, a cylinder is a stage in the reciprocating compressor. The centrifugal and the other dynamic compressors to be discussed have the problem of a dual vernacular, one used by the machine design engineer and the one used by the process engineer. To the machine builder, a stage is an impeller-diffuser pair, whereas the process designer tends to think of a stage as a process block that equates to an uncooled section of one or more impeller and diffuser sets. There is no problem with the single impeller machine as the two are synonymous. The confusion comes with the use of the multiple impeller machine. To make everyone equally happy or unhappy, as the case may be, hereinafter, a process compression stage will be referred to as an *uncooled section*. Whenever the term *stage* must be used in the process connotation, it will be called a *process stage*. The multiwheeled machine will retain the name of multistage, and the individual impeller and diffuser pairs will be called a stage.

With the foregoing discussion as an entreé to the types of centrifugal compressors, it seems redundant to classify them as single and multistage. A cross classification can be established by the manner in which the machine casing is constructed, whether it has an axial or radial joint. More commonly, this type of construction is referred to as horizontal and vertical split. For simplicity, the second terminology will be used. The *overhung* style of single stage is an example of the vertical split type of compressor (see Figure 5-1). An example of the horizontally split compressor is the common multistage. Maintenance of the horizontally split compressor is very simple and straightforward, as the rotor may be removed without disturbing the impellers. When the pressure is too high to maintain a proper joint seal or for low molecular weight service, another style commonly used is referred to as a barrel compressor (see Figure 5-2). The barrel uses a vertical split construction. In the multistage configuration, it is constructed with a removable, horizontally split, inner barrel that permits the removal of the rotor without removing the impellers. Many overhung compressors do not permit the removal of the rotor without first removing the impeller.

Another common type of compressor is manufactured in an integrally geared configuration. It is basically an overhung style machine mounted on a gear box and uses the gear pinion shaft extension to mount an

Figure 5-1. Single-stage, vertically split, overhung style centrifugal compressor. (*Courtesy of Elliott Company*)

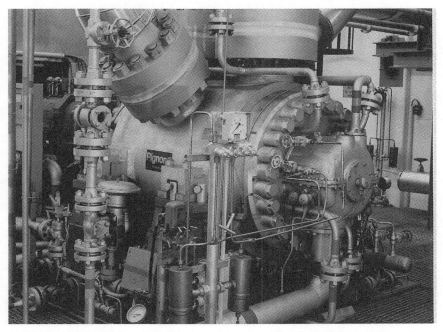

Figure 5-2. Multistage barrel type compressor. (*Courtesy of Nuovo Pignone*)

impeller (see Figure 5-3). The casing is also attached to the gear box. This style is built in both the single and multistage configuration. The most common form of multistage is the plant air compressor, which also has intercoolers included as part of the machine package.

Figure 5-3. Overhung, gearbox mounted centrifugal compressor. (*Courtesy of Atlas Copco Comptec Inc.*)

Arrangement

The single stage can be arranged, as has been discussed in the previous paragraphs, in the overhung style. Figure 5-4 shows a schematic of the compressor. Note that the flow enters axially and exits in a tangential direction. For a comprehensive discussion, it should be mentioned that the overhung style is, on very rare occasions, constructed in the multi-stage form, usually overhanging no more than two impellers. The over-hung compressor is generally more competitively priced than the between-bearing design. Careful application must be made because the overhung impeller configuration is more sensitive to unbalance than the between-bearing design. If impeller fouling is anticipated, this design may not be acceptable.

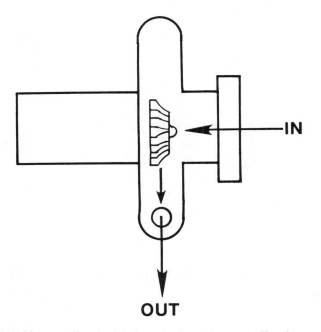

IN

OUT

Figure 5-4. Diagram of a single-stage overhung type centrifugal compressor.

A less common form of the single stage is shown in Figure 5-5. In this form, the impeller is located between two bearings, as is the multistage. This type of compressor is sometimes referred to as a *beam type* single stage. The flow enters and leaves in a tangential direction with the nozzles located in the horizontal plane. The between-bearing single stage is found most commonly in pipe line booster service where the inherent rigidity of the two outboard bearings is desirable.

Figure 5-6 is a flow diagram and schematic layout of the integrally geared compressor, and Figure 5-7 shows exploded view. It consists of three impellers, the first located on one pinion, which would have a lower speed than the other pinion that has mounted the remaining two impellers. This arrangement is common to the plant air compressor. Configurations such as this are used in process air and gas services, with the number of stages set to match the application.

Figure 5-8 shows the multistage arrangement. The flow path is straight through the compressor, moving through each impeller in turn. This type of centrifugal compressor is probably the most common of any found in process service, with applications ranging from air to gas. The latter includes various process gases and basic refrigeration service.

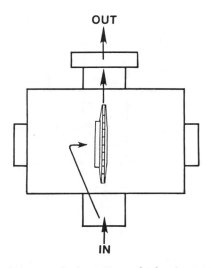

Figure 5-5. Diagram of a beam type single-stage compressor.

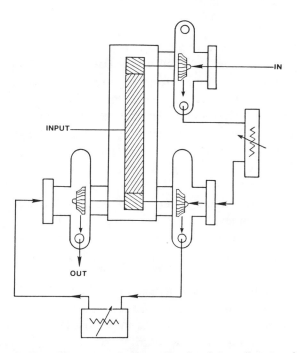

Figure 5-6. Flow diagram and schematic of an integrally geared compressor.

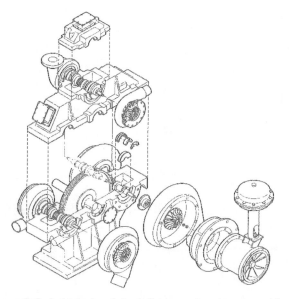

Figure 5-7. An exploded view of an integrally geared compressor. (*Courtesy of Cooper Turbocompressor*)

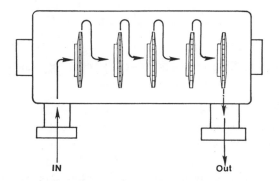

IN

Out

Figure 5-8. Diagram of a multistage centrifugal compressor with a straight-through flow path.

Figures 5-9 and 5-10 depict the two most common forms of in-out arrangements. This arrangement is also referred to as a compound compressor. In these applications, the flow out of the compressor is taken through an intercooler and back to the compressor. The arrangement is not limited to cooling because some services use this arrangement to remove and scrub the gas stream at a particular pressure level. Provision for liquid removal must be made if one of the gas components reaches its saturation

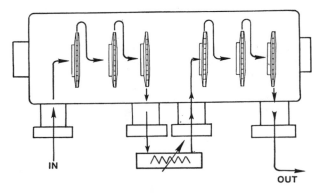

Figure 5-9. Diagram of an in-out arrangement with intercooling.

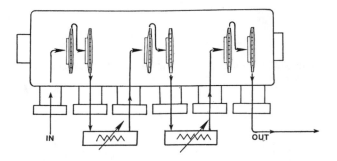

Figure 5-10. Diagram of a double-cooled centrifugal compressor.

temperature in the process of cooling. Figure 5-10 shows a double-cooled or double compound compressor. This arrangement is used mostly when the gas being compressed has a temperature limit. The limit may be imposed by the materials of construction or where the gas becomes more reactive with an increase in temperature and thus sets the limit in a given application. Polymer formation is generally related to temperature and may form the basis for an upper temperature limit. However, with the external cooling, the amount of compression needed can be accomplished in a single case. The physical space needed to locate the multiple nozzles normally limits the number of in-out points to the two shown.

The arrangement shown in Figure 5-11 is referred to as a double-flow compressor (see also Figure 5-12). As indicated in the figure, the flow enters the case at two points, is compressed by one or more stages at each end, and then enters the double-flow impeller. The flow passes through each individual section of the double-flow impeller and joins at

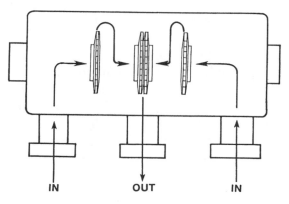

Figure 5-11. Diagram of a double-flow compressor with two inlets.

Figure 5-12. A double-flow compressor with inlets on each end and a common center discharge. (*Courtesy of Elliott Company*)

the diffuser. There are various physical arrangements to accomplish the double-flow compression. One variation is to use two back-to-back stages for the final compression and join the flow either internally, prior to leaving the case, or join two separate outlet nozzles outside the case.

From a process point of view, the flow should be joined prior to exiting the discharge nozzle.

Another variation of this arrangement is to use it in the single-stage configuration, where only a single inlet and outlet nozzle is used. The flow enters the case and is divided to each side of the double-flow impeller and then joins at the impeller exit prior to entering the diffuser. Figure 5-13 shows a schematic diagram of the flow in this machine. The advantage of the double-flow arrangement is, of course, that in the same casing size, it doubles the flow. However, the realization of the advantage is more complex. The losses in the flow paths through the double-flow impeller must, in theory, be identical. In practice, of course, this is not possible. The sensitivity is a function of the total head level. The lower the levels, the more nearly the paths must be the same.

The single-stage configuration, the lower head compressor, will exhibit the highest degree of sensitivity to the flow imbalance and have its performance most adversely affected. The multistage configuration, while not as sensitive to the flow anomalies because of the higher head generated, will benefit from careful flow path design to keep the flow balanced to each section of the double flow inlets. If a number of options are open for a given application, the double-flow option should not be the first choice; although, it should be evaluated because successful applications in service indicate that with careful design the compressor will perform satisfactorily.

The arrangement in Figure 5-14, generally called "back to back," is normally considered useful in solving difficult thrust balance problems where the conventional thrust bearing and balance drum size are inade-

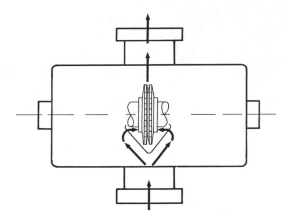

Figure 5-13. Diagram of a double-flow compressor with flow split internally.

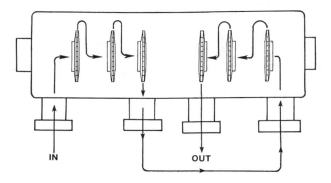

Figure 5-14. Diagram of an arrangement used to overcome a thrust balance problem.

quate or become excessively large. The balance drum will be described in detail in a following section. The flow is removed part way through the compressor and reintroduced at the opposite end, then allowed to exit at the center. Because centrifugal impellers inherently exhibit a unidirectional thrust, this arrangement can be used to reduce the net rotor thrust. The obvious use is for applications generating high thrusts, higher than can be readily controlled by a normal size thrust bearing and balance drum. An evaluation of the cross leakage between the two discharge nozzles must be made and compared to the balance drum leakage to determine the desirability of the "back to back." It can be combined with the sidestream modes, discussed in the next paragraph, to possibly help sway a close evaluation. In some rare cases, this design has been used for two different services. Unfortunately, it is difficult to totally isolate the two streams because of the potential cross leakage. In cases where the two services may have a common source, or the mixing of the streams does not cause a problem, it is possible to generate savings by using only one compressor case.

A very common compressor design used in the chemical industry, particularly in large refrigeration systems, is the *sidestream* compressor (see Figure 5-15). Gas enters the first impeller and passes through two impellers. As the main stream approaches the third impeller, it is joined by a second stream of gas, mixed, and then sent through the third impeller. The properties of the gas stream are modified at the mixing point, as the sidestream is rarely at the same temperature as the stream from the second impeller. In refrigeration service, this stream is taken from an exchanger where it is flashed to a vapor, resulting in a stream temperature near saturation. As such, the sidestream would act to cool

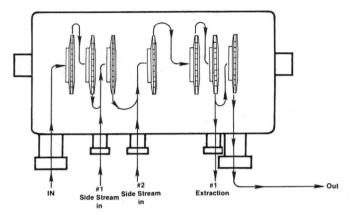

Figure 5-15. Diagram of flow path through a sidestream compressor.

the total stream. The weight flow to the third impeller is the combined weight flow of the two streams.

The second sidestream follows the same logic. To show the flexibility of the arrangement, the last sidestream is indicated as an extraction. This stream could be used where heated gas at less than discharge pressure is required. Using the extraction saves the energy needed to compress this quantity of gas to the full discharge pressure and then throttling for the heating service. One potential application of an extraction stream is for use in a reboiler. The arrangement shown was arbitrarily chosen to illustrate the available options. The total number of sidestream nozzles is limited only by the physical space required to locate them on the case. Three nozzles are not uncommon.

When applications are more complex than can be accommodated by a single-case compressor, multiple cases can be used. The most frequently used is the tandem-driven series flow arrangement using a common driver (see Figure 5-16). A gear unit may be included in the compressor train, either between cases or between the driver and the compressors. The individual compressor cases may take the form of any of the types described before. The maximum number of compressors is generally limited to three. Longer, tandem-driven series-connected compressor trains tend to encounter specific speed problems. In the longer trains, the double-flow arrangement can be useful in permitting more compressors to run at the same speed. At the inlet, where flow is the highest, the gas stream is divided into parallel streams and the volume is reduced by compression to a value within the specific speed capability of a single-flow compressor. The

Figure 5-16. A tandem driven multi-body centrifugal compressor train with a steam turbine driver. (*Courtesy of Demag Delaval Turbomachinery Corp.*)

alternative to the double-flow arrangement is the use of a speed increasing gear between compressor bodies to permit the flow matching of downstream stages. This is one case where the double-flow compressor should be considered first. When longer trains are needed, the cases are grouped with several individual drivers, maintaining the series flow concept. One installation that can be recalled used nine individual cases, separately driven and series connected, for a very high pressure air application.

Drive Methods

Historically, the most popular driver for the centrifugal compressor has been the steam turbine. Steam turbines can readily be speed matched to the compressor. Prior to the upsurge in energy costs, reliability, simplicity, and operational convenience were the primary factors in driver selection. The steam turbine, with its ability to operate over a relatively wide speed range, was ideal for the centrifugal compressor, which could be matched to the process load by speed modulation.

With the advent of energy as a more significant consideration in driver selection, the electric motor received a higher degree of attention. While motors were probably second to the steam turbine in general industry usage, the limitations imposed by a constant speed driver tended to discourage their use in process plants. But because fossil fuel can be more efficiently converted to electricity in large central generating stations, the cost of electrical energy for motors became such that they began to displace the more convenient steam turbines. Local steam generation cannot be accomplished at a competitive energy cost in many instances. While large electric drivers using variable frequency conversion to provide for variable speed are relatively new, they provide an alternative to the steam turbine. Two primary factors that have prevented universal acceptance of the variable frequency system are cost and experience. As more units are furnished, and with the passing of time, the negative factors will undoubtedly begin to diminish.

Electric motors, whether speed controlled or not, are either *induction* or *synchronous* in design. Size and plant electric system requirements set the parameters for motor selection. Synchronous motors normally receive consideration only for the larger drives, with the individual plant setting the minimum size at which the synchronous machine is used. Regardless of which motor type is selected, a speed increasing gear will be needed, because motor speed is rarely high enough to match the necessary centrifugal compressor speed.

As an alternate to the drivers mentioned, a gas turbine may be selected as the driver. If exhaust heat recovery or regeneration is used, the efficiency of the gas turbine is quite attractive. Unfortunately, the gas turbine is expensive and in some cases has demonstrated high maintenance cost. It should be understood that gas turbines are relatively standardized even though they cover a wide range of power and speed. They are not custom engineered to the specific application for a power and speed as is customary with steam turbines. In many applications, a speed matching gear must be included, which adds the complication of another piece of equipment, subsequently higher capital cost, and potentially decreased reliability. This gear also inherently has a high pitch-line velocity making for one of the more difficult applications. Despite some of the hurdles just mentioned, the gas turbine is widely used in offshore installations because of its superior power-to-weight ratio over other drivers. It is quite popular for use in remote locations where the package concept minimizes the need for support equipment. As an example, the north slope of Alaska is estimated to have in excess of 1.5 million horsepower in gas turbine powered compressors.

The remaining driver is the *gas expander,* which can only be considered if the process stream has the potential for energy recovery. The expander can be either *cryogenic* or *hot gas* in design depending on the application. Normally the cryogenic expanders are relatively small in size and may be integral with the compressor. These are relatively special purpose and do not have a wide range of application. The hot-gas expander tends to be a larger machine and makes an excellent driver in that it can be speed matched to the compressor and may have variable speed capability. The expander must operate at high temperatures to have sufficient energy for a reasonable output power level. The high temperature does make the supply piping design somewhat complex and also makes the cost of the expander higher than a comparably sized steam turbine. Alignment maintenance is more difficult than with other drivers. It would seem fairly obvious that the economic return of this driver would have to be quite favorable to entice someone to consider it. There are numerous successful installations using the expander, so it is a viable alternative to consider under proper circumstances.

Performance

Compression Cycle

Figure 5-17 is a section of a typical multistage compressor, which should aid the reader in following the flow path through the machine.

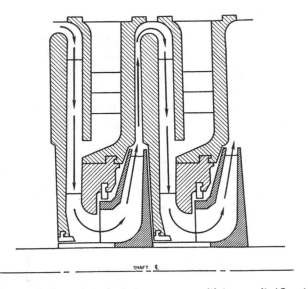

Figure 5-17. Flow path through typical stages on a multistage unit. (*Courtesy of Elliott Company*)

Gas enters the impeller from one of several sources. In the case of the first impeller of a multistage, the flow has moved through an inlet nozzle and is collected in a plenum from which it is then directed into the first impeller. Another possible path occurs when the flow has passed through one or more stages and approaches the impeller through a channel referred to as a return passage. In the return passage, the flow stream passes through a set of vanes. The vanes are called *straightener vanes,* if the flow is directed axially at the impeller entrance (eye), or *guide vanes,* if the flow is modified by the addition of prerotation. The final possible path occurs when the flow comes into the compressor from a sidestream nozzle. This stream is directed into the flow stream to mix and be direct- ed into the impeller eye using one of two alternative methods as shown on Figure 5-18. One method is by way of a blank section between the stages where the stream mixing point is immediately ahead of the impeller inlet. This method is used if the sidestream flow is large in com- parison to the through flow. The alternative is used when the flow is small compared to the through flow, and consists of injecting the flow into the return passage from the previous stage. The latter has better mix- ing, and takes less axial space, but has a higher pressure drop. For the former, the opposite is true. It has a lower pressure drop, but exhibits somewhat poorer mixing and uses more axial space, normally at least a

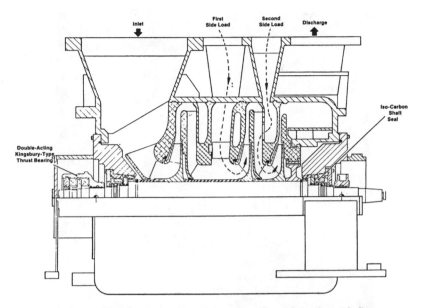

Figure 5-18. Two methods of directing sidestream flows into through flows. (*Courtesy of Elliott Company*)

full-stage pitch in length. A *stage pitch* is defined as the axial distance measured from the entrance of one impeller to the same location on the following impeller. Stage pitch may be a constant, as on low-volume ratio staging, or variable, as may be found in higher-volume ratio stages. The variable stage pitch is commonly used on higher flow coefficient stages using the 3D impeller designs. The importance of physical length will become apparent as the entire compressor is explored, but at this point, it will suffice to say that there never seems to be enough.

Generally, there are no vanes in the inlet of an axial entry compressor (see Figure 5-19). Normally there is no more than the plenum divider vane in the inlet section of the typical multistage compressor, although there are designs that use vanes in this area. These are externally movable and are used to provide flow control for constant speed machines. The use of these vanes will be explored further in the section on capacity control.

After the flow has been introduced into the compressor and has been acted on by one or more stages, it must be extracted. Because there is a relatively large amount of velocity head available in the stream, care must be used when designing the discharge section to keep the head loss low and maintain overall efficiency. The flow from the last stage is gath-

Figure 5-19. The impeller blades can be seen in this view through the inlet of a single-stage compressor. (*Courtesy of Atlas Copco Comptec, Inc.*)

ered in some form of collector, normally a scroll, in an effort to convert as much of the remaining velocity head as possible into pressure. With intermediate extraction, or for some of the in-out designs, a compromise must be made, reducing large passages to preserve axial length.

Having gotten the flow in and out of the machine, a closer examination of just how the compression takes place is needed. An important concept to maintain throughout the following discussion is that all work done to the gas must be done by the active element, the impeller. The stationary element is passive, that is, it cannot contribute any additional energy to the stage. It can only convert the energy and unfortunately contribute to the losses. Figure 5-20 is a schematic diagram of an impeller and the basic inlet and outlet flow vector triangles.

The impeller will be covered in detail in the following sections; therefore, a brief review of the various impeller components is in order. The

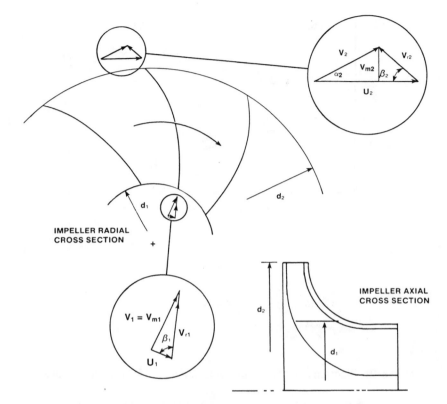

Figure 5-20. Impeller inlet and outlet flow vector triangles.

impeller consists of a set of vanes radially oriented on a hub. The vanes are enclosed either by a rotating or stationary front and rear shroud. If both front and rear shroud are stationary, the impeller is referred to as an *open impeller.* If the rear shroud is attached to the vanes and rotates as a part of the impeller assembly, it is referred to as *semi-open.* If the front shroud is also attached to the vanes and rotates with the assembly, it is referred to as a *closed impeller.* The vanes may be forward curved, radial, or backward curved, as shown diagramatically in Figure 5-21. Forward curved vanes are normally only used in fans or blowers, and rarely, if ever, used in centrifugal compressors.

Figure 5-21 includes an outlet velocity vector triangle for the various vane shapes. Figure 5-20 shows a backward curved impeller that includes the inlet and outlet velocity vector triangle. Because most of the compressors used in process applications are either backward curved or radial, only these two types will be covered in detail.

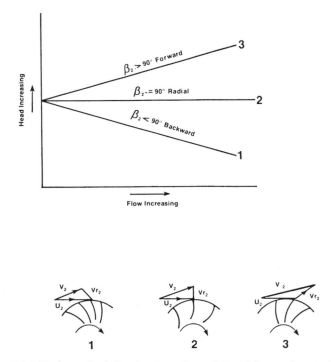

Figure 5-21. Diagram depicting backwards, radial and forward curved blades.

Vector Triangles

Gas enters the impeller vanes at the diameter d_1. The absolute gas velocity approaching the vanes is V_1. As shown in Figure 5-20, the gas approaches the vane in a radial direction after entering the impeller in an axial direction and makes the turn to a radial direction inside the impeller. The vane leading edge velocity is represented by the velocity vector u_1. The net velocity is the relative velocity V_{r1}. It should be noted for this basic example that the relative velocity vector aligns itself with the vane angle β_1, resulting in zero incidence. In this idealized case, the *meridional flow vector* V_{m1} is aligned with and equal to the absolute velocity. After passing between the vanes, the gas exits the impeller at the diameter d_2. The velocity of the gas just prior to leaving the impeller is the relative velocity V_{r2} and leaves at the vane angle β_2 in the idealized example. By the addition of the impeller tip velocity vector u_2, the absolute leaving velocity V_2 is generated. The angle of the absolute flow vector is α_2. This is the velocity and direction which the gas assumes as it leaves the impeller and enters the diffuser. The meridional velocity V_{m2} is shown by the radial vector passing through the apex of the outlet velocity triangle. If the vane was radial, rather than backward leaning, $\beta_2 = 90°$, the relative velocity and the meridional velocity would be equal and aligned.

Slip

In real world application, the gas leaving the impeller will not follow the vane exit angle. The deviation from the geometric angle is referred to as *slip*. The leaving angle will be referred to as the gas angle β'_2. Figure 5-22 shows the discharge velocity vector triangle, including the effect of slip. The terms on the ideal triangle are the same as those used in Figure 5-20. Superimposed over the ideal triangle is the velocity triangle, including the effect of slip. Note that the terms are indicated with the prime (′) symbol. While there are numerous papers written on the subject of slip, none seem to present a complete answer. One of the better papers, which summarizes the field and brings the subject into focus, is the one by Wiesner [7]. In this book, for the purpose of understanding the workings of the centrifugal compressor, the Stodola slip equation will be used. It is probably one of the oldest and has been used in practical design prior to the advent of some of the more sophisticated methods available now. Returning to the triangle under discussion, the gas angle, β'_2, is always less than the geometric angle, β_2. In Figure 5-22, projections are

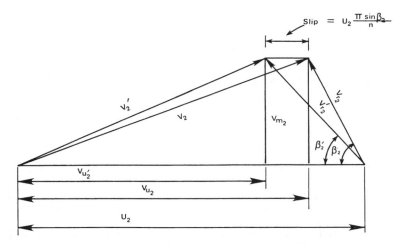

Figure 5-22. Discharge velocity vector triangle showing the effect of slip.

made onto the tip velocity vector from the absolute gas vectors, V_2 and V_2'. These are labeled as V_{u2} and V_{u2}', respectively, and have the designation of tangential component of the absolute velocity. From these vectors, some simple relationships can be presented that will give a reasonable explanation of how the centrifugal compressor geometry relates to its ability to compress gas. The ideal work input coefficient, ζ_i, is given by the following expression:

$$\zeta_i = \frac{V_{u2}}{u_2} \tag{5.1}$$

where

    V_{u2} = tangential component of the absolute velocity
    u_2 = impeller tip velocity

The ideal head input to the stage is given by

$$H_{in\ ideal} = (1/g)\zeta_i\, u_2^2 \tag{5.2}$$

The Stodola slip factor is defined as

$$Slip = u_2\, \frac{\pi \sin\beta_2}{n_v} \tag{5.3}$$

where

    β_2 = geometric vane exit angle
    n_v = number of vanes in the impeller

The slip factor SF follows.

$$SF = \frac{V_{u2}'}{V_{u2}} \tag{5.4}$$

Reference is made to Figure 5-22, where

$$V_{u2}' = V_{u2} - slip \tag{5.5}$$

Substituting into Equation 5.4 yields the following slip factor equation:

$$SF = 1 - \frac{u_2}{V_{u2}} \left(\frac{\pi \sin \beta_2}{n_v} \right) \tag{5.6}$$

The actual work input coefficient, ζ, is written by taking the ideal work input coefficient, Equation 5.1, and modifying by the addition of the slip factor, SF. The geometric relationship of the Stodola slip function is shown in Figure 5-23.

$$\zeta = \frac{V_{u2}}{u_2} (SF) \tag{5.7}$$

By replacing the ideal work input coefficient with actual work input coefficient, the actual head input can be written as

$$H_{in} = (1/g)\zeta u_2^2 \tag{5.8}$$

If the head coefficient is written as

$$\mu = \eta \zeta \tag{5.9}$$

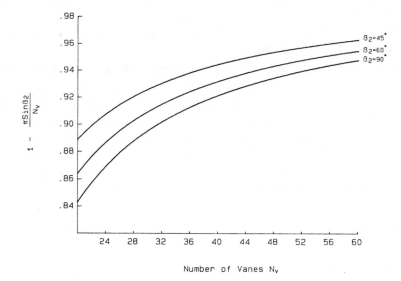

Figure 5-23. Geometric relationship of Stodola slip function.

where

η = stage efficiency, then

$$H_{out} = (1/g)\mu u_2^2 \tag{5.10}$$

For adiabatic head, the head coefficient is defined as μ_a and Equation 2.70 is recalled. The geometric and the thermodynamic head relationships for a stage may be equated.

$$H_a = \frac{\mu_a u_2^2}{g} = Z_{avg} RT_1 \frac{k}{k-1} (r_p^{\frac{k-1}{k}} - 1) \tag{5.11}$$

Similarly, for polytropic head, the head coefficient is defined as μ_p and Equation 2.73 is recalled, the geometric and thermodynamic head relationships, on a per-stage basis, may be equated as above.

$$H_p = \frac{\mu_p u_2^2}{g} = Z_{avg} RT_1 \frac{n}{n-1} (r_p^{\frac{n-1}{n}} - 1) \tag{5.12}$$

In the previous paragraphs, the term *specific speed* has been used. This is a generalized turbomachinery term used quite successfully with pumps and to some extent with turbines. It can be used with turbocompressors to help delimit the various kinds of machines. It is also used as a general term to describe the need for a correction on multistage machines when the wheel geometry at the current speed will no longer support a reasonable efficiency. For compressors, specific speed is paired with specific diameter to include the geometric factors. In centrifugal compressors, attempts have been made to correlate efficiency directly to these parameters. Most designers feel the relationships, while satisfactory to set bounds, are not adequate for describing impeller efficiency with good resolution. Definitions for specific speed, N_s, and specific diameter, D_s, are

$$N_s = \frac{NQ_1^{1/2}}{H_a^{3/4}} \tag{5.13}$$

$$D_s = \frac{DH_a^{1/4}}{Q_1^{1/2}} \tag{5.14}$$

Reaction

The outlet vane angle for the normal centrifugal compressor varies from radial to a backward leaning angle. An ideal vector tip triangle, with no slip, is shown in Figure 5-24. Three angles are illustrated to show the effect of varying the vane outlet angle.

Reaction is defined as the ratio of the static head converted in the impeller to the total head produced by the stage. Restating in a more philosophical sense, the object of the compressor stage is to increase the pressure of the gas stream, and reaction gives the relationship of the division of effort between the impeller and the diffuser.

Ideal reaction, R_i, is defined as

$$R_i = \frac{2 + \cot \beta_2}{4} \tag{5.15}$$

One of the practical aspects of reaction is that for a well-proportioned stage, the higher the reaction, the higher the efficiency. Again, using a philosophical approach to explain, for a given stage the impeller is more efficient than the diffuser. This is particularly true for the typical process compressor that uses a simple vaneless diffuser. If the radial vane impeller is used for the reference, it will have an ideal reaction of 50%, as calculated using Equation 5.15. Because the static head conversion is evenly divided between the impeller and the diffuser, the net stage effi-

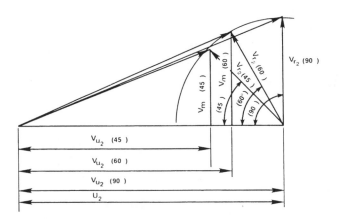

Figure 5-24. Vector tip triangle without slip, showing the effect of different exit vane angles.

ciency is the numeric average of the impeller efficiency and the diffuser efficiency. Figure 5-25 shows that as the vane angle decreases, the reaction increases. If the efficiency is evaluated for the lower angle, the net stage efficiency is now the weighted average of the two component individual efficiencies, with the higher impeller efficiency contributing a greater influence. A numeric example may help to illustrate the idea.

Example 5-1

Assume

Impeller efficiency = .90

Diffuser efficiency = .60

Calculate an ideal stage efficiency for a radial and a 45° backward leaning impeller.

For the radial impeller, using Equation 5.15,

$R_i = .50$

$$.50 \times .90 = .45$$

$$.50 \times .60 = .30$$

$.45 + .30 = .75$ net stage efficiency

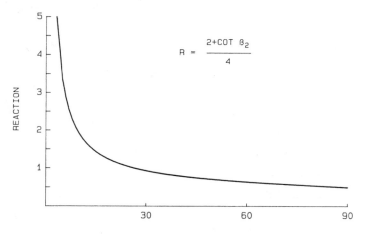

Figure 5-25. Theoretical reaction without slip.

For a 45° backward leaning impeller,

$R_i = .75$

The diffuser then converts $1 - R_i$ or .25

$$.75 \times .90 = .68$$

$$.25 \times .60 = .15$$

$.68 + .15 = .83$ net stage efficiency

The example indicates that an improvement of seven percentage points was achieved by backward leaning the vanes 45°. The obvious question arises. Why not make all impellers high reaction? Maybe this can be put into a good/bad analogy. The good is better efficiency. The bad is a lower head produced by the stage. To see why the head is less, review Figure 5-24. It can be seen that as the outlet angle, β_2, is decreased, the tangential component of the absolute velocity, V_{u2}, is decreased. If Equation 5.1 is recalled, it should be noted that a decrease in V_{u2} will decrease the value of the head input coefficient, ζ_i. By carrying a lower value of ζ_i into Equation 5.9, the head coefficient, μ, is decreased. In Equation 5.10, it is obvious that for a lower μ the output head is decreased. There is some relief in that in Equation 5.9, the stage efficiency η increases to offset the lowered ζ. However, in real life, this is not enough to make up the difference and the output head of a higher reaction stage is indeed lower. There are several effects that influence a commercial design and, again, the designer is faced with trade-offs. Equation 5.10 indicates that increase in the tip velocity u_2 would offset the loss in μ. Impeller stresses and rotor dynamics must also be considered and may act to limit the amount of correction that can be made. Another possibility is using additional stages. A well-proportioned stage is assumed, which brings to light the fact that the high reaction stage tends to use more axial length. This tends to counter the addition of extra stages, especially where the length of the rotor is beginning to cause critical speed problems. Despite the conflicts, changing reaction can sometimes aid the designer in achieving a higher efficiency. Another benefit is a steeper head-capacity curve. Also in some cases, the higher reaction stage seems to perform better where fouling is evident.

Sizing

Many of the steps used in sizing estimates are also useful for checking bids or evaluating existing equipment. In the latter two endeavors, there is one advantage: someone else has established the initial evaluation criteria. When working from a material balance flow sheet as a starting point, it is sometimes difficult to envision what the compressor should look like. Except for the addition of a few rules of thumb, most of the tools needed have already been established. The method outlined is based on the more conventional multistage compressors used in process service. Earlier, integrally geared, as well as direct expander driven compressors, were briefly described. These compressors may also be sized by the method outlined, but because they are tailored for higher head service, modifications to the method regarding the head per stage and the head coefficient are necessary.

To start, convert the flow to values estimated to be the compressor inlet conditions. Initially, the polytropic head equation (Equation 2.73) will be used with n as the polytropic compression exponent. If prior knowledge of the gas indicates a substantial nonlinear tendency, the real gas compression exponent (Equation 2.76) should be substituted. As discussed in Chapter 2, an approximation may be made by using the linear average of the inlet and outlet k values as the exponent or for the determination of the polytropic exponent. If only the inlet value of k is known, don't be too concerned. The calculations will be repeated several times as knowledge of the process for the compression cycle is developed. After selecting the k value, use Equation 2.71 and an estimated stage efficiency of 75% to develop the polytropic compression exponent n.

The molecular weight, inlet temperature, and inlet pressure are combined with the compressibility and discharge pressure in Equation 2.73 to estimate the polytropic head. The average of inlet and outlet compressibility should be used, using the polytropic discharge temperature calculated by the following equation to evaluate the discharge compressibility.

$$T_2 = T_1 \, r_p^{\frac{n-1}{n}} \tag{5.16}$$

where

T_2 = absolute discharge temperature of the uncooled section
T_1 = absolute inlet temperature of the uncooled section

To determine the number of stages, using the impeller and diffuser defined as the stage, assume 10,000 ft-lb/lb of head per stage. This value can be used if the molecular weight is in the range of 28 to 30. For other molecular weights, this initial value must be modified. As a rule of thumb, lower the head per stage by 100 ft-lb/lb for each unit increase in molecular weight. Conversely, raise the allowable head per stage 200 ft lb/lb for a unit decrease in molecular weight. The rule of thumb gives the best results for a molecular weight range of 2 through 70. Because this sizing procedure is being used only to establish the rough size of the compressor, the upper range may be extended with some loss in accuracy.

Once the head per stage has been established, the number of stages can be estimated by taking the total head, as calculated by the head equation, and dividing by the head per stage value. A fraction is usually rounded to the next whole number. However, if the fraction is less than .2, it may be dropped. The stage number should be used to calculate a new head value per stage. This method assumes an uncooled or no sidestream compressor. If either of the two are involved, the uncooled sections can be estimated, taken one at a time. Assumptions for between-section pressure drop or sidestream mixing can be added to the calculation as appropriate to account for all facets of the process. When all calculations are completed, the compressor sections can be arranged to form a complete unit.

Before proceeding, a few limits need to be considered. The temperature, if not limited by any other consideration, should not exceed 475°F. This limitation is arbitrary, as centrifugals may be built to higher limits, but the estimator is cautioned not to venture too far into this region without additional considerations. The number of stages per casing should not exceed 8 for rotor dynamics considerations. Also, knowledge of auxiliary nozzle stage pitch would be needed to evaluate exactly how far to venture in this direction. Vendor literature advertises the availability of as many as ten stages; however, an estimate should never go to the edge without a background of considerable experience. These limits can also be used to evaluate proposals and help to determine a series of questions for the vendor skirting the upper limits.

The next step begins by assuming a head coefficient equal to .48. Equation 5.12 can be used to calculate the tip speed, u_2. Figure 5-26 can be used to get an impeller diameter estimate from the inlet volume calculated earlier. The diagonal line on the diagram marks the right extremity of each impeller's flow range to guide the user in making the first selection. The tip speed and diameter can be used to calculate an approximate speed, N, by

$$N = \frac{u_2}{\pi d_2} \tag{5.17}$$

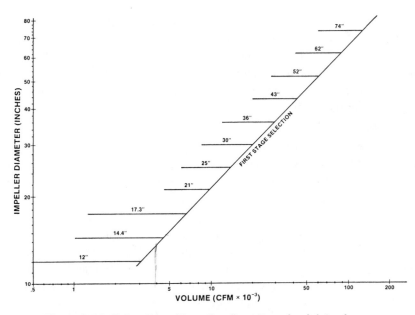

Figure 5-26. Estimation of impeller diameter using inlet volume.

where

d_2 = impeller outside diameter

To summarize the sizing to this point: the inlet volume, an overall head, number of stages, head per stage, impeller tip speed, and impeller diameter have been established. The one parameter of interest still missing is the efficiency. To obtain an estimate of efficiency without empirical data, a generalized form may be used. As in the previous chapters, where estimates were involved, the data presented is just one way to approach the problem, and any other reasonable source such as specific vendor data may be used. To use the generalized curve, Figure 5-26, the volume for the first and last stage must be developed. The volume for the first stage is the inlet volume. The volume for the last stage, Q_{ls}, can be estimated by

$$Q_{ls} = \frac{Q_{in}}{\left(r_p^{1-\frac{1}{z}}\right)^{\frac{1}{n}}} \qquad (5.18)$$

where

Q_{in} = inlet volume
r_p = pressure ratio for an uncooled section
z = number of stages in the uncooled section

Use the inlet and the last stage volume for the uncooled section and use the following equation to calculate the inlet flow coefficient δ.

$$\delta = 700 \, \frac{Q_i}{Nd_2^3}$$ (5.19)

where

Q_i = volumetric flow, ft³/min
N = rotational speed, rpm
d_2 = impeller diameter, in.

Note: This equation is not in the primitive form. While δ is basically dimensionless, the constant 700 is not easily derived; therefore, units were assigned.

The value for the first-stage flow coefficient should not exceed .1 for a 2D type impeller and for a 3D design, the upper value can be as high as .15. The value for the last stage should be no less than .01. If the flow coefficients should fall outside these limits, another impeller diameter should be selected. It may be necessary to interpolate to obtain a reasonable diameter from Figure 5-26. This can be done because this is an estimate and not bound to an arbitrary line of compressor frames. The diagram was set up to give the user an idea of how a compressor line might be organized. A vendor may quote values outside the guidelines due to the constraints of his available frame sizes. For estimates, values as close as possible to the given guidelines are recommended. At the time of a proposal, the benefit of stages beyond either extreme value of flow coefficient can be evaluated. It should be noted at this point that not all vendors report their flow coefficients on the same basis. If necessary, the parameters for flow coefficient should be obtained to permit evaluation with Equation 5.19. An average of efficiency can be calculated from two efficiencies selected from Figure 5-27. The figure includes efficiency values for 2D and 3D impeller designs. While it would appear obvious that only 3D impellers should be used, there is a caveat. Generally, 3D impellers require more space, that is, the axial stage spacing (stage pitch)

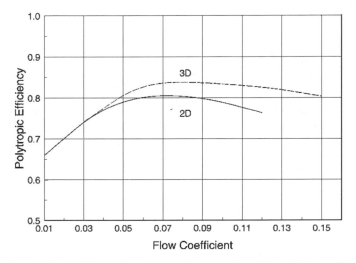

Figure 5-27. Centrifugal stage efficiency vs. flow coefficienct for 2D and 3D blading.

is longer. This will result in a longer compressor, which makes for possible rotor dynamic problems and does also increase cost. Also, it should be pointed out that the increase in efficiency begins above flow coefficients of .04. The increase in stage pitch can vary from approximately 1.1 to 1.3 times a 2D stage pitch with the values increasing with increased flow coefficient. For the 2D impeller, it should be noted that the peak efficiency occurs at a flow coefficient of approximately .07. The 3D impeller peak efficiency value curve is broader and occurs in the range of from .07 to values as high as 1.3.

At this point, after a first pass through the calculation, a new polytropic exponent should be calculated. All values calculated to this point should be rechecked to see if original estimates were reasonable. If the deviation appears significant, a second pass should be made to improve the accuracy. Equation 2.78 can be used to calculate the power for the uncooled section. For an estimate, use a value of 1% for the mechanical losses.

If time permits and a more accurate estimate is desired, particularly if the compressor is intercooled or has sidestreams, the velocity head losses through the nozzles can be estimated using the values from Table 2-2. This is possible where the nozzle sizes are available or can readily be estimated. When coolers are involved, the drop through the cooler should be included. Subtract the pressure drop from the inlet pressure (of the stage following the element) and recalculate a modified pressure ratio for the section. The cooler pressure drop can be approximated by using 2%

of the absolute pressure at the entrance to the cooler. Because the percentage gives unrealistic values at the lower pressures, a lower limit of 2 psi should be used. Compressors with in-out nozzles used to take gas from the compressor for external cooling and return to the compressor can experience some temperature crossover in the internal sections of the machine. Unless the design has specifically provided for a heat barrier, heating of the return gas can be expected. For a first estimate, a 10°F rise should be used. Balance pistons will be described in the mechanical section of this chapter. Briefly, the balance piston contributes a parasitic loss to the compressor not accounted for in the stage efficiency. The weight flow passing from the balance piston area, normally the discharge, and entering the suction must be added to the flow entering the first stage or the stage receiving the balance piston flow. Unfortunately, the flow is not the only problem, as the return flow also acts to heat the inlet gas. For discharge pressures of 150 psia or less, a value of 1% can be used. For pressures higher than 150 psia but under 1,000 psia, a value of 2% is a reasonable starting point. An equation for the heating is

$$t_w = \frac{t_i + t_d BP}{1 + BP} \qquad (5.20)$$

where

t_w = impeller inlet temperature

t_i = nozzle inlet temperature

t_d = temperature at the balance piston

BP = balance piston leakage fraction

The relationships are given to help the user size a compressor from scratch. The same relationships can be used in the bid evaluation process. The vendor-provided geometry and performance values can be compared to the original sizing, which should have been performed prior to going out for the bid. The vendor's results can be evaluated using some of the rules of thumb or guidelines provided. Any deviations can be used as a focus for additional discussions. Also, some insight can be gained into the vendor's sizing techniques, particularly the way the vendor trims out a selection. Incremental wheel sizing is fairly universal. Some vendors also offer fixed guide vane sections as part of a stage to aid in the achievement of a particular performance specification.

Example 5-2

Using the results of Example 2-2, size a centrifugal air compressor using the sizing procedure. A summary of the results is:

$Q1 = 6,171$ cfrn inlet volume

$w_m = 437.5$ lbs/min

$mw = 28.46$ molecular weight

$P_1 = 14.7$ psia inlet pressure

$t_1 = 90.0°F$ inlet temperature

$T_1 = 550°R$ absolute inlet temperature

$Rm = 54.29$ specific gas constant

Add the following conditions to complete the application:

$k = 1.395$ isentropic exponent for air

$P_2 = 40$ psia discharge pressure

Assumed polytropic efficiency

$\eta_p = .75$

Step 1. Calculate the polytropic exponent using Equation 2.71.

$$\frac{n-1}{n} = \frac{1.395-1}{1.395} \times \frac{1}{.75}$$

$$\frac{n-1}{n} = .378$$

$$\frac{n}{n-1} = 2.646$$

$$n = 1.608$$

From Equation 2.64,

$r_p = 40/14.7$

$r_p = 2.721$ pressure ratio

Step 2. Calculate the total required polytropic head using Equation 2.73, assuming a value for $Z_{avg} = 1$:

$H_p = 1 \times 54.29 \times 550 \times 2.646 \, (2.721^{.378} - 1)$

$H_p = 36,338.4$ ft-lb/lb overall polytropic head

Step 3. Determine the number of stages, z, required using the recommended 10,000 ft-lb/lb head per stage.

$z = 36,338.4/10,000$

$z = 3.63$ stages, round off to 4

Calculate a new head per stage using four stages:

$H_p = 36,338.4/4$

$H_p = 9,085$ ft-lb/lb

Step 4. Use the geometric form of Equation 5.12 to calculate a tip speed to produce the head per stage just calculated. Also, use the recommended head coefficient $\mu = .48$ in the equation.

$u_2 = (9,085 \times 32.2/.48)^{.5}$

$u_2 = 780.7$ fps impeller tip speed

From Figure 5-26 and the inlet volume, select an initial impeller diameter.

$d_2 = 17.3$ inches impeller diameter

Use Equation 5.17 to calculate the initial speed, N, and use the conversion factors of 12 in/ft and 60 secs/min.

$$N = \frac{60 \times 12 \times 780.7}{\pi \times 17.3}$$

$$= 10,342 \text{ rpm shaft speed}$$

Step 5. The volume into the last impeller, in this example stage 4 inlet, is calculated using the Equation 5.18.

$$Q_{ls} = \frac{6,171}{(2.721^{1-1/4})^{1/1.608}}$$

$$Q_{ls} = 3,869 \text{ cfm volume at last stage}$$

To obtain an efficiency for the geometry selected, the value of the flow coefficient must be calculated using Equation 5.19 for the first inlet and the last stage flow.

$$\delta = (700 \times 6,171)/(10,342 \times 17.3^3)$$

$$\delta = .081 \text{ first stage flow coefficient}$$

$$\delta = (700 \times 3,869)/(10,342 \times 17.3^3)$$

$$\delta = .051 \text{ last stage flow coefficient}$$

Using the flow coefficients just calculated and Figure 5-26, the corresponding efficiencies may be looked up:

$$\delta = .081, \eta_p = .79$$

$$\delta = .051, \eta_p = .79$$

The average is rather easy to calculate.

$$\eta_p = .79 \text{ the average efficiency}$$

Step 6. Recalculate the polytropic exponent using Equation 2.71 and the new efficiency.

$$\frac{n-1}{n} = \frac{1.395 - 1}{1.395} \times \frac{1}{.79}$$

$$\frac{n-1}{n} = .359$$

$$\frac{n}{n-1} = 2.787$$

$$n = 1.559$$

Using the new polytropic exponent, calculate the discharge temperature using Equation 5.16.

$$T_2 = 550 \times 2.721^{.359}$$

$$T_2 = 787.8°R$$

$$t_2 = 787.8 - 460$$

$$t_2 = 327.8°F \text{ discharge temperature}$$

Calculate the power required using Equation 2.78 and the recommended 1% for mechanical losses.

$$W_p = \frac{437.5 \times 36,338.4}{33,000 \times .79} + .01W_p$$

$$W_p = 609.8 + 6.1$$

$$W_p = 615.9 \text{ hp total for the compressor}$$

Note, the polytropic head was not recalculated as the change in efficiency only made an approximate 1% difference in original value and is well within the accuracy of an estimate.

Example 5-3

For a sample problem that will include some of the additional losses that are normally encountered in an actual situation, size a compressor to the following given conditions for a hydrocarbon gas:

$mw = 53.0$

$k_1 = 1.23$

$Z_1 = 0.97$

$t_1 = 85°F$

$P_1 = 40$ psia

$P_2 = 120$ psia

$w = 2,050$ lb/min

Step 1. Use Equation 2.5 to calculate the specific gas constant.

$R = 1,545/53$

$R = 29.15$

Step 2. Convert the inlet temperature to absolute.

$T_1 = 85 + 460$

$T_1 = 545°R$

Step 3. Calculate the polytropic exponent using Equation 2.71. Assume an efficiency of $\eta_p = .75$. Use as $k_{avg} = k_1 = 1.23$.

$$\frac{n-1}{n} = \frac{1.23-1}{1.23} \times \frac{1}{.75}$$

$$\frac{n-1}{n} = .249$$

$$\frac{n}{n-1} = 4.011$$

$$n = 1.332$$

Step 4. From Equation 2.64,

$r_p = 120/40$

$r_p = 3.0$ pressure ratio

Step 5. Calculate the estimated discharge temperature using Equation 5.14.

$T_2 = 545 \times 3^{.249}$

$T_2 = 716.7°R$ absolute discharge temperature estimate

Convert to °F:

$t_2 = 716.7 - 460$

$t_2 = 256.7°F$

Correct for the balance piston leakage using 1% for pressures of 150 psia and under. The weight flow into the impeller must be increased to account for the leakage.

$w = 1.01 \times 2,050$

$w = 2,070.5$ lb/min net flow to the impeller.

The temperature at the entrance to the impeller is increased because of the hot leakage. Calculate the corrected impeller inlet temperature using Equation 5.20.

$$t_w = \frac{85 + 256.7(.01)}{1.01}$$

$t_w = 86.7°F$ corrected impeller inlet temperature

Convert to absolute:

$T_w = 86.7 + 460$

$T_w = 546.7°R$

Step 6. Substitute into Equation 2.10 and using 144 in²/ft².

$$Q_1 = \frac{.97 \times 29.15 \times 546.7}{40 \times 144} \times 2,070.5$$

$Q_1 = 5,557$ cfm inlet flow to the impeller

Step 7. Calculate the total required polytropic head using Equation 2.73, assuming the average value of $Z_{avg} = .97$.

$H_p = 0.97 \times 29.15 \times 546.7 \times 4.011(3^{.249} - 1)$

$H_p = 19,508$ ft-lb/lb total polytropic head required

Step 8. Determine the number of stages required using the modified rule of thumb on head per stage, H_{stg}.

$H_{stg} = 10,000 - ((53 - 29)100)$

$H_{stg} = 7,600.$ ft-lb/lb

$z = \dfrac{19,508}{7,600}$

$z = 2.57$ stages, round off to 3

Calculate a new head per stage using three stages.

$H_p = 19,508/3$

$H_p = 6,502.7$ ft-lb/lb head per stage

Step 9. Use the geometric portion of Equation 5.12 to calculate a required tip speed, which will produce the head per stage. Use the recommended head coefficient $\mu = .48$ for the calculation.

$u_2 = (6,502.7 \times 32.2/.48)^{1/2}$

$u_2 = 660.5$ fps impeller tip speed

Step 10. From Figure 5-26 and the inlet volume, select an initial impeller diameter.

$d_2 = 17.3$ in. initial impeller diameter

Use Equation 5.17 to calculate the initial speed, N.

$N = \dfrac{60 \times 12 \times 660.5}{\pi \times 17.3}$

$N = 8,750$ rpm compressor shaft speed

Step 11. The volume into the last impeller is calculated with the use of Equation 5.18.

$$Q_{ls} = \frac{5,557}{(3^{1-1/3})^{1/1.332}}$$

$Q_{ls} = 3,206$ cfm volume into last stage

With the volumes just calculated, calculate the inlet flow coefficient for each of the two stages using Equation 5.19.

$\delta = (700 \times 5,557)/(8,750 \times 17.3)^3$

$\delta = .086$ first stage flow coefficient

$\delta = (700 \times 3,206)/(8,750 \times 17.3)^3$

$\delta = .050$ last stage flow coefficient

Look up the efficiencies for the two flow coefficients on Figure 5-27.

$\eta_p = .79$ first stage efficiency

$\eta_p = .79$ last stage efficiency

$\eta_p = .79$ average of the two efficiencies

Step 12. Recalculate the polytropic exponent using Equation 2.71 and the new average efficiency.

$$\frac{n-1}{n} = \frac{1.23-1}{1.23} \times \frac{1}{.793}$$

$$\frac{n-1}{n} = .236$$

$$\frac{n}{n-1} = 4.24$$

With the new polytropic exponent, calculate the discharge temperature by substituting into Equation 5.16.

$T_2 = 546.7 \times 3.236$

$T_2 = 708.5°R$ absolute discharge temperature

$t_2 = 708.5 - 460$

$t_2 = 248.5°F$ discharge temperature

Step 13. Calculate the power required using Equation 2.78, allowing 1% for the mechanical losses. Use the conversion 33,000 ft-lb/min/hp.

$$W_p = \frac{2,070.5 \times 19,508}{33,000 \times .793} + .01W_p$$

$$W_p = 1,543.1 + 15.4$$

$$W_p = 1,558.5 \text{ hp shaft horsepower}$$

There is no need to recalculate the polytropic head for the changed efficiency because the head difference from the original value is negligible. Another item to note is that the horsepower is 1.5% higher than if the balance piston had been neglected. The interesting part is not the value itself, but the fact that the slight temperature addition at the impeller inlet is responsible for .5% of the increase and the remainder is the 1% weight flow increase through the compressor. As the small, but significant "real life" items are included, the actual efficiency is being eroded. If the calculation had been made with only the original weight flow, the equivalent efficiency would prorate to .781.

Example 5-4

This example presents a gas with a temperature limit and is typically found in a halogen mixture. A multi-section compressor is required to accommodate the limit. This example illustrates one approach for the division of work between the sections to achieve a discharge temperature within the specified bound.

$$mw = 69$$

$$k_1 = 1.35$$

$$k_2 = 1.33$$

$$Z_1 = .98$$

$$Z_2 = .96$$

$$t_1 = 80°F$$

$$P_1 = 24 \text{ psia}$$

$$P_2 = 105 \text{ psia}$$

$$w = 3,200 \text{ lbs/min}$$

The temperature, t_2, is limited to a value of 265°F.

Step 1. Use Equation 2.5 to calculate the specific gas constant.

R = 1,545/69

R = 22.39

Step 2. Convert the inlet temperature to absolute.

$T_1 = 80 + 460$

$T_1 = 540°R$

Substitute into Equation 2.10 and using the conversion constant of 144 in.^{2}/ft^2, calculate the inlet volume.

$$Q_1 = \frac{.98 \times 22.39 \times 540}{24 \times 144} \times 3200$$

$Q_1 = 10,971$ cfm inlet flow

Step 3. Calculate the overall poltropic exponent using Equation 2.71 and an assumed polytropic efficiency of $\eta_p = .75$.

$k_{avg} = (1.35 + 1.33)/2$

$k_{avg} = 1.34$

The average was used in evaluating k because the values were not all that different.

$$\frac{n-1}{n} = \frac{1.34-1}{1.34} \times \frac{1}{.75}$$

$$\frac{n-1}{n} = .338$$

$$\frac{n}{n-1} = 2.956$$

$$n = 1.511$$

Step 4. From Equation 2.64,

$r_p = 105/24$

$r_p = 4.375$ overall pressure ratio

Step 5. Calculate the discharge temperature for the total pressure ratio to check against the stated temperature limit, using the assumed efficiency, $\eta p = .75$ and the polytropic exponent. Apply Equation 5.14.

$T_2 = 540 \times 4.375^{.338}$

$T_2 = 889.3°R$

$t_2 = 889.3 - 460$

$t_2 = 429.3°F$ discharge temperature

Since the limit is 265°F and the overall temperature is obviously in excess of this limit, intercooling is required.

Intercooler outlet temperature must be determined. If cooling water at 90°F and an approach temperature of 15°F are assumed, the gas outlet from the cooler returning to the compressor will be 105°F.

If Equation 3.12 is borrowed from the reciprocating compressor chapter and used for an uncooled section, the pressure ratio per section may be calculated assuming an approximate equal-work division. For the first trial, assume the limit of temperature may be achieved in two sections.

$r_p = 4.375^{1/2}$ pressure ratio per section

$r_p = 2.092$

$P_2 = 2.092 \times 24$

$P_2 = 50.2$ psia first section discharge pressure

From the rule of thumb given for estimating intercooler pressure drop, a value of 2 psi is used because it is larger than 2% of the absolute pressure at the cooler. The pressure drop must be made up by the compressor by additional head, and can be added to the first or second section pressure ratio. By applying a little experience, the guessing can be improved. The front section has a lower inlet temperature and is generally more efficient, so the best location for additional pressure would be in the first section. The first section discharge pressure is $50.2 + 2 = 52.2$ psia. A new pressure ratio for the first section must be evaluated.

$r_p = 52.2/24$

$r_p = 2.175$

Step 6. Evaluate the discharge temperature, continuing the use of the previously calculated polytropic exponent.

$T_2 = 540(2.175)^{.338}$

$T_2 = 702.2°R$

$t_2 = 702.2 - 460$

$t_2 = 242.2°F$ first section discharge temperature

This temperature is within the limit.

Intercooler outlet pressure is 50.2 psia. Calculate the second section pressure ratio.

$r_p = 105/50.2$

$r_p = 2.092$

Evaluate the Section 2 discharge temperature.

$T_2 = 565(2.092)^{.338}$

$T_2 = 725°R$

$t_2 = 725 - 460$

$t_2 = 265°F$ discharge temperature

Because the temperature just calculated is right on the temperature limit and there is margin in the Section 1 temperature, the pressure may be arbitrarily adjusted to the first section to better balance the temperatures. A Section 1 discharge pressure of 54.5 psia is selected, which results in a new pressure ratio.

$r_p = 54.5/24$

$r_p = 2.271$

Now calculate a new Section 1 discharge temperature for the pressure just assumed.

$$T_2 = 540(2.271)^{.338}$$

$$T_2 = 712.5°R$$

$$t_2 = 712.5 - 460$$

$$t_2 = 252.5 °F$$

The temperature is still within the required limit. Correct the cooler outlet pressure and evaluate a new ratio for Section 2. The corrected cooler outlet pressure is 52.5 psia.

$$r_p = 105/52.5$$

$$r_p = 2.0$$

Recalculate the discharge temperature for Section 2, using the previous cooler outlet temperature.

$$T_2 = 565(2.0)^{.338}$$

$$T_2 = 714.2°R$$

$$t_2 = 714.2 - 460$$

$$t_2 = 254.2°F$$

The temperature is now below the 265°F limit and consistent with the Section 1 temperature. At this point, the initial assumption for 2 sections can be considered a firm value.

Step 7. Calculate the polytropic head for each section, using the overall average compressibility of $Z_{2avg} = .97$.

Section 1

$$H_p = .97 \times 22.39 \times 540 \times 2.956(2.271)^{.338}$$

$$H_p = 11,074 \text{ ft-lb/lb}$$

Section 2

$$H_p = .97 \times 22.39 \times 565 \times 2.956(2.0)^{.338}$$

$$H_p = 9,576 \text{ ft-lb/lb}$$

Step 8. Develop the allowable head per stage by the use of one of the rules of thumb.

$H_{stg} = 10,000 - ((69 - 29)(100))$

$H_{stg} = 6,000$ ft-lb/lb

Divide the total head per section by the allowable head per stage to develop the number of stages required in each section.

Section 1

$z = 11,074/6,000$
$z = 1.84$ stages, round off to 2

Section 2

$z = 9,576/6,000$
$z = 1.6$ stages, round off to 2

Step 9. Calculate a head per stage for each section based on two stages each.

Section 1

$H_p = 11,074/2$
$H_p = 5,537$ ft-lb/lb head per stage, Section 1

Section 2

$H_p = 9,576/2$

$H_p = 4,788$ ft-lb/lb head per stage, Section 2

Use the geometric portion of Equation 5.12 to calculate the tip speed. Assume $\mu_p = .48$ for the pressure coefficient.

Section 1

$u_2 = (5,537 \times 32.2/.48)^{.5}$

$u_2 = 609.5$ fps tip speed first two stages

Section 2

$u_2 = (4,788 \times 32.2/.48)^{.5}$

$u_2 = 566.7$ fps tip speed last two stages

Step 10. From Figure 5-26 and the inlet volume to the first section, select an initial impeller diameter.

$d_2 = 25$ in.

Because the second section shares a common shaft with the first section, it is not necessary to look up a new impeller size. Apply the Section 1 impeller diameter, Equation 5.15, and the conversion constants of 12 in./ft and 60 sec/min. to calculate a shaft speed.

$$N = \frac{12 \times 60 \times 566.7}{\pi \times 5,588}$$

$N = 5,588$ rpm

With the shaft speed and the tip speed calculated in Step 9 for the Section 2 stages, calculate an impeller diameter using Equation 5.15.

$$d_2 = \frac{12 \times 60 \times 566.7}{\pi \times 25}$$

$d_2 = 23.24$ in. Section 2 impeller diameter

Step 11. Calculate the inlet volume into Section 2. Use $Z_{avg} = 97$, $P_1 = 52.5$ psia, and $t_1 = 105°F$. Substitute into Equation 2.10 as was done in Step 2.

$$Q_1 = \frac{.97 \times 22.39 \times 565}{52.5 \times 144} \times 3,200$$

$Q_1 = 5,194$ cfm inlet volume into Section 2

Calculate the last impeller volume for each section using Equation 5.18.

Section 1

$$Q_{ls} = \frac{10,971}{(2.271^{1/2})^{1/1.511}}$$

$Q_{ls} = 8,363.4$ cfm last stage volume, Section 1

Section 2

$$Q_{ls} = \frac{5,194}{(2.0^{1/2})^{1/1.511}}$$

$Q_{ls} = 4,129.4$ cfm last stage volume, Section 2

Use Equation 5.19 to evaluate the flow coefficient for the first and last impeller of each section.

Section 1

$\delta = (700 \times 10,971)/(5,588 \times 25^3)$

$\delta = .088$ flow coefficient, first stage

$\delta = (700 \times 8,364.4)/(5,588 \times 25^3)$

$\delta = .067$ flow coefficient, last stage

Section 2

$\delta = (700 \times 5,194)/(5,588 \times 23.24^3)$

$\delta = .052$ flow coefficient, first stage

$\delta = (700 \times 4,129.4)/(5,588 \times 23.24^3)$

$\delta = .041$ flow coefficient, last stage

Step 12. Use Figure 5-27 and the flow coefficients to determine the efficiencies for the stages.

Section 1

$\delta = .088, \eta_p = .79$

$\delta = .067, \eta_p = .80$

The average is

$\eta_p = .795$

Section 2

$\delta = .052, \eta_p = .793$

$\delta = .041, \eta_p = .78$

The average is

$\eta_p = .787$

Step 13. Recalculate the polytropic exponent.

Section 1

Use $k_{avg} = 1.345$

$$\frac{n-1}{n} = \frac{1.345-1}{1.345} \times \frac{1}{.795}$$

$$\frac{n-1}{n} = .323$$

$$\frac{n}{n-1} = 3.1$$

Section 2

Use $k_{avg} = 1.335$

$$\frac{n-1}{n} = \frac{1.335-1}{1.335} \times \frac{1}{.787}$$

$$\frac{n-1}{n} = .319$$

$$\frac{n}{n-1} = 3.136$$

Step 14. Use the polytropic exponents calculated in the previous step and recalculate the discharge temperature of each section to correct for the average stage efficiency.

Section 1

$T_2 = 540(2.271)^{.323}$

$T_2 = 703.8°R$

$t_2 = 703.8 - 460$

$t_2 = 243.8°F$ final Section 1 discharge temperature.

Section 2

$T_2 = 565(2.0)^{.319}$

$T_2 = 704.8°R$

$t_2 = 704.8 - 460$

$t_2 = 244.8°F$ final Section 2 discharge temperature.

The temperature is approximately 20°F below the 265°F temperature limit. The sections differ by less than 1°F. This is probably just luck because that good a balance is not really necessary. Also, it should be noted that to maintain simplicity the additional factors were ignored, such as the 10°F temperature pickup in the return stream due to internal wall heat transfer. Also, nozzle pressure drops for the exit and return were not used. Balance piston leakage was not used as it was in Example 5-3. When all the factors are used, the pressures for each section would undoubtedly need additional adjustment as would the efficiency. However, for the actual compression process, the values are quite realistic, and for doing an estimate, this simpler approach may be quite adequate.

Step 15. To complete the estimate, calculate the shaft power, using the conversion of 33,000 ft-lb/min/hp.

Section 1

$$W_p = \frac{3,200 \times 11,074}{33,000 \times .795}$$

$W_p = 1,350.8$ hp gas horsepower, Section 1

Section 2

$$W_p = \frac{3,200 \times 9,576}{33,000 \times .787}$$

$W_p = 1,179.9$ hp gas horsepower, Section 2

Combine the two gas horsepower values and add 1% for the mechanical losses.

$W_p = 1,350.8 + 1,179.9 + 25.3$

$W_p = 2,556.0$ hp compressor shaft power

Fan Laws

These relationships were actually developed for pumps instead of compressors, but they are very useful in rating compressors that are being considered for reapplication. The equations used to this point are adequate to perform any rerate calculation; however, looking at the fan laws may help establish another perspective. The following relationships are a statement of the fan laws.

$$Q_i \, \alpha \, N \tag{5.21}$$

$$H_p \, \alpha \, N^2 \tag{5.22}$$

$$W_p \, \alpha \, N^3 \tag{5.23}$$

The equations have been expressed as proportionals; however, they can be used by simply "ratioing" an old to a new value. To add credibility to fan law adaptation, recall the flow coefficient, Equation 5.19. The term Q_i/N is used which shows a direct proportion between volume Q_i and speed N. Equation 5.12 indicates the head, H_p, to be a function of the tip speed, u_2 squared. The tip speed is, in turn, a direct function of speed making head proportional to speed. Finally, the power, W_p, is a function of head multiplied by flow, from which the deduction of power, proportional to the speed cubed, may be made.

Curve Shape

Figure 1-3 presented a general form performance curve for each of the compressors. The centrifugal compressor exhibited a relatively flat curve compared to the other machines. Flat is defined as a relatively low head rise for a volume change. Translated to pressure terms, it means a relatively low pressure change for a given volume change. It is important to understand some of the basics that contribute to the curve shape.

Figure 5-24 shows that if the flow is reduced for the radial wheel, a reduction occurs in the vector, V_{r2}, but there is no influence on the tangential component of the absolute velocity, V_{u2}. In fact, the ratio of $V_{u2}/u_2 = 1$. In this case, the ideal curve would be flat, something that really does not happen due to the effect of slip and efficiency. Looking at the 60° curve, the V_{u2} vector will increase with a decrease in flow. This is shown as decrease in the length of the V_{r2} vector, raising the work input coefficient and putting a slope into the curve. Then, if the 45° vector triangle is examined, the same thing will happen: V_{u2} will increase for a decrease in the flow. Because the angle β_2 is less, the V_{u2} increases faster for 45° than for 60°, making for a steeper curve. This is consistent with the earlier statement about the higher reaction wheel having a steeper curve.

Flow passing through an impeller is constantly changing in volume because of the compressible nature of the gas. If an impeller is operated first with a light molecular weight gas and then a heavy gas, the curve will be steeper with the light gas because the volume ratio is higher for the heavy gas. An examination of Equation 5.12 shows that head for a given geometry is fixed, within reasonable limits. Therefore, substituting different molecular weights in the head equation will indicate a higher pressure ratio directly proportional to the molecular weight. The volume ratio, then, is directly proportional to the pressure ratio making it also directly proportional to the molecular weight. Since the geometry was not modified to match the different volume ratio, the vectors, V_{r2}, are shorter for the lower outlet volume. As such, the change to the vector V_{u2} is not as great and the curve is not as steep.

The compressibility of the gas going through the impeller causes some problems. The assumption in the use of the fan law, when speeding up an impeller, is that the inlet volume follows the speed in a proportional manner. At the same time, the head is increased as a function of the speed squared. Just as the head increases with a given gas, so does the pressure ratio and therefore the volume ratio. It wasn't pointed out, but the alert reader may have noticed that the outlet triangle, not the inlet triangle, was used to discuss the curve shape. The problem is that the outlet volume is not exactly proportional to the inlet volume. For a 10% speed change, the compressor does not truly respond with a 21% head change. For small speed changes the problem is not serious; however, the basics should be remembered if a compressor is being rerated.

One last item should be noted regarding the shape of the curve. As stages are put together, the overall flow range of the combined stages is never larger and, in most cases, is less than the smallest flow range of the individual stages. Because of the compounding effect, as the volume is changed, the combined curve is always steeper.

Surge

Notice that the left end of a centrifugal compressor pressure volume curve does not reach zero flow. The minimum flow point is labeled as the *surge limit* and is the lowest flow at which stable operation can be achieved. Attempted operation to the left of that point moves the compressor into surge. In full surge the compressor exhibits an extreme instability; it backflows to a point and then temporarily exhibits forward flow. This oscillating flow is accompanied by a large variety of noises, depending on the geometry and nature of the installation. Sometimes it is a deep low frequency booming sound and for other machines it is a squeal. The pressure is highly unsteady and the temperature at the inlet rises relatively fast. The latter is caused by the same gas backing up in the machine and then recompressing until the next backflow. Each pass through the compressor adds additional heat of compression. Mechanically, the thrust bearing takes the brunt of the action and, if not left in surge indefinitely, most compressors do survive. In fact, most compressors that have operated for any period have experienced surge at one time or another. If left unchecked, and assuming the thrust bearing is well-designed, the compressor will more than likely destroy itself from the temperature rise.

Surge is due to a stalling of the gas somewhere in the flow path, although opinions seem to differ as to exactly where. For the process plant type low head compressor, it would appear to start in the diffuser. It can also take place at one of several points in an impeller depending on the geometry. For compressors designed for higher heads, the primary stall point appears to move into the impeller. Compressors exhibit a phenomenon referred to as *incipient surge* or *stall*. This is where one element stalls but not severely enough to take the stage into a complete stall. An experienced listener can readily hear and identify the stall. If the flow is not further reduced, it can remain in this condition without further stalling. It is very close to the limit, however, and only a minor flow disturbance can trigger a full-stage surge, which may then spread through the whole compressor. Stall is a flow separation. It may be compared to an airplane wing that produces lift until the angle of attack exceeds a limiting value at which point separation becomes great and the ability to continue producing lift is lost.

Choke

The right side of the curve tends to slope in an orderly manner and then falls off quite rapidly. If taken far enough, the compressor begins to choke or experience the effect of "stonewall." If the internal Mach numbers are near 1 and/or the incidence angle on the inlet vane becomes high enough to reduce the entrance flow area and force the Mach number high enough, the compressor will *choke*. At this point, no more flow will pass through the compressor. The effect is much greater on high molecular weight gas, particularly at a low temperature and with the k value on the low side. The problem is that the compressor reaches the "stonewall" limit in flow before the designer had intended. If compressors are rerated, this effect must be kept in mind, particularly when the new conditions are for a lower molecular weight gas. It is possible to choke the front-end stages and starve the downstream stages, causing these stages to be in surge.

Normally, operation of a compressor in choke flow is relatively benign, particularly for compressors operating at nominal pressures of less than 2,000 psig. As the application pressurs are raised, with the higher resulting density, there is the possibility that the off-design differential pressures could become high enough to increase the stresses to a level of concern. It would be wise if in the application of very high density compressor, due to the nature of the operation, the supplier be advised if prolonged operation in choke flow is anticipated. The supplier should review past experience with similar installations and critique the design to avoid potential problems.

Application Notes

As with the reciprocating compressor, care must be exercised when liquids are present in the gas stream. Unlike the reciprocator, the centrifugal is somewhat more forgiving if the liquid is in the form of mist. Small droplets can pass through the machine without problem if the duration is short. The problems with liquid have both short- and long-term characteristics. Short term, the biggest problem with liquid is the ingesting of slugs of water. The compressor is in danger of severe mechanical damage if suddenly deluged with a great quantity of liquid. For this reason, compressors taking suction from vessels containing either liquid and vapor or vapors near saturation should have suction drums to trap any potential liquids. The suction drum is also a good idea where the possibility exists that condensation could take place in the suction lines, forming a slug on its way to the compressor. The long-term problem is with mist or small droplets. With time, any liquid will start an erosion of the moving parts, particularly the impeller vanes. As the tips of the vanes erode, the effective diameter of the impeller is reduced. The foregoing equations showed that the head-producing capability of the impeller is a function of the tip speed squared.

Interestingly enough, centrifugals can be washed "on stream" to counteract the effects of fouling. In some cases, where fouling is continuous and severe, a liquid wash may be used continuously. Care must be used and a certain amount of trial-and-error steps taken to ensure the proper quantity: enough to do the job and not enough to cause significant erosion. In the same manner, when a compatible liquid is available, liquid can be injected into the machine to provide auxiliary cooling. In all liquid injection applications, it is important not to inject the liquid so that it impinges on any of the surfaces. It is better to use tangential sprays to the degree practical to have the liquid flash in the gas stream.

One question that arises quite often is the orientation of inlet piping and its influence on compressor performance. The flow into the impeller has been assumed axial or radial, depending on the impeller geometry, which means there is no pre- or antirotation and it is free from random flow distortions. While centrifugals are somewhat more forgiving than other machines like axials, there are limits. If the flow has rotation or distortion as it enters the impeller, the compressor performance will be influenced in a negative manner. Correct piping practices at the compressor inlet will help ensure the proper performance of the compressor. Figure 5-28 includes a set of curves that may be used as guidelines to establish a minimum length of straight pipe to use ahead of the inlet. The base case is shown in Figure 5-29 and consists of an elbow turned in the plane of the rotor. While the sketches are shown as multistage compressors, they may be used for axial entry single-stage compressors by obtaining the multiplier for the base case and taking the final result and multiplying by 1.25. The higher multiplier accounts for the more sensitive nature of the axial inlet. These sketches and pipe lengths are conservative, but should a vendor recommend a longer length, the vendor's recommendation should receive the first consideration. When there are problems achieving some of the minimum lengths, vaned elbows and straighteners can be used. Figures 5-30 and 5-31 offer suggestions for those not experienced in these areas. Again, these are methods that have been used, but are not the only, or necessarily the best, solutions for any and all applications.

Mechanical Design

Introduction

The centrifugal compressor is composed of a casing containing a rotating element, *rotor,* which is supported by a set of bearings. For most multistage compressors, shaft end seals are located in-board of the bearings.

(text continued on page 192)

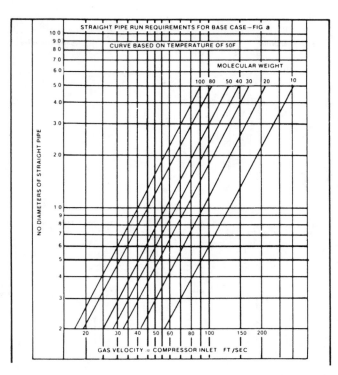

COMPRESSOR INLET CORRECTION FACTORS FOR VARIOUS PIPING ARRANGEMENTS

	Factor	Figure
1. One long radius elbow (plane parallel to rotor)	1.0	a
2. One long radius elbow (plane normal to rotor)	1.50	b
3. Two elbows at 90 to each other with second elbow plane parallel to rotor	1.75	c*
4. Two elbows at 90″ to each other with second elbow plane normal to rotor	2.0	d
5. Butterfly valve before an elbow		
a. valve axis normal to compressor inlet	1.5	e*
b. valve axis parallel to compressor inlet	2.0	—
6. Butterfly valve in straight run entering compressor inlet		
a. valve axis normal to rotor	1.5	—*
b. valve axis parallel to rotor	2.0	—
7. Two elbows in same plane (parallel to rotor)	1.15	f*
8. Two elbows in same plane (normal to rotor)	1.75	—
9. Gate valve (wide open)	1.0	—*
10. Swing check valve (balanced)	1.25	—*

*Factors also apply to single stage, axial inlet compressors.

Note:

1. Factors are applied to the base straight run requirements from the chart.

2. Factors for butterfly valves assume minimum throttling at design conditions. If heavy throttling is required, factors should be doubled.

3. For axial inlets, use 1.25 with appropriate figure.

Figure 5-28. Chart for minimum straight inlet piping. Use this chart and the given factors in conjunction with Figure 5-29. (*Courtesy of Elliott Company*)

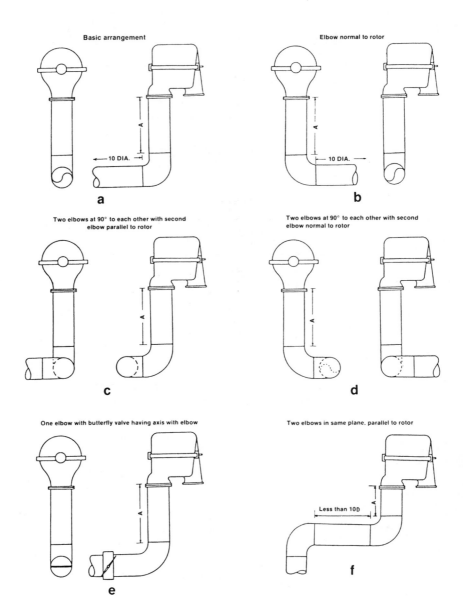

Figure 5-29. Methods of piping. To find "A," multiply the number of diameters of straight pipe from the chart in Figure 5-28 with the appropriate correction factor. (*Courtesy of Elliott Company*)

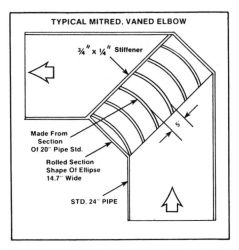

Figure 5-30. Elbow straightening vanes. (*Courtesy of Elliott Company*)

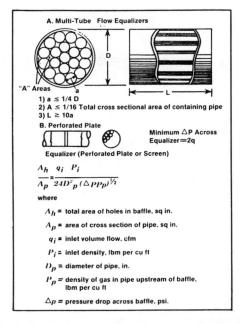

Figure 5-31. Straightening vanes. (*Courtesy of Elliott Company*)

(text continued from page 188)

The internal passages are formed by a set of diaphragms. Figure 5-32 depicts a typical multistage barrel compressor. Refer to this figure to locate the relative position of the various parts that are described in the following section.

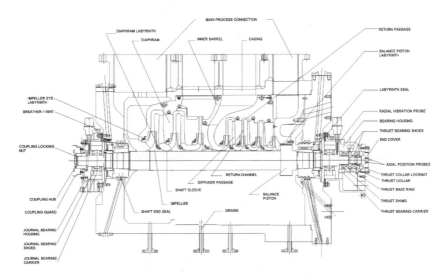

Figure 5-32. Centrifugal compressor nomenclature. (*Courtesy of A-C Compressor Corporation*)

Casings

All centrifugal compressor casings were initially of cast construction, and this method is still used on many casings today, particularly in the smaller sizes. In the past 15 years, some manufacturers have fabricated steel casings, generally converting their line of steel casings beginning with the larger frame sizes. The reason for this was economics; while the fabricated casings cost more to make on a per-unit weight basis, the net cost was less. Two factors were responsible. Quality of large steel castings was hard to control, with much time spent repairing the casing after inspection. Secondly, fabrication techniques and costs have improved significantly. There also came a side benefit of flexibility, once the manufacturer discovered he was no longer tied to a set of patterns or was bound by the time consumption and cost of pattern changes.

Casing materials are, in most cases, cast iron, nodular iron, or cast steel. Fabricated casings are generally made of carbon or alloy steel. Casings are, on occasion, made of austenitic stainless steel or one of the high nickel alloys. For low temperature inlet conditions, a low nickel alloy may be used. API Standard 617 [12] includes material guidelines in its appendix. The standard also mandates steel for all flammable and toxic gases, for air or nonflammable gas at pressures in excess of 400 psig, and for air or nonflammable gas with operating temperatures anywhere in the operating range in excess of 500°F.

The casing construction and materials covered to this point have generally applied to all kinds of compressors, including both horizontally split and vertically split. The vertically split, multistage barrel compressor is somewhat different. It is generally constructed of steel or steel alloy. It may be cast, fabricated, or, for very high pressure service, it may be forged. It should always be used when the gas contains hydrogen at or above a partial pressure of 200 psig. It may also be required in those services where the overall pressure is too high for the horizontally split compressor. This occurs when the horizontally split joint deforms too much at the operating pressure to maintain a gas-tight seal.

Diaphragms

The stationary members located inside a multistage casing are referred to as diaphragms. The function of the diaphragm is to act as a diffuser for the impeller and a channel to redirect the gas into the following stage. The diaphragm also acts as the carrier for the impeller eye seal and the interstage shaft seal. Diaphragms are either cast or fabricated. Most cast diaphragms are made of iron. Fabricated diaphragms are steel or composite steel and cast iron, with straightener or guide vanes of cast iron. Diaphragms are normally not highly stressed, with some exceptions. On compressors with out-in streams, if the differential pressure is relatively high from the outlet to the return nozzle, then the differential is taken across the diaphragm at the two nozzles. This diaphragm should be made of steel. The diaphragms are split, located with matching grooves in the upper and lower half casing and pinned to the upper half for maintenance ease. The diaphragms are hand-fitted to center them to the rotating element. It is important for the horizontal joint to match well, to keep the joint leakage to a minimum. On barrel compressors, the diaphragm assembly makes up an inner barrel (see Figure 5-33). The assembly and rotor are removed from the barrel casing as a unit using a special fixture. The diaphragm assembly is split to permit the removal of the rotor, and the diaphragms are generally constructed in the same way as those of the horizontally split compressor.

Figure 5-33. Inner barrel assembly. (*Courtesy of A-C Compressor Corporation*)

Casing Connections

Casing inlet and outlet nozzles are normally flanged. General preference, in process service, is for all casing connections to be flanged or machined and studded. On steel-cased machines, this normally is not a problem. On the smaller, refrigeration compressors that are highly standardized, constructed of cast iron, and originally designed for other than process service, connections will generally have flanged inlet and outlet nozzles. However, most of the auxiliary connections on these machines will be screwed. It is desirable to use standard flanges throughout the connections on the casing. However, for space reasons, on rare occasions, a nonstandard flange arrangement may become necessary. It is quite important to have the equipment vendor furnish all nonstandard mating flanges and associated hardware.

Forces and moments which the compressor can accept without causing misalignment to the machine are to be specified by the vendor. Many factors go into this determination, and as one may guess, the limits are determined quite arbitrarily in most cases. With all the many configurations a compressor can take, a single set of rules cannot fit all. Despite this, NEMA SM-23[13] for mechanical drive steam turbines is used as a basis. API 617 has adapted the NEMA nozzle criteria to centrifugal compressors. This works on larger steel-cased multistage compressors, but is not good for the overhung style. Moreover, the user or piping designers want a higher number to simplify piping design, while the manufacturer wants a small number to assure good alignment and fewer customer complaints. From a user's point of view, where long-term reliability is a must, the vote must go to the manufacturer. Experience shows that the lower the piping loads on the nozzles, the easier coupling alignment can be maintained. This seems reasonable since most compressors are equipped with plates called wobble feet to provide flexibility for thermal growth. The feet will flex from pipe loads as well as from the temperature. The piping loads tend not to align themselves as well with the shaft as the temperature gradients. Even when guides and keys are used, as is customary on the larger machines, they may bind despite the fact that they are stout enough to carry the load.

Impellers

Impeller construction was covered in the performance section and need not be repeated here. The impeller is the most highly stressed compressor component, and generally becomes the limiting item when it comes to establishing the rotating element performance limit. Impellers are made of low alloy steel for most compressors in process service, either chrome-moly or chrome-moly-nickel. Because of the high strength-to-weight ratio, many of the high head, integrally geared units use aluminum. Austenitic stainless, monel, and titanium are some of the other materials used for impellers in certain special applications, generally with corrosive gases involved. Stress levels must be adjusted for the materials involved. Some of the precipitation hardening steels in the 12 chrome alloy have been used and found to provide a good alternate material with moderately good corrosion resistance and very good physical properties.

Impeller construction for the cover-disk style impeller historically has been by built-up construction and welding. The traditional method uses die formed blades (see Figure 5-34). More recently, with the increased use of 5-axis milling, blades have been milled integrally with the hub disk. This alternate construction method is somewhat more costly because of the machining but produces a more accurate and repeatable gas path, which offsets the added expense (see Figure 5-35). Cover disks are welded to the blades to complete the milled impeller. Physical prop-

Figure 5-34. A fabricated centrifugal compressor impeller. (*Courtesy of A-C Compressor Corporation*)

Figure 5-35. A centrifugal compressor impeller during manufacture. The blading was milled with a five-axis milling machine. The blading is integral with the back plate. (*Courtesy of Dresser-Rand*)

erties are derived by heat treating and stress relieving. Some small sizes are cast. In the semi-open construction, casting is quite common, though fabricated impellers are used. Fully open impellers, which are not as common, can be either fabricated or cast. Impeller shaft attachment for multistage applications is by shrinking the hub to the shaft either with or without a key, depending on the vendor philosophy. There are numerous other methods used, each peculiar to the individual vendor.

Although not universal, on most multistage compressors, the impellers are axially located by shaft sleeves. The sleeves form a part of the interstage seal and are shrunk onto the shaft with a shrink level less than the impeller.

Shafts

Shafts are made of material ranging from medium carbon to low alloy steel and are usually heat treated. Shafts were originally made of forgings for the compressors in process service. But because of the availability of high quality material, hot rolled bar stock has been used for shafts up to 8 inches in diameter. Bar stock shafts are given the same heat treatment and quality control as forgings. Many of the process users prefer a low alloy, chrome-moly-nickel material for shafting, particularly for compressors in critical service.

Shafts require a good finish that can be achieved by machining. Honing, or sometimes grinding, is used to improve the finish in selected areas. Since proximity probes are used with most process compressors, the probe area must receive extra attention to minimize mechanical and electrical runout. On the whole, the shaft is the foundation for good mechanical performance to keep the rotor dynamics in control and maintain good balance. The requirements are that the shaft must be round and all turns must be concentric to the journals. As simple as it sounds, it is not easy to accomplish. The tighter the tolerance, the closer to perfection, the more expensive that particular manufacturing step. However, some added expense at this point will save time in subsequent rotor balancing providing the user with a rotor that can be more easily maintained. By using CNC machine tools for manufacturing shafting, the cost should come down, quality improve, and the product should become more consistent.

Radial Bearings

Radial bearings or journal bearings are usually pressure-lubricated. Most compressors use two bearings on opposite ends of the rotor assembly or on the overhung design, located adjacent to each other between the drive coupling and the impeller. It is highly desirable for ease of maintenance to have the bearings horizontally split. On centrifugal compressors, the bearings size is not a function of the load but rather it is dictated by critical speed considerations. Rotors in centrifugal compressors are by nature not very heavy; therefore, the bearings are lightly loaded. Because of the light loading, there are potential bearing-induced rotor dynamics problems.

Straight cylindrical bearings, as shown in Figure 5-36, are the most simple in concept. Because of low resistance to bearing-induced prob-

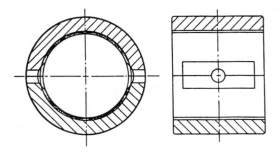

Figure 5-36. Straight cylindrical bearing. (*Courtesy of Turbocare, a Division of Demag Delavel Turbomachinery Corp., Houston facility*)

lems, application of this bearing is limited in centrifugal compressors. It is found normally in very large compressors with relatively heavy rotors and low compressor operating speeds. They are also used in fluorocarbon refrigerant compressors, where speeds are low because of high molecular weight and where relatively short rotors are used. As a minimum, most compressors with sleeve bearings use a modified sleeve bearing, such as the dam type shown in Figure 5-37. A relief groove is cut in the upper half of the bearing. The groove is stopped near the center of the upper portion of the bearing in a square, sharp-edged dam. As the shaft rotates, oil is carried through the groove to the end where the oil velocity is suddenly brought to a halt thereby converting it to pressure. A stabilizing force is formed on the top of the journal by the pressure. The maintenance of the sharp edge at the end of the bearing is very important. In service, if the groove ends become rounded, the bearing will cease to function as intended and can become unstable.

To facilitate maintenance and avoid the tedious scraping and other fitting steps required in early forms of plain journal bearings, replaceable

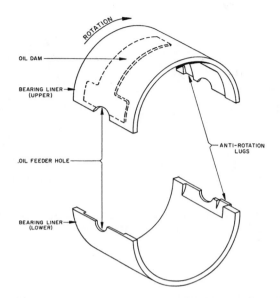

Figure 5-37. Dam type sleeve bearing. (*Courtesy of Elliott Company*)

inserts are used. The inserts are lined with a thin layer of babbitt on a steel backing. Precise manufacturing assures interchangeability. Babbitt thickness is a compromise, balancing enough depth for particle imbedability against keeping the strength up by staying close to the steel liner. This form of journal bearing is also referred to as a *liner bearing.*

The bearing most often found in centrifugal compressors is the *tilting pad bearing,* shown in Figure 5-38, which is inherently stable. The individual pads break up the rotating oil film and discourage the tendency for the oil to whirl. Each pad also acts as a separate force to keep the bearing loaded and thereby stabilized. The bearing, also known as the tilting shoe bearing, has grown in popularity in recent years and is found in most process compressors in critical service. The bearing can be furnished with various numbers of pads, with five being the most common. Bearing dynamics can be altered by a variety of configuration changes, such as load on or between pads. The number of pads can be changed for alternative dynamic parameters, with the four-pad bearing the most common alternative. Bearing clearance for a journal bearing is on the order of 1 to 1.5 mils per inch of journal diameter and is generally the same value for both the liner and the tilting pad bearings.

The pads are fabricated of steel with a babbitt coating, the thickness determined by the same argument as stated for the liner bearing. The

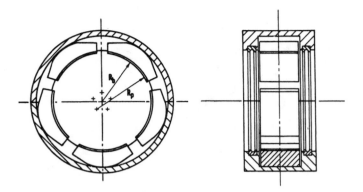

Figure 5-38. Five-pad tilting pad bearing. (*Courtesy of Turbocare, a Division of Demag Delavel Turbomachinery Corp., Houston facility*)

backside of the pad is fitted with some form of rocker, the exact shape varying from one maker to the next. The pads are contained by a horizontally split base ring assembly.

Thrust Bearings

Centrifugal compressor impellers, with the exception of the open impeller, are thrust unbalanced. The machine also has a requirement for a location device to maintain axial clearances. For these reasons, all centrifugal compressors use some form of thrust bearing.

API 617 recognizes the need for the compressor thrust design to take into account peripheral factors such as the coupling. Gear couplings can transmit thrust to the compressor because of tooth friction. The standard uses an arbitrary friction coefficient of .25, which can be a design basis. Flexible element couplings transmit less thrust because of the lower flexing element axial stiffness.

The basic type of thrust bearing consists of a thrust collar attached to the shaft running against a *flat land* (see Figure 5-39). The land is normally a steel ring with a babbitted surface. The load-carrying capacity of this bearing is quite limited, making the bearing suitable only for locating purposes. This bearing is commonly used with double helical gear units and is not normally found in centrifugal compressors.

A thrust bearing that physically resembles the flat land bearing is the *tapered land bearing*. The modification is the construction of the land. The land is grooved radially, dividing the land into segments that are individu-

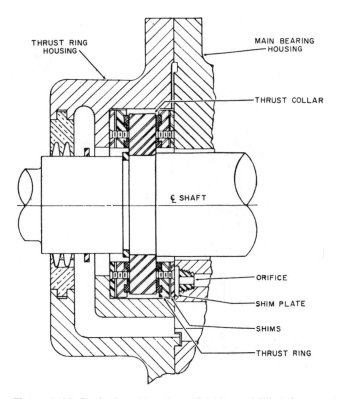

Figure 5-39. Basic thrust bearing. (*Courtesy of Elliott Company*)

ally tapered to form a wedge. As the collar rotates, relative to the land, oil is carried past the tapered wedges, developing pressure in the oil film resulting in an outward force. This force generates a load-carrying capacity. Theoretically, the tapered land bearing is capable of handling large axial loads, but it can only do so at a limited speed range. The bearing also requires good perpendicular alignment between the shaft and the land to maintain a uniform face gap. This bearing is used only for limited applications in the centrifugal compressor and when used is highly derated.

The tilting pad thrust bearing is available in two forms. The first form is named alternately for one or the other of the inventors, as it was developed by Albert Kingsbury in the United States and A.G.M. Michell in Australia working independently [14]. The bearing consists of a collar or thrust runner, attached to the shaft with the collar either integral or separable, and the stationary carrier in which the pads reside. Various numbers of pads are used, with six or eight being the most common. The pad consists of a babbitted segment, normally made of steel. The load is transmitted to the carrier by way of a button at the back of the pad, which also acts as the pivot point. The button may be centered or offset. The bearing is suitable for variable speeds because the pivoting feature allows the pad to adjust to the differing velocity of the oil film. The basic tilting pad thrust bearing is shown in Figure 5-40.

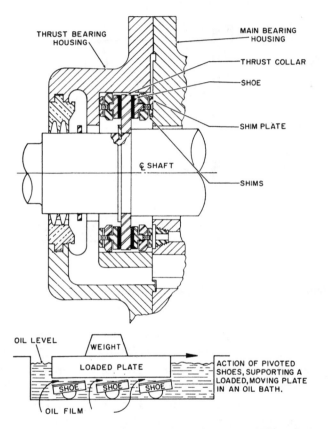

Figure 5-40. Tilting pad thrust bearing. (*Courtesy of Elliott Company*)

The second form of the multiple segment or tilting pad thrust bearing retains all the features of the first type, but includes a further refinement. This bearing is referred to as a *self-equalizing bearing*. Instead of a simple carrier to house the pads, each pad rides against an equalizing bar. Between each pad's equalizing bar is a secondary bar that carries the ends of two adjacent pad bars and transmits the load to the carrier ring. All bars are free to rock and thereby adjust themselves until all pads carry an equal share of the load (see Figure 5-41). The advantage of the self-

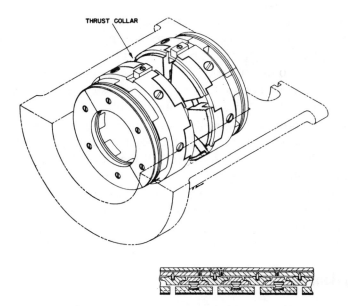

Figure 5-41. Self-equalizing tilting pad type thrust bearing. (*Courtesy of Elliott Company*)

equalizing bearing is obvious—it can adjust for minor irregularities of the rotor-to-bearing position.

Each bearing described requires a certain amount of axial space, with the simple thrust ring using the least and the self-equalizing bearing the greater amount. The thrust carrying capability of the latter two bearings is theoretically the same, and proponents of the Michell bearing cite deflections in the carrier ring as providing sufficient adjustment to achieve full potential load within the practical limits of bearing misalignment.

API 617 mandates the self-equalizing feature. Steady loads as high as 500 psi can be accommodated with transient loads going higher. However, conservative design practices and some encouragement from API tend to keep the loads on the thrust bearing in the range of 150 psi to 350 psi.

Pads for the multiple pad bearings may be made of a higher heat conducting material such as copper. Load-carrying capability can be increased by use of the copper pads. This option is a good alternate for difficult applications where size limits the use of a standard bearing. Many users do not permit the use of the alternate materials in a new compressor, using the argument that the option should be available for the solution of field problems. It should be mentioned that while the reference to copper is the common usage, in reality the material is a chromium copper alloy.

Compressors built to the API standard are required to have equal thrust capability in both directions; that is, the bearing is to be symmetrically constructed. However, the thrust bearings can be combined using various numbers of pads on each side, or a tapered land can be combined with the multiple-segment design. Other combinations of the four thrust bearings discussed are found in certain isolated applications.

The thrust bearing is responsible for large portions of the mechanical horsepower losses; however, it is the sophisticated multiple-segment bearing that has the highest losses of all the thrust bearing types. The power consumption is due to the churning of the oil in the essentially flooded bearing; therefore, care must be used in sizing the bearing to maintain margins for reliability. Alternative lubrication methods, such as the directed lubrication that uses a spray or jet to apply oil to the pads and eliminate the need for flooding, are available on some designs.

Bearing Housings

The bearing support system is normally separable from the casing, as mandated by API 617 and should be made of steel, particularly when used with a steel case. Provision should be made to maintain alignment of the rotor to the casing. The housing should be horizontally split and nonpressurized with provision for circulation of bearing lubrication. Care should be taken to prevent foaming of the lubricant. The housing is the desirable place to locate radial vibration probes as required by API 617.

The preceding paragraph assumes the bearings are located outboard of the seals on a multistage compressor, and also applies to most of the overhung types with the exception of the integrally geared machine. For the multistage, which has the seals outboard of the bearings, it is recognized that the housing will assume some pressure level used in the compressor. Therefore, provision should be made to minimize the amount of lubricant entering the gas stream. Also, for maintenance purposes, a port access to the bearing should be furnished. This type of compressor is generally limited to fluorocarbon refrigerant service and is not recommended for general gas service.

Magnetic Bearings

With the advent of magnetic bearings, the dream of an all-dry compressor can now be realized. This is to say that no external lube system is needed. Not all compressor applications at this point can qualify, because control oil is generally required for steam and gas turbine drivers. Gear bearing loads at present are higher than can be carried by current magnetic bearing designs.

The magnetic bearing is made up of a series of electromagnets located circumferentially around the shaft to form the radial bearing. The electromagnets (Figure 5-42) are laminated to limit the eddy current losses. The shaft must be fitted with a laminated sleeve (see Figure 5-43) for the

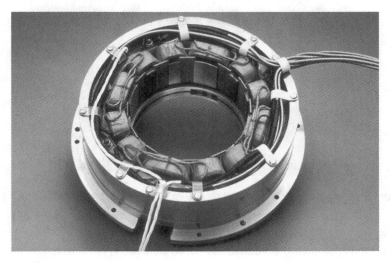

Figure 5-42. Radial magnetic bearing with a view of the circumferential electromagnets. (*Courtesy of Mafi-Trench Corp.*)

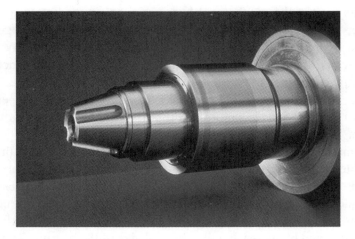

Figure 5-43. Radial magnetic bearing rotor sleeve. (*Courtesy of Mafi-Trench Corp.*)

same reason. The thrust is carried by a single-acting or dual-acting set of electromagnets (see Figure 5-44) depending on the need for a unidirectional or bidirectional thrust load. The magnetic bearing operates with a fixed air gap so there is no contact under operating conditions.

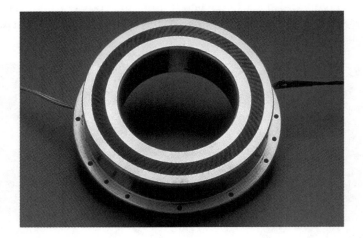

Figure 5-44. Magnetic bearing thrust electro-magnets. (*Courtesy of Mafi-Trench Corp.*)

Sensors are incorporated in the bearing assemblies to sense position of the rotor relative to the bearing. A servo control system uses the position information provided by the sensors to increase or decrease the bearing force on the rotor as needed to keep the rotor properly positioned. Magnetic bearings react differently than hydrodynamic or rolling element bearings in that the mechanical bearings react immediately to a load change, while the electromagnet in line with the load change must increase its force at the same rate to maintain rotor position. The actual rate at which the servo amplifier can increase the force is a function of volt ampere product of the amplifier. If the rate at which the load is applied exceeds the capability of the servo control, a temporary perturbation will be experienced before the shaft is brought back to its normal position.

The magnetic bearing load capacity on a per-unit basis is less than that available from hydrodynamic bearings. The specific load limit is at approximately 80 psi with typical design values of 60 psi. Higher values can be achieved with special magnetic materials, but these are not normally used in compressor applications. The load carried by bearing may be compensated by increasing the physical size of the bearing. The heaviest compressor rotor weight has been approximately 4,000 pounds.

Speed is not limited by a surface speed as in the hydrodynamic bearing. Unfortunately, however, the stresses in the thrust collar pose one limit and the other is caused by need for an auxiliary bearing, which does have a limitation based on the type of bearing being used.

The auxiliary bearing may be of the rolling element type, which is currently most common, or the dry lubricated bushing. The auxiliary bearing, which normally does not contact the shaft, is used to protect the rotating components from loss of the servo amplifiers (see Figure 5-45). The aux-

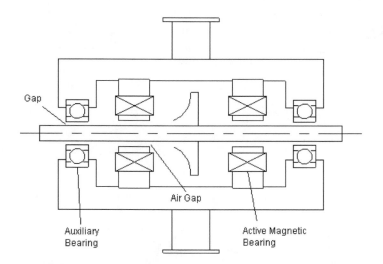

Figure 5-45. Schematic illustration of magnetic bearings and auxiliary rolling element bearings.

iliary bearing gap is approximately one half the air gap. Displacement due to momentary overload would also cause the auxiliary bearings to be pressed into service on a transient basis. It is critical that the compressor be tripped off line should the power to the magnetic bearings fail, because the auxiliary bearings have a limited life and are primarily intended for coastdown use. The life of the auxiliary bearings in general is considered to be five coastdowns from full speed. In some cases, the life has proven to be somewhat longer, particularly with the dry bushing design. Continued development in this area will no doubt increase this value in time.

An interesting aspect of the magnetic bearing is that in pre-startup the bearing servos are energized and the rotor levitated. It remains suspended as the startup begins. There is no minimum oil film type phenomena to pass through. On shutdown, the rotor is allowed to cease rotating and to remain in the levitated position until the power is removed should a full shutdown be required.

Balance Piston

It is desirable to have additional axial-load control on the multistage compressor. A *balance piston,* also referred to as the *balance drum,* can be located at the discharge end (see Figure 5-46). The balance piston consists of a rotating element that has a specified diameter and an extended rim for sealing. The area adjacent to the balance piston (opposite the last stage location) is vented, normally to suction pressure. The differential pressure across the balance piston acts on the balance piston area to develop a thrust force opposite that generated by the impellers. The pressure on the

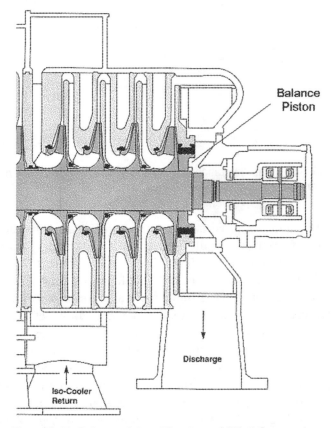

Figure 5-46. Balance piston. (*Courtesy of Elliott Company*)

low pressure side of the balance piston is higher than the reference pressure by an amount equal to the resistance of the balance line, the line taking the flow from the low pressure cavity to the reference point. Line resistance, of course, is a function of the flow in the line. To permit efficient balance, an effective seal must be used at the rim of the balance piston because the leakage also represents parasitic power loss. In the earlier paragraph on sizing, a target value of 1% was used as a base value, recognizing that for higher pressure applications this value would tend to be greater. While full control of the thrust can be developed by controlling the diameter, limits are in order. Generally, the balance force is kept less than that developed by the impellers, with the thrust bearing taking the remainder of the load. This keeps the rotor on one face of the thrust bearing for all load conditions and is the recommended practice. An alternative philosophy overbalances the thrust with the balance piston, arguing that balance piston seal deterioration will unload the thrust bearing for more conservative design. The problem with this approach is that the rotor will tend to shift its operating position from one side of the thrust bearing to the other for varying loads and conditions. Because the thrust bearing has .012 to .015 inches of float, the rotor will not be in a fixed position, making instrumentation for rotor position difficult to judge. Also, oversizing the balance piston means a larger seal diameter, making the potential seal leakage greater. Besides the ramifications of the higher leakage, the method tends to be somewhat self-fulfilling in that the deterioration will tend to increase at a higher rate.

Interstage Seals

Interstage and balance piston seals of the labyrinth type are universally used in centrifugal compressor service. Multistage compressors are equipped with impeller eye seal and interstage shaft seals to isolate the stages. Figure 5-47 shows various labyrinth configurations. Labyrinth seals consist of a tooth-like form with spaces in between. Leakage is a function of both the tooth or fin clearance and the spacing. As shown in the figures, the fins can be stationary or rotating. The basic labyrinth design is the straight seal, where the teeth are at the same height. Another, rarely used for the interstage seal but frequently used for the balance piston, is the staggered or stepped form. When rotating seals are used, they can be machined integral on a sleeve or into the rim of the balance piston. Another type of rotating seal is constructed of a strip material with one edge rolled. The rolled edge is then caulked into a groove in the rotating

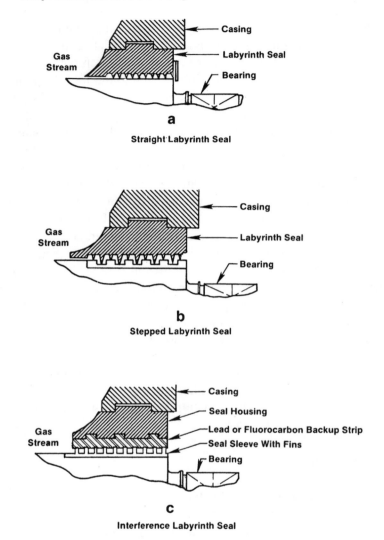

Figure 5-47. Three types of labyrinth seals. (*Courtesy of Elliott Company*)

element. The shape of the rolled edge gives the name *J-Strip,* which is sometimes used by the manufacturers who use the seal (see Figure 5-48). For use with the rotating seal, a soft backing surface is provided on the stationary surface opposite the seal. The backing can be lead, babbitt, or a stabilized fluorocarbon material. This arrangement allows the seal clearance to be set to a smaller value compared to the stationary finned seal with the objective of running the seal fins into the soft material to cut run-

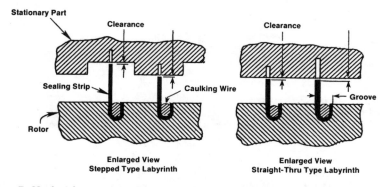

Figure 5-48. J-strip type labyrinth seal. (*Courtesty of A-C Compressor Corporation*)

ning grooves. While somewhat more expensive, the method does tend to keep leakage down and is particularly well-suited to fouling service; whereas, the stationary teeth would tend to fill and lose effectiveness.

When teeth are stationary, the material chosen for the labyrinth must be a relatively soft nongalling material, because the teeth tend to touch the shaft during upsets, startup, or shutdown. The clearance chosen must be large enough to avoid excessive rubbing yet close enough to control the leakage. If set too tight, the extra rubbing may cause the edges to roll and affect the performance. Overall, the stationary seal is simple and relatively easy to replace.

Shaft End Seals

Restrictive Seals

In controlling gas leakage, shaft end seals are either restrictive or positive in nature. The labyrinth seal is one form of restrictive seal. The reasoning for the labyrinth end seal is generally the same as discussed in the interstage seal section. A procedure for the calculation of restrictive seal leakage is given in Appendix D.

Another common form of restrictive seal is the *carbon ring seal* (see Figure 5-49). This seal consists of a series of carbon rings, using either solid or segmented rings. The segmented rings are enclosed with a retaining spring, called a garter spring. This seal, while somewhat more complex, is easier to replace than its solid counterpart. The carbon ring seal is able to operate with a close clearance, closer than bearing clearances, because the rings can move radially and the carbon acts to self-lubricate

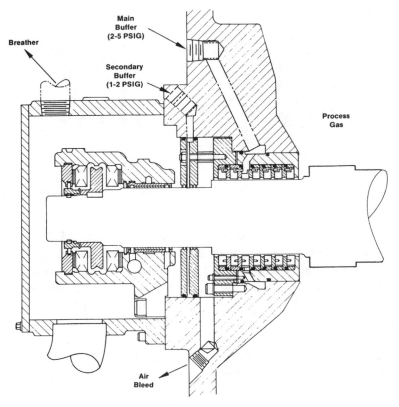

DRY CARBON RING SEAL

Figure 5-49. Dry carbon ring seal. The carbon rings on this seal are buffered by dry air. (*Courtesy of Elliott Company*)

when the seal rubs. Because rubbing does take place from time to time, the carbon ring tends to need more frequent replacement than the labyrinth. But for equal axial length, the carbon ring seal can be designed for leakage an order of magnitude less.

Liquid Buffered Seals

The positive seals are positive in the sense that the process gas is completely controlled, and in most applications, can be designed to avoid the loss of any gas, if the process gas and the sealing fluid are compatible to permit safe separation. In any event, the gas taken from the process is orders of magnitude lower than is the case for the restrictive seal. The positive seals take on the form of a *liquid film seal* or a *contact seal,* also known as the *mechanical seal.* The buffer fluid aids in the sealing process in the liquid film type and acts as coolant in both types. Each manufacturer generally has a proprietary form for one or both types of seal. Figures 5-50, 5-51, and 5-52 show the various seals available. The liquid film type operates with a close clearance and is used for high pressure applications. One modification of the liquid film seal uses a pumping bushing to control gas side leakage and, therefore, operates at bearing clearances (see Figure 5-52).

The contact seal can be used under 1,000 psig. It is more complex, but has the advantage of not leaking while shut down. The contact seal is used extensively in refrigeration service where the compressor is part of a closed loop, and the shutdown feature is desirable. As mentioned, the seals must have a source of cooling and buffer fluid. In many cases, this fluid is lubricating oil. If contamination is not a problem, a combined lube and seal system can be used.

Positive seals have been used in flammable and some toxic services. In toxic applications, an isolating seal must be included in the seal configu-

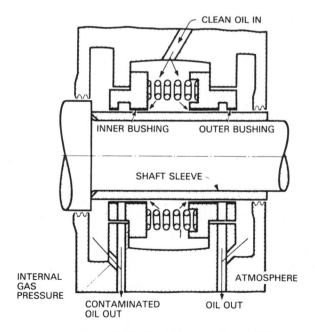

Figure 5-50. Liquid film shaft seal [12].

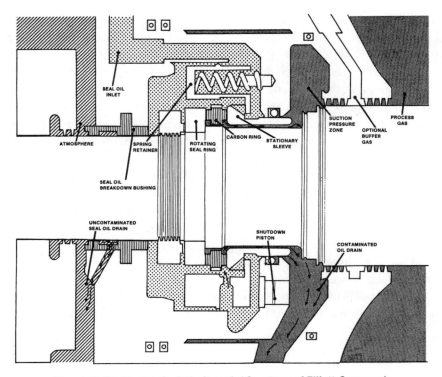

Figure 5-51. Mechanical shaft seal. (*Courtesy of Elliott Company*)

ration. By careful application, the isolating seal can also act as a backup to the primary seal.

In all situations, seals must function over the entire operating range, including startup and shutdown. If a compressor shuts down and is to be restarted hot after being down only a short time, the possibility exists of differential growth of the various components, closing the clearances to the point of seizure of the parts. The seal should be selected well inside its operating pressure range. With the liquid buffered seals, a value for the allowable leakage toward the gas side must be determined. This liquid is removed from the compressor by traps, referred to as *sour oil pots,* even when the fluid can be recycled. On small to intermediate compressors, the leakage flow should not be more than three to five gallons per day (gpd). Large compressors can have larger leakages, but should not average more than ten gpd per seal.

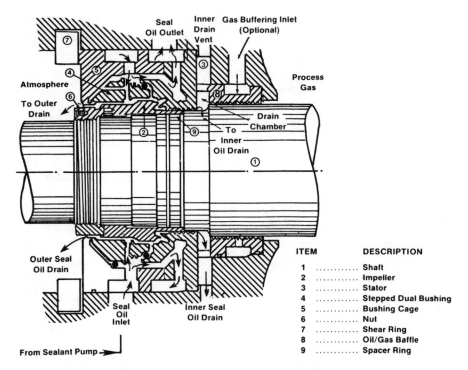

Figure 5-52. Liquid film type seal with pumping bushing. (*Courtesy of A-C Compressor Corporation*)

ITEM	DESCRIPTION
1	Shaft
2	Impeller
3	Stator
4	Stepped Dual Bushing
5	Bushing Cage
6	Nut
7	Shear Ring
8	Oil/Gas Baffle
9	Spacer Ring

Dry Gas Seals

Dry gas seals are in the positive seal class and have the same basic design features as mechanical face seals with one significant difference. The dry gas seal has shallow grooves cut in the rotating seal face located part way across the face. The grooves may be in a spiral pattern; the exact location and pattern vary from one manufacturer to another. Lubrication and separation is effected by a microscopically thin film of gas. This implies some finite amount of leakage, which is quite small but must be accounted for in the design.

The seal unit located at each end of the multistage compressor rotor is installed as a cartridge. The cartridge has positive locating features to permit proper placement on installation. It normally includes a provision to ensure that cartridges are not interchanged from the intended end. This is to prevent reverse rotation on the unidirectional configurations.

Gas leakages range from less than 1 Scfm to 1 Scfm. The maximum rubbing speed is considered to be 590 fps. Operating pressures may range up to 3,000 psi. The temperature range using elastomers range from −40°F to 450°F. By using non-elastomers in the seal design, the temperature range is widened to −250°F to 650°F. From these values, it can be seen that the dry gas seal has a wide application range potential.

The dry gas seal has numerous advantages, but, as with most things in life, it also comes with some disadvantages. It is fair to state that for most of the applications, the good outweighs the bad and as such these seals are used extensively in the industry. However, each application should be evaluated on its own merits.

Probably the biggest single advantage of dry gas seals is getting rid of the seal oil. The seal oil system, even when part of a combined lube and seal system, is a complex assembly. With the dry gas seal, the lubrication oil system is all that is needed to service the compressor train bearings and, on turbine driven units, to also supply turbine control oil. As an aside, it makes feasible the dream long held by the compressor vendors of having a standardized lube system line.

Eliminating the oil gets rid of the disposal problem of the contaminated oil, which must be properly disposed of or cleaned up and recycled. It also eliminates the fouling problems in components downstream of the compressor. Despite all efforts to the contrary, oil from liquid buffered seals finds its way into the gas stream.

In most applications, the net loss of gas is less. The oil buffered seal loses gas both with the contaminated oil due to gas in solution and through the gas leakoff required to keep the various differential pressures in the proper orientation.

In those application where the cross-coupling effects from the oil seal were detrimental to the rotor dynamics, the use of the gas seal is a distinct advantage. However, the down side is that should the oil seal have provided a good measure of damping, the impact on the rotor dynamics is reversed. None of this is irreversible, but certainly must be kept in mind at the time of design.

As stated, the dry gas seal does come with its own set of disadvantages. The biggest of these is that the buffer gas must be reliable. Loss of buffer gas in some cases will reverse the differential pressure across the seal faces, which will damage the seal in short order. The seals will operate at a zero differential pressure level, but when possible, even a small differential in the proper direction is recommended by the manufacturers. Another disadvantage is the requirement for clean and dry gas at the seal faces. The issue of providing a dry gas supply to the seal is covered in Chapter 8. For dirty gas applications, a sidestream from the compressor discharge will have to be filtered and injected on the process side of the seal. Of course, all buffer gas must be filtered. A 2-micron nominal level is considered sufficient. While the requirement for cleanliness of the gas is a disadvantage, it is not unique to the gas seal as the liquid buffered seal, particularly the mechanical type, also has a relatively stringent cleanliness requirement.

One final negative comment: some of the dry gas seals are unidirectional. This is a problem for compressors that are subject to reverse rotation. It is a problem for using a common spare seal for a compressor, because the rotation makes a seal rotor end specific. For compressors prone to reverse rotation and for the spare parts concern, seals that are bidirectional are available. There may be a small leakage penalty. Other considerations are that the compressor bearings may not tolerate reverse rotation, making the seal limitation not the only factor. Also, though definitely not recommended, unidirectional seals have rotated in the reverse direction for short periods of time without any major problem. The best solution is to address the reverse direction problem itself. The negatives were pointed out only as a caution to the user. The dry gas seal advantages definitely outweigh the negatives and are a significant addition to compressor shaft sealing.

Seal configurations are single, tandem, and double opposed (shown in Figures 5-53 A, B, and C, respectively). The single configuration, as the name implies, is a single set of sealing faces with the leakage either flared or vented. The tandem seal, which is probably the most common, consists of two single seals oriented in the same direction. The first seal is considered a primary seal and handles full pressure, while the second seal, which is referred to as secondary, operates at near zero differential and acts as a backup to the primary. Figure 5-54 shows a tandem seal. The leakage is removed from between the seals, and either flared, vented, or recovered if the recovery system can maintain a relatively low pressure. In applications where it is undesirable to permit the primary gas to leak through the secondary seal, such as with hazardous gas, a baffle can be installed between the primary and secondary seal. An additional port is added to permit the injection of a secondary gas with inert properties. This secondary gas then flows through the secondary seal. A variation of the tandem seal is referred to as the triple seal, which uses a two-seal arrangement to break down the pressure. By design, the two seals divide the pressure drop approximately in half and use the third seal as a backup.

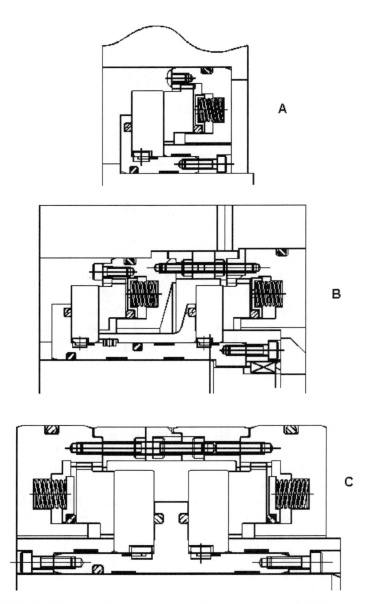

Figure 5-53. Section drawings of non-contacting dry gas seals: A. single seal arrangement, B. tandem seal arrangement, C. double opposed seal arrangement. (*Courtesy of John Crane International*)

Figure 5-54. Cutaway of a tandem arrangement non-contacting dry gas seal. (*Courtesy of John Crane International*)

The double opposed seal is used in applications where a zero process leakage is mandated. The seal consists of two seal faces, with the process side seal reversed. An inert gas is injected between the two seals at a positive differential over the process gas pressure. A small amount of the inert gas leaks into the process. The process must be able to accept the contamination of the buffer gas for this seal to be used.

Dry gas seals use a separation seal on the bearing side of the seal as a barrier. The purpose of the barrier seal is to prevent lubricating oil from migrating along the shaft and into the dry gas seal. This seal also serves the purpose of preventing any gas leakage from the dry gas seal from leaking into the bearing cavity.

The barrier seals come in two basic forms. One is a labyrinth design, which is probably the most common. It has the features of a conventional labyrinth discussed earlier. The alternative is the carbon ring seal. The carbon ring is used either as a single ring or, in some cases, it is of a multiple ring configuration. In the latter, it is normally a double ring. The carbon ring may be split and use a garter spring around the outside segments or it may be one piece. The carbon ring has the advantage of being a lower leakage seal and uses less barrier gas. It also provides a more effective seal against oil migration. The carbon seal features in general were covered in the earlier discussions on carbon ring seals.

Capacity Control

Probably the most widely used capacity control for the centrifugal compressor is speed control. The capacity curve when used with speed control covers a wide range. While electric variable speed motors offer a continuation to the speed control practice, there are some other alternatives available. Suction throttling has been widely used and offers a reasonable control range for a relatively low cost.

A more efficient control available on some centrifugals is the *movable inlet guide vane.* The movable inlet guide vane adds pre-whirl to the gas stream entering the impeller, which, in turn, reduces the axial component of the absolute velocity, which controls the capacity to the impeller as discussed earlier. By modifying the inlet whirl component, the capacity is reduced with little loss in efficiency. It is obvious that this method would be most effective on the single-stage compressor (see Figure 5-55). It can be used on the multistage compressor, however, it can only be installed in front of the first impeller. If the compressor has more than a few stages or is more complex in arrangement the idea is not practical.

The guide vanes in a single-stage are generally pie shaped and center pivoted. They are located directly in the flow path immediately in front of the impeller. The shanks of the vanes extend through the inlet housing and connect to an external linkage. The linkage is connected to a power operator to supply the motive power to position the vanes. Control for the vanes can be by a remote manual station or connected to an automatic control as the final element.

In the multistage compressor, the vanes are rectangular and located in a radial position ahead of the first impeller, with a linkage connecting the vanes to a power positioner. From that point, the control is affected in the same manner as the single-stage.

The largest problem with the use of movable inlet guide vanes is the danger of the vanes or the mechanism sticking. Obviously, the vanes are not suitable for dirty or fouling gas service. The vane bearings and linkage should be buffered with clean, dry gas and exercised regularly. While this does take extra effort, the vanes will work well and give efficient control.

Figure 5-55. Single-stage centrifugal with movable inlet guide vanes. (*Courtesy of A-C Compressor Corporation*)

Maintenance

At the risk of misleading the reader, this section will just touch a few points concerning maintenance. One frequently asked question is, what is critical and what needs special consideration when performing maintenance on the centrifugal compressor? While there are many areas that must be carefully reviewed, the clearances should be restored as close as practical to the new machine values. Probably the single most important consideration, therefore, concerns concentricity. Interstage seal clearances should be concentric to the rotating element. It is better to allow larger than desired clearances in the machine than to leave seals in an eccentric condition. Leakages approach 2½ times the concentric values for eccentric seals, with the same average clearance.

While the axial position of multistage impellers to their diffusers is not critical, they should line up reasonably well. Impellers are not extremely sensitive to leading-edge dings and minor damage, but anything, such as erosion on the exit tips, that tends to decrease the effective diameter of the impeller is more serious. Front shroud clearance on open impellers should be maintained close to the design values to minimize capacity loss.

Bearings normally have a specified clearance range. Allowing clearances to exceed the specified maximum clearance may encourage the onset of rotor dynamics problems. Dams in dam type bearings are very critical. The edge of the relief must be square and sharp, not rounded. The clearance of this bearing is also quite sensitive and must remain inside the specified limits for stability.

Care must be taken in the assembly of the buffered shaft end seals, particularly in the area of the secondary *o-ring* seals. A cut or damaged ring can allow more oil to be bypassed than from a damaged main seal.

References

1. Balje, O. E., "Study on Design Criteria and Matching of Turbomachines, Part B: Compressor and Pump Performance and Matching of Turbocomponents," *ASME Paper No. 60-WA-231,* ASME Transactions, Vol. 84, *Journal of Engineering for Power,* January 1962, p. 107.

2. Boyce, Meherwan P., *Gas Turbine Engineering Handbook,* Houston, TX: Gulf Publishing Company, 1982.

3. Durham, F. P., *Aircraft Jet Power Plates,* Englewood Cliffs, NJ: Prentice-Hall, Inc., 1951.

4. Lapina, R. P., *Escalating Centrifugal Compressor Performance, Process Compressor Technology,* Vol. I, Houston, TX: Gulf Publishing Company, 1982.

5. Scheel, Lyman F., *Gas Machinery,* Houston, TX: Gulf Publishing Company, 1972.

6. Sheppard, D. G., *Principles of Turbomachines,* The MacMillan Co., 1956, pp. 60, 67, 9th Printing 1969, pp. 238–244.

7. Wiesner, F. J., "A Review of Slip Factors for Centrifugal Impellers," *ASME 66-WA/FE-18,* American Society of Mechanical Engineers, New York, NY, 1966.

8. Hallock, D. C., "Centrifugal Compressor, the Cause of the Curve," *Air & Gas Engineering,* January 1968.

9. Boyce, Meherwan P., *et al., Practical Aspects of Centrifugal Compressor Surge and Surge Control,* Proceedings of the 12th Turbomachinery Symposium, Purdue University, West Lafayette, IN, 1983, pp. 147–173.

10. Hackel, R. A. and King, R. F., "Centrifugal Compressor Inlet Piping—A Practical Guide," *Compressed Air & Gas Institute,* Vol. 4, No. 2.

11. Brown, Royce N., "Design Considerations for Maintenance Clearance Change Affecting Machine Operation," Dow Chemical USA, Houston, TX, 1976.

12. API Standard 617, *Centrifugal Compressors for General Refinery Services,* Sixth Edition, 1995, Washington, DC: American Petroleum Institute, 1979.

13. NEMA Standards Publication No. SM 23-1979, *Steam Turbines for Mechanical Drive Service,* National Electrical Manufacturers Association, Washington, DC, 1979.

14. Elliott Company, *Compressor Refresher,* Elliott Company, Houston, TX., pp. 3–29 (other pages)

15. Dugas, J. R., Southcott, J. F. and Tran, B. X., *Adaptation of a Propylene Refrigeration Compressor With Dry Gas Seals,* Proceedings of the 20th Turbomachinery Symposium, Texas A&M University, College Station, TX, 1991, pp. 57–61.

16. Feltman, P. L., Southcott, J. F. and Sweeney, J. M., *Dry Gas Seal Retrofit,* Proceedings of the 24th Turbomachinery Symposium, Texas A&M University, College Station, TX, 1995, pp. 221–229.

17. Bornstein, K. R. *et al., Applications of Active Magnetic Bearings to High Speed Turbomachinery with Aerodynamic Rotor Disturbance,* Proceedings of MAG '95 Magnetic Bearings, Magnetic Drives and Dry Gas Seals Conference, The Center for Magnetic Bearings, A Technology Development Center of the Center for Innovative Technology and the University of Virginia, Charlottesville, VA, August 1995.

6

Axial Compressors

Historical Background

The basic concepts of multistage axial compressors have been known for approximately 130 years, being initially presented to the French Academie des Sciences in 1853 by Tournaire. One of the earliest experimental axial compressors was a multistage reaction type turbine operating in reverse. This work was performed by C. A. Parsons in 1885. Needless to say, the efficiency was not good, primarily because the blading was not designed for the condition of a pressure rise in the direction of flow. Around the turn of the century, a few axial compressors were built using blading based on propeller theory. The efficiency was better but still marginal, achieving levels of 50 to 60%. Further development of the axial compressor was retarded by ignorance of the underlying fluid mechanics principles.

World War I and the interest in aviation gave rise to a rapid development of fluid mechanics and aerodynamics. This, in turn, gave a renewed impetus to axial compressor research. The performance of the compressor was considerably improved by the isolated air foil theory. As long as the pressure ratio per stage was moderate, the axial compressors were capable of quite high efficiencies. The compressor began to see commercial service in ventilating fans, air-conditioning units, and steam-generator fans.

Beginning in the 1930s, interest was increased in axial compressors as a result of the quest for air superiority. Efficient superchargers were necessary for reciprocating engines in order to increase power output and improve aircraft high altitude performance. With the development of efficient compressor and turbine components, turbojet engines for aircraft also began receiving attention. In 1936, the Royal Aircraft Establishment in England began the development of axial compressors for jet propulsion. A series of high performance compressors was developed in 1941 [1]. In this same period, Germany was doing similar research that ultimately produced several jet aircraft. In the United States, research was directed by the National Advisory Committee for Aeronautics (NACA). This was the forerunner of the National Aeronautics and Space Administration (NASA).

In the development of all these units, increased stage pressures were sought by using high blade cambers and closer blade spacing. Under these conditions the blades began to affect each other, and it became apparent that the isolated airfoil approach was not adequate. Aerodynamic theory was, therefore, developed specifically for the case of cascaded airfoils. In addition to the theoretical studies, systematic experimental investigations of airfoils in cascade were conducted to provide the required empirical design information.

While the aircraft-oriented research was going on in the mid '30s, commercial axial compressors were being built and installed in various process plants. The technology from the aircraft industry did not penetrate the commercial compressor business until 1958 when many of the NACA reports were declassified. Today, much of the commercial compressor design worldwide is based on the published NACA reports. An interesting comparison is to take similar applications from different time periods and look at the number of stages required to perform a given pressure ratio on air. For example, the '30s vintage compressor sized to compress air from 14.7 to 45 psia required 21 stages. If the '50s technology is applied to the same application, only 11 stages are needed.

Description

Axial compressors are high speed, large volume compressors but are smaller and somewhat more efficient than comparable centrifugal compressors. The axial compressor's capital cost is higher than that of a centrifugal but may well be justified by energy cost in an overall evaluation. The pressure ratio per stage is less than that of the centrifugal. In a general comparison, it takes approximately twice as many stages to perform the same pressure ratio as would be required by a centrifugal. The characteristic feature of this compressor, as its name implies, is the axial direction of the flow through the machine.

The energy from the rotor is transferred to the gas by rotating blades—typically, rows of unshrouded blades. Before and after each rotor row is a stationary (stator) row. The first stator blade row is called the guide vane.

The volume range of the axial starts at approximately 30,000 cfm. One of the largest sizes built is 1,000,000 cfm, though this size is certainly not common. The common upper range is 300,000 cfm. The axial compressor, because of a low pressure rise per stage, is exclusively manufactured as a multistage machine.

By far, the largest application of the axial compressor is the aircraft jet engine. The second most common usage is the land-based gas turbine, either the aircraft derivative or generic designed type. In last place of the applications comes the process axial compressor. All principles of operation are exactly the same. About the most obvious difference is that the gas turbine compressor is a higher pressure ratio machine and therefore has more stages.

Performance

Blades

While the process engineer does not need to understand the aerodynamic theory that makes the axial compressor perform, a few basics may be helpful. Acquaintance with the blading nomenclature and the fundamental velocity triangle may also keep the terms used around these compressors from seeming like a foreign language. The airfoil in most axials is of the NACA type, generally a 65 series. A typical designation is 65-(18) 10. The 65 defines the profile shape. The 18 represents a lift coefficient of 1.8 and the 10 designates a thickness/chord ratio of 10%. Figure 6-1 is a sketch of a blade cascade with various angles and profiles defined. The blade angles are designated as β. The *chord,* b, is the length of the straight line connecting the leading and the trailing edge. The *pitch,* s, is the measure of the circumferential spacing. *Solidity,* σ, is the relative interference, obtained by the ratio of b/s. *Camber,* is the curved mean line of the blade as is defined as the angular difference between the

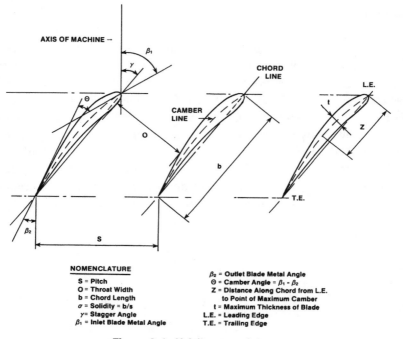

Figure 6-1. Airfoil nomenclature.

inlet blade angle β_1 and outlet blade angle β_2. *Aspect ratio,* AR, is the blade height, h, divided by the chord. *Stagger angle,* γ, is the angle between the chord line and machine axis. The gas angles are designated as α. Figure 6-2 gives a comparison of the various 65 series profiles. Note the increase in camber as the lift coefficient increases.

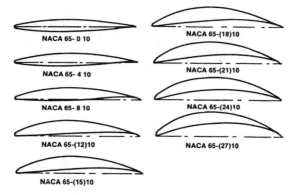

Figure 6-2. The NACA 65 series of cascade airfoils.

Compression Cycle

The gas enters the axial compressor through a guide vane row, then is acted on by the rotor, and moves on through the first stator row. From the stator, the gas again enters the rotor and exits to the next stator row. The gas continues its axial path until it has been acted on by all the rotor-stator pairs (stages). After the last stator, the gas may encounter an additional stator row or two, referred to as straightener rows. These stator blades take the whirl component out of the gas, so entry into the final diffuser or collector may be accomplished with a minimum of loss. Figure 6-3 shows a diagram of the blade path and the resulting vector triangles. Because of the finite length of the blades, the blade velocity, u, for the

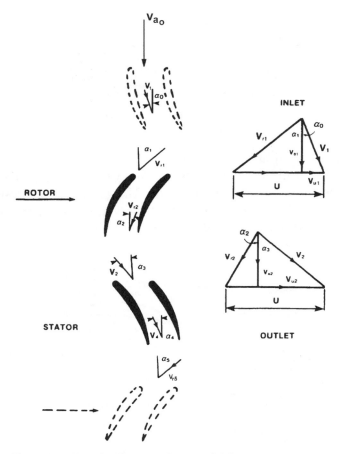

Figure 6-3. Velocity diagrams for an axial-flow compressor stage.

diagrams will be assumed to apply to the mean height of the blade. This simplifying assumption will be carried through most of the discussion because the three-dimensional flow path becomes quite complex. The absolute velocity is designated as V_1 and V_2. The relative velocity is given by V_{r1} and V_{r2}. The axial velocity is V_a.

Figure 6-4 is a diagram of the gain in energy of the gas shown by the enthalpy plot and the increase and decrease in velocity as the pressure is increased. While energy can only be added at the rotor blades, shown by

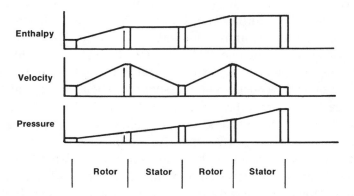

Figure 6-4. Variation of enthalpy, velocity, and pressure through an axial-flow compressor. (Modified from [4])

a horizontal line at the stator, the gain in pressure can be divided between the rotor and stator, as indicated by the steady increase in pressure as the gas moves from rotor through the stator.

Reaction

The degree of *reaction,* R, in the axial compressor is defined as the ratio of the static differential pressure in the rotor to the static differential pressure developed across the stage.

$$R = \frac{\Delta P_{\text{static rotor}}}{\Delta P_{\text{static stage}}} \tag{6.1}$$

The reaction can also be derived in terms of velocity components as given in the following relationship:

$$R = \frac{1}{u}\left(\frac{V_{ru1} + V_{ru2}}{2}\right) \qquad (6.2)$$

$$R = \frac{1}{u}\left(V_{rum}\right) \qquad (6.3)$$

Equation 6.3 shows that the degree of reaction can be seen directly on the velocity diagram as proportional to the whirl component of the mean relative velocity. These diagrams are helpful in comparing blade configurations, and with a further knowledge of the actual blade angles, a considerable amount of information can be obtained. Figure 6-5 shows a series of blade arrangements. These are all for the same blade speed, axial velocity, and change of whirl velocity. This, in effect, says that they have the same flow coefficient, pressure coefficient, and energy transfer.

Figure 6-5(a) shows the special case of axial outflow, associated with a single-stage fan with a stator row preceding the rotor. This case has no residual whirl velocity at the exit. As a multistage design, it offers the advantage of acceleration in the stator, since R > 1, which has the effect of smoothing out the flow and providing the best possible conditions for the rotor. However, it has the disadvantage of having a very high relative velocity, V_{r1}, and possibly a high Mach number. It is, therefore, unsuited for the first stage of the compressor, where V_a and u are high and the temperature is at its lowest, but may be more suited for the later stages where the Mach number may be lower.

Figure 6-5(b) is a case for a reaction of unity, that is, all the pressure rise is in the rotor, with the stator blades acting only as guide vanes to deflect the gas. A reaction of unity is aerodynamically the equivalent of R = 0 or impulse, as shown at Figure 6-5(f) since it corresponds to an interchange of the moving and fixed blade row.

Figure 6-5(c) shows the diagram for R between 0.5 and 1.0 for the special case of axial inflow associated with a single-stage fan with a stator following the rotor to remove the residual whirl velocity. This is necessary when the gas is drawn straight into the rotor. As a multistage design, this arrangement offers no special advantages or disadvantages. It will be seen again that from the aerodynamic viewpoint, it is the equivalent of R between 0 and 0.5, Figure 6-5(e).

Figure 6-5(d) shows R = 0.5 or the symmetrical case, in which rotor and stator are similar. $V_1 = V_{r2}$ and $V_2 = V_{r1}$. The degree of reaction is defined as

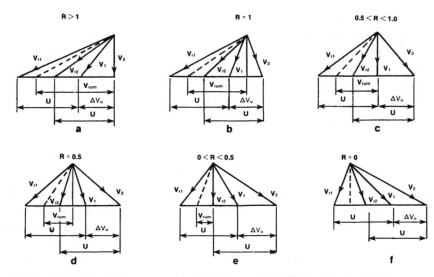

Figure 6-5. Velocity diagrams for various degrees of reaction in an axial-flow compressor stage. (*Reprinted with permission of Macmillan Publishing Company from* Principles of Turbomachinery *by D. G. Shepherd, Macmillan Publishing Company, 1956*)

$$R = 1 - \frac{(V_{u1} + V_{u2})}{2u} \tag{6.4}$$

From the symmetry $V_{u1} = V_{ru2}$, hence

$$R = 1 - \frac{V_{ru2} + V_{u2}}{2u} \tag{6.5}$$

substituting $u = V_{ru2} + V_{u2}$

$$R = 1 - \frac{u}{2u}$$

$$R = .5$$

Any effect of Mach number is experienced by rotor and stator equally and thus neither (or both) are limiting, and this Mach number will be lower than for other degrees of reaction under the conditions stated. If equal lift and drag are assumed in both rotor and stator, then optimum efficiency is obtained with $R = 0.5$ and $V_a/u = 0.5$. Although the latter is not always true, it does provide a useful criterion. Furthermore, the blade angles are similar in rotor and stator, which may be an advantage in the manufacturing phases in relation to tooling and inspection facilities. These factors help make the 50% reaction design a common choice by the compressor manufacturers when using the NACA 65 series blading.

It would seem that a high-reaction design would be desirable for the highest stage pressure rise and, therefore, the fewest number of stages for a given overall pressure ratio. Another factor must be considered however, and that is Mach number. The diagram for R = 1.0 shows that the relative inlet velocity to the rotor is considerably higher than for R = 0.5. The cascade data are for low speed (exclusive of Mach number). Therefore, the high reaction stage might be penalized by poor efficiency due to high Mach number, or alternatively, u and V_a would have to be reduced and thus the gain in energy transfer might be nullified. It is not possible to generalize further on degree of reaction, because the choice rests on the selection of velocities, Mach numbers, and effect of Mach number on efficiency. It may be reasonable, however, to state that for a high performance compressor, one in which high velocities and high efficiencies are required, a controlled diffusion airfoil may be used (see Figure 6-6). This type of airfoil has been made possible with the development of computational fluid dynamics (CFD). By using the controlled diffusion airfoil, the optimum reaction can be raised to a range of .6 to .7.

Stagger

For a given degree of reaction and value of V_a or flow rate, a choice can be made of the stagger angle or setting of the blades. From the consideration of two defining equations:

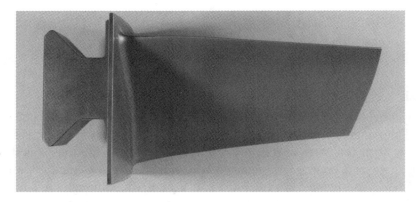

Figure 6-6. An axial compressor controlled diffusion airfoil. (*Courtesy of Elliott Company*)

$$R = \frac{V_a}{2u} (\tan \alpha_1 + \tan \alpha_2) \tag{6.6}$$

$$E = \frac{uV_2}{g} (\tan \alpha_1 - \tan \alpha_2) \tag{6.7}$$

where

E = energy transfer

For fixed values of R and V_a, a selection of a value of α_2 (corresponding to a value of stagger angle) fixes a limiting value of α_1 from cascade data. Thus the blade speed, u, is determined and then the energy transfer E is found.

To show the general effect of outlet angle or stagger, an approximate relation is used,

$$\tan \alpha_1 - \tan \alpha_2 = \frac{1.55}{1 + 1.5\,s/b} \tag{6.8}$$

where

s = blade pitch
b = blade chord

For the conditions of $s/b = 1$ and $R = 0.5$, the result is

$$\tan \alpha_1 - \tan \alpha_2 = 0.62$$

and from Equation 6.6,

$$\frac{V_a}{u} = \frac{1}{\tan \alpha_1 + \tan \alpha_2} \tag{6.9}$$

substituting the previous calculated value for $\tan \alpha$,

$$\frac{V_a}{u} = \frac{1}{0.62 + 2 \tan \alpha_2} \tag{6.10}$$

As the stagger angle increases, for higher values of α_2, the optimum V_a/u decreases and thus for a given value of V_a, the blade speed, u, must increase. Also as α_2 increases, $(\alpha_1 - \alpha_2)$ decreases, hence high stagger implies high rpm and blades of low camber.

Another important factor in design is the steepness of the characteristic curve, that is, the variation of pressure ratio with mass flow (see Figure 1-3). From consideration of the velocity diagram for 50% reaction, such as (d) of Figure 6-5, it can be shown that the symmetrical arrangement gives

$$\tan \alpha_1 = \frac{u - V_a \tan \alpha_2}{V_a} \tag{6.11}$$

and, substituting in Equation 6.7, this produces

$$E = \frac{u}{g}(u - 2V_a \tan \alpha_2) \tag{6.12}$$

Differentiating with respect to V_a, and selecting the proportional variables

$$\left(\frac{\partial E}{\partial V_a}\right) u \propto (-\tan \alpha_2) \tag{6.13}$$

that show the energy transfer, E, which is head or pressure ratio change with respect to axial velocity, V_a, which is mass flow at a greater rate as α_2 increases. Therefore, high stagger blades tend to have a steeper characteristic curve. However, another feature of increased stagger that cannot be demonstrated simply, but requires use of cascade data, is that the design point or point of maximum efficiency on the characteristic curve is at a value of mass flow somewhat greater than that for maximum pressure ratio. As a result, the design point is further away from the surge mass flow and allows more flexibility on that part of the characteristic curve. Low stagger, on the other hand, tends to place the design point closer to the surge point. While this approach offers more flexibility for increased mass flow, some efficiency may be sacrificed to avoid operating too near surge.

Since for a given value of V_a the blade speed must be greater for high stagger, the energy transfer is increased because

$$E \propto uV_a \tag{6.14}$$

and thus fewer stages may be required. Because the relative velocity is higher, the Mach number may be prohibitive, or alternatively thinner blades may have to be used in order to obtain a higher critical Mach number for the drag value. Thinner blades would require larger chords in order to maintain reasonable levels of bending stress. As a net result, the overall length would not decrease in proportion to the decrease in number of stages.

Earlier, it was stated, on the basis of simplifying assumptions, that the maximum efficiency for 50% reaction blading was obtained at a value of $V_a/u = 0.5$, requiring mean gas angles of 45°. The assumption for this result was that the drag-lift ratio was constant. In actual practice, cascade data indicate that drag-lift is not constant but increases as α_2 increases. It would appear that the maximum efficiency may be close to $\alpha_2 = 30°$. However, the reduction in efficiency is not severe because for values of α_2 of 15° and 45°, the drop is only about 1%.

Compressor performance can be changed by alteration of blade stagger. This is usually done by changes of stator stagger in preference of changes to the rotor. This can be accomplished on both process compressors and gas turbines, including the aircraft engine by use of variable stator vane control. While the mechanism is somewhat complex, it gives the axial compressor, with its inherently steep pressure-volume curve, the ability to be matched to a changing load without changing the speed. Figure 6-7 is a pressure-volume chart for an axial compressor with a partial stator vane control (only a portion of the stator blades is movable).

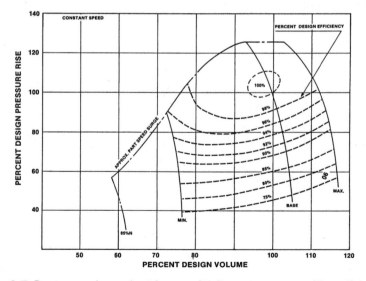

Figure 6-7. Pressure-volume chart for an axial-flow compressor with partial stator vane control. (*Courtesy of A-C Compressor Corporation*)

Curve Shape

The role of stagger, in developing the curve shape, was just covered; however, the shape deserves a few comments on comparison with other compressors. Even though it is a dynamic compressor, the axial's inherently steep pressure-volume curve makes it more akin to the positive displacement compressor shown in Figure 1-3. The steep curve contrasts with the flat curve of the centrifugal compressor. The horsepower characteristic also contrasts the shape of the centrifugal horsepower curve. Note that while the centrifugal's required horsepower increases with volume, the axial compressor's required horsepower does just the opposite (see Figure 6-8). This unique characteristic should be kept in mind when starting an axial. The axial is unloaded by opening a discharge bypass or otherwise removing the downstream load restriction during startup. This means that the lowest gas load is away from surge, compared to the centrifugal which may be unloaded by discharge throttling, which tends to bring the machine up to speed in surge.

The steep pressure-volume curve permits the axial compressor to operate very well in parallel with other axial compressors. The pressures do not have to match precisely to permit load sharing, as the steepness of the curve allows for adjustment without danger of going into surge or taking wild load swings as sometimes happens when attempts are made to operate centrifugal compressors in parallel.

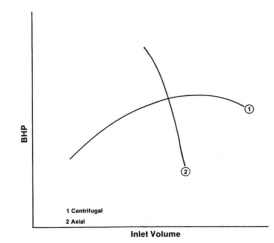

Figure 6-8. Typical brake horsepower vs. inlet volume curves for a centrifugal and an axial-flow compressor.

The axial compressor matches up well with a centrifugal compressor in a tandem-driven series connected arrangement, when higher process pressures are needed. The interstage matching is quite easy as the steep curve will provide sufficient changes in discharge volume to easily accommodate the requirement of the centrifugal. It can be thought of as stacking a constant ratio compressor on top of the axial's more vertical curve. With the axial compressor's high volume and speed attribute, the two machines will match in speed and volumes very neatly without the use of interstage gears.

Surge

In Chapter 5, this characteristic was applied to centrifugal compressors. The airplane wing analogy of stall was used, which is very directly applicable to the axial's airfoil-shaped blades. The incidence angle, described earlier, defines the onset of surge by stating that when the incidence exceeds the stall point, as developed from the cascade data, the foil ceases to produce a forward motion to the gas. When the gas cannot move forward, it moves in reverse, opposing the incoming flow. When the two collide, there is a noise, sometimes very loud. Recompression of the gas causes the temperature to rise very high very quickly. There have been cases where, when the blades were sufficiently strong not to break from the unsteady forces, they melted. It is more normal with prolonged surge to experience catastrophic blade breakage. The axial can also exhibit a phenomenon referred to as *rotating stall*. Rotating stall (propagating stall) is generally encountered when the axial compressor is started or operated too near the surge limit. This is especially true for compressor with adjustable vanes with the vanes in their extreme open or closed position. Vane movement is limited in some cases to minimize this problem. A flow perturbation causes one blade to reach a stalled condition before the other blades. This stalled blade does not produce a sufficient pressure rise to maintain the flow around it, and an effective flow blockage or a zone of reduced flow develops. This retarded flow diverts the flow around it so that the angle of attack increases or decreases on adjacent blades. These blades, with the increased angle of attack, stall and stay in a cell-like form. The cell then propagates around the stage or possibly two in which it occurs and at some fraction (40–75%) of rotor speed. Once begun, the cells continue to generate, causing inefficient performance, and if not terminated, may continue until a blade failure occurs. This is especially true if the cell's rotating speed coincides with the blade's natural frequency. Rotating stall is sometimes accompanied by audible pressure pulsation. Momentarily venting the compressor can inhibit cell formation. Besides the fact that the compressor tends to unload when taken to the maximum flow condition when starting, the additional problems are avoided.

Sizing

A short procedure will enable the reader to size an axial compressor. The complete and rigorous sizing of the axial is quite tedious and requires a comprehensive computer program. By making a few simplifying assumptions, a reasonable first approximation can be derived. While axials have been used rather extensively for gas service in their history, the bulk of the applications has been in air service. This is the first limitation to the sizing method, as it is only good for gases in the air molecular weight range eliminating the Mach number considerations. The head is assumed to be reasonably well divided over the various stages, and axial velocity is assumed constant throughout the machine. The numerical values are tabulated as follows:

Hub/tip ratio, $d_h/d_t = 0.7$ minimum, 0.9 maximum

Adiabatic efficiency, $\eta_a = 0.85$

Pressure coefficient, $\mu = 0.29$

Mean blade velocity, $u_m = 720$ fps

Because the constants and frame data include units, the relations presented here will depart from the primitive form used elsewhere in the book and will incorporate the necessary units.

Calculate an inlet volume, correct for moisture, if necessary, as outlined in Chapter 2. Select the frame size using Figure 6-9. It's probably a good idea to select the smallest frame that seems to accommodate the volume at the start. The frame size has been made easy as it is the hub diameter in inches. There have been five sizes presented, which should cover most ranges encountered, although the commercially available size range extends over a somewhat broader volume range. Select the number of stages, z.

Calculate the overall required head from the pressure ratio and the inlet temperature using Equation 2.70 from Chapter 2. It is repeated here for convenience.

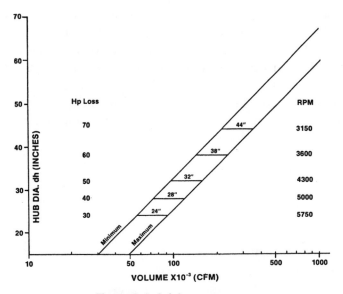

Figure 6-9. Axial compressor.

$$H_a = Z_{avg}\, RT_1\, \frac{k}{k-1}\, (r_p^{\frac{k-1}{k}} - 1) \qquad (2.70)$$

$$z = \frac{gH_a}{\mu\,(u_m)^2} \qquad (6.15)$$

Round off the number of stages to the nearest whole number, then recalculate the pressure coefficient, μ, using

$$\mu = \frac{gH_a}{z\,(u_m)^2} \qquad (6.16)$$

If the pressure coefficient just calculated is within 5 % of the original value of 0.29, then proceed using the calculated pressure coefficient and the assumed value of mean blade velocity as the final value. Continue to the speed calculation. If the pressure coefficient is higher than 5 %, add an additional stage to the compressor and again calculate the pressure coefficient using Equation 6.16.

If the pressure coefficient is now, or was in an earlier step, 5% under the 0.29 value, calculate a new mean blade velocity using the rounded-off number of stages and the original pressure coefficient, 0.29. Use the calculated blade velocity in the subsequent step for compressor speed. Calculate the speed.

In order to calculate the speed, a mean blade diameter must be established, and to calculate the mean diameter, a tip diameter is needed. The first step is to calculate the tip diameter.

$$d_t = [(6.05 \, Q_1/u_m) + d_h^2]^{1/2} \tag{6.17}$$

where

d_t = tip diameter, inches
d_h = hub diameter, inches
Q_1 = inlet flow, cfm
u_m = mean blade velocity, fps

then calculate a mean diameter, d_m in inches:

$$d_m = \frac{d_t + d_h}{2} \tag{6.18}$$

Before proceeding, make sure the hub tip ratio is within the minimum limit of 0.67. If satisfactory, continue with the speed calculation. If the value is unsatisfactory, repeat the previous steps with an alternate frame choice:

$$N = \frac{229 \, u_m}{d_m} \tag{6.19}$$

where

N = shaft speed, rpm

The speed must not exceed the speed given in Figure 6-9 for the selected hub diameter. Calculate the last stage volume using the following:

$$Q_{1s} = \frac{Q_1}{(r_p^{z-1/z})^{1/k}} \tag{6.20}$$

where

Q_{ls} = last stage inlet volume, cfm
r_p = pressure ratio across the compressor
k = isentropic compression exponent

Using Equation 6.16, calculate a stage tip diameter. Then check the hub tip ratio against the maximum value, 0.9. If the value is greater, the last stage blading is getting too short and probably the only solution is to use a smaller frame.

The guidelines presented are simplified and may not be sufficient for all applications. This does not mean that an axial cannot be used, because the vendors can perform a much more complex analysis and change factors that this simplified method chose to hold constant. Undoing some of these values is probably beyond the scope of most of the users. The best way to interpret a potential application is that an extra measure of care might be exercised when going out for bid. This can generate additional questions concerning the vendor's proposal.

To complete the sizing, calculate the discharge temperature using the Equation 4.6 from Chapter 4.

$$t_2 = t_1 + \frac{T_1 \, (r_p^{\frac{k-1}{k}} - 1)}{\eta_a} \times \eta_t \qquad (4.6)$$

where

t_2 = discharge temperature, °F
T_1 = inlet absolute temperature, °R
t_1 = inlet temperature, °F
η_a = adiabatic efficiency

Calculate the shaft horsepower, using Equation 4.7 from Chapter 4. Read the mechanical losses from Figure 6-9.

$$W_s = \frac{w \times H_a}{33,000 \, \eta_a} + \text{Mech losses} \qquad (4.7)$$

where

w = weight flow of the gas in the compressor, lb/min.
H_a = total adiabatic head, ft-lb/lb

Example 6-1

Size an axial compressor for air service using the procedures outlined in the chapter. The following conditions are given:

Molecular weight: 28.65

Isentropic exponent: 1.395

Compressibility: 1.0

Inlet temperature: 80.0°F

Inlet pressure: 23.0 psia

Discharge pressure: 60.0 psia

Weight flow: 28,433.7 lb/min.

Step 1. Use Equation 2.5 to calculate the specific gas constant.

R = 1,545/28.65

r = 53.93 specific gas constant

Convert temperature to absolute.

$T_1 = 460 + 80$

$T_1 = 540°R$

k/(k − 1) = 1.395/.395

k/(k − 1) = 3.53

(k − 1)/k = .395/1.395

(k − 1)/k = .283

Step 2. Substitute into Equation 2.10 and, using the conversion constant of 144 in²/ft², calculate the inlet volume.

$$Q_1 = \frac{1.0 \times 53.93 \times 540}{23 \times 144} \times 28,433.7$$

$Q_1 = 250,000$ cfm inlet flow

Select the frame size (hub diameter, d_h) using Figure 6-9. At the inlet volume value, a 44 frame is selected with a maximum speed of 3,150 rpm. This frame has a 44-inch hub diameter.

Step 3. To calculate the number of stages, the overall head is required. The head is calculated using Equation 2.70 and $r_p = 60/23 = 2.61$ for the pressure ratio.

$H_a = 1 \times 53.93 \times 540 \times 3.53 \, (2.61^{.283} - 1)$

$H_a = 32,080.2$ ft-lb/lb total adiabatic head

Then using Equation 6.12 and the pressure coefficient, $\mu = .29$ and $u_m = 720$ fps given in the chapter.

$z = 32.2 \times 32,080.2/(.29 \times 720^2)$

$z = 6.87$

This value is rounded to the next whole number, 7. Recalculate μ using Equation 6.13 and the number of stages.

$\mu = 32.2 \times 32,080.2/(7 \times 720^2)$

$\mu = .285$

Since the μ is within 5% of the target value of .29, then use the mean blade velocity, $u_m = 720$ fps, as a final value and proceed with the sizing.

Step 4. Calculate the tip diameter using Equation 6.17.

$d_t = [(6.05 \times 250,000/720) + 44^2]^{1/2}$

$d_t = 63.53$ in. first stage tip diameter

Check the hub-to-tip ratio, d_h/d_t. If greater than .67, continue. If not, go back to Step 2 and try another frame size.

$$d_h/d_t = 44/63.53$$

$$d_h/d_t = .69$$

This is greater than the stated limit, proceed to the next step.

Step 5. Calculate a mean diameter in preparation for calculating the compressor speed. Use Equation 6.18.

$$d_m = (63.53 + 44)/2$$

$$d_m = 53.77 \text{ in. mean blade diameter}$$

Calculate the speed using Equation 6.18.

$$N = 229 \times 720/53.77$$

$$N = 3066 \text{ rpm compressor speed}$$

The speed just calculated is less than the maximum speed of 3150 given for the frame and is therefore acceptable.

Step 6. Calculate the last stage volume using Equation 6.20.

$$Q_{ls} = \frac{250,000}{(2.61^{7-1/7})^{1/1.395}}$$

$$Q_{ls} = 138,698.9 \text{ cfm last stage volume}$$

Using Equation 6.14 calculate the last stage tip diameter.

$$d_t = [(605 \times 138,698.9/720) + 44^2]^{1/2}$$

$$d_t = 50.9 \text{ in. last stage tip diameter}$$

Check the last stage hub-to-tip ratio. It should be less than 0.9. If a problem is encountered with meeting this ratio, select another frame. If one frame exceeds the lower limit and the alternative choice exceeds the higher limit, multiple cases may be needed. The casing passing the lower hub-to-tip limit of .67 should be selected, except with the pressure ratio varied until the high limit of .9 can be met. The balance of the compression could be completed with a centrifugal compressor. In sizing a centrifugal compressor by the procedure outlined in Chapter 5, the speed of the axial can be assumed to be the centrifugal speed. This would permit a tandem drive arrangement. Cooling can be added, depending on the discharge temperature of the axial.

Calculate the last stage hub-to-tip ratio.

$d_h/d_t = 44/50.9$

$d_h/d_t = .864$

This value is less than the limit of 0.9. Proceed to the next step.

Step 7. Calculate the discharge temperature using the efficiency stated of .85 and Equation 4.6.

$$t_2 = 80 + \frac{540\,(2.61^{.283} - 1)}{.85}$$

$t_2 = 278.2°F$ discharge temperature

Step 8. Calculate the shaft horsepower using Equation 4.7 and the mechanical losses from Figure 6-9 at the frame selection. Use the efficiency $\eta_a = .85$ as recommended.

$$W_s = \frac{28,433.7 \times 32,080.2}{33,000 \times .85} + 70$$

$W_s = 32,589\text{hp}$ shaftpower

Application Notes

The axial compressor is a highly refined, sophisticated compressor. It is capable of very high efficiency, to the point that some of the designers feel there is no area of improvement left. As efficiency gets higher, the margin left between the ideal and current design makes each point much more difficult to achieve. Most of the development activity has centered on higher velocities, and the development of cascade data at the higher Mach numbers. The developments are more significant in aircraft engines where power-to-weight (size) ratio has a greater impact. The technology, however, is being applied to the land-based axial compressor. While the cost of the machine will somewhat follow the number of stages, and cost is probably one of the more significant factors retarding the application growth. The higher Mach number stages are more expensive to manufacture, somewhat offsetting the savings of having fewer stages. Gas turbines seem to be using the newer technologies as their size capabilities are increased.

Because the axial is a sophisticated compressor, it tends to show its "blueblood" at times, in lack of ability to cope with common plant problems such as fouling. The sophisticated airfoils, while capable of such nice high efficiency performance, have a real problem with dirt. It does not have to be polymers or other chemical reactions of the kind that cause problems with the centrifugal, but rather it can be ordinary atmospheric air. Some of the tendency to foul can be averted by changing the reaction at the expense of efficiency. This has not been completely successful, however, due to the complex modes in which fouling takes place. The best solution is filtration, which is attended by an increase in inlet pressure drop. The filtration should be of the dry type. Moisture, even a high humidity, can make whatever dirt does pass through the filter stick to the blading. On-stream washing has been successful in some cases, but must be carefully done and is somewhat of a trial-and-error method, until an operable mode is established. An alternative to washing is the use of organic abrasives. These have been reported as an effective and low-cost method of cleaning up this type of build-up [2].

Larger axial compressors have a physical space problem with the inlet nozzle, requiring a departure from the conventional round flanged nozzle customarily used in centrifugal application. This means either custom engineered rectangular duct work supplied by the user or an off-machine transition piece. For atmospheric suction compressors, where the inlet is connected to a nearby filter housing, this is not a serious problem.

Inlet startup screens have been recommended for other compressors covered in the earlier chapters. If the point has not been made yet, it should be with the axial compressors. Considering that most of the cost of the compressor is in the hundreds of vulnerable blades just waiting to be hit by some foreign object, it should be obvious that some protection is needed until the piping has been proven to be clear and clean.

Mechanical Design

Casings

The casings on axial compressors are somewhat unusual, because of the disproportionately large inlet and outlet nozzles. This makes the compressors appear to be only nozzles connected by a long tube. Casings can be fabricated or cast, with the fabricated obviously being steel, while the castings can be cast iron or cast steel. In some designs, the casing is an outer shell containing an inner shell, which acts as the stator vane carrier. In other designs, the stators are directly carried on the casing, which are of a one-part construction. With this latter design, the casing is made up of three distinct parts, bolted at two vertical joints. The parts are the inlet section, the center body with the stators, and the discharge section. The three sections are also split horizontally for maintenance. With the three piece-bolted construction, a mixture of fabrications and castings may be used. The mounting feet are attached to the outside casing and so located as to provide a more or less centerline support. As mentioned earlier, some designs use a rectangular inlet section to provide more axial clearance. The reason for using an entire separate casing is that it forms a separate pressure casing that can readily be hydrotested. The disadvantage is that there is more material involved making the cost higher. The compressors that use the integral stator section or single case approach have somewhat of a cost advantage, and in general, may have a slight advantage in being able to keep the stator carriers round, because of the end-bolting to the other casing components. The disadvantage is that there are more joints to seal and maintain. Checking out the entire casing for strength and leakage in a hydro and gas test becomes somewhat more complex.

Stators

As already described, the stators may be carried in a separate inner casing or may be carried by the outer, center section of the main casing. When movable stator vanes are used, the vanes pass through the wall of the carrier. The outer side of the casings exposes the shank ends, which are used as shafts to connect to the linkage that will control the movement. Because of leakage at the mounting bushings in the stator liner or carrier, the single-case unit has some sealing problems, which are inherently taken care of in the double-case construction. The single-case construction uses a lagging over the linkage, which can act as a collector.

For non-movable construction, some vendors use a conical shank and bolt arrangement. The stator vanes are set to a gauge at the factory and locked in place by tightening the hold-down nut and then staking. This construction permits flexibility for capacity adjustment at relatively low cost because the vane stagger can be reset without the manufacture of new parts or a complete machine disassembly. Other bolting arrangements are manufactured to give the same flexibility as the conical shanks. Some designs use a more permanent fixture for the stator vanes, setting the vanes in a diaphragm similar to the steam turbine or fixing the vane to the stator casing by dovetailing.

The stator vanes are usually not shrouded. This is not a hard and fast rule. In practice, often vanes are mixed with some shrouded and some unshrouded. When a separate stator inner case is used, it is normally of cast iron. If the temperature is expected to exceed 500°F, the liner will be multipart with the discharge end being made of steel.

Casing Connections

The casing inlet and outlet are flanged, for connection to the user's piping system, with the exception of some rectangular inlet connections as already mentioned. Communication with the manufacturer must take place concerning allowable forces and moments. While this is recommended for all types of compressors, the axial is a relatively small machine with disproportionately large piping, which results in the inherent potential for large forces and moments.

The balance chamber leakoff line, while recommended to be held within the confines of the compressor casing may well turn out to require some user piping. There are some situations where the desire for keeping open space around the compressor for maintenance may require compromise on the part of the user. The balance of the connections on the axial are for lube oil and other auxiliary equipment not different from that found on other compressors.

Rotor

Rotor construction tends to vary from vendor to vendor. The blades are attached to the outer surface of the rotor. The rotor may be of basic disc or drum type construction, with the disc type having some variations. The two most common disc construction modes are shrunk discs on a shaft and stacked discs, normally through-bolted together. A final method is the solid rotor construction.

When the disc construction method is used, the blades are attached by a single-lobe dovetail root design. The slots are broached into the rotor and the blade roots fitted into the slots and keyed in place. Figure 6-10 shows a bladed disc type rotor.

When the discs are of the shrunk-on design, they are made up individually and stacked onto the shaft by first heating the disc to dilate the bore. They are then allowed to cool and thus attach themselves to the shaft. Keys are normally not used. When the discs are of the stacked design, the discs are equipped with rabbet fit to radially lock the discs to maintain concentricity of assembly. The through-bolts are usually tensioned by stretching hydraulically to a precise value to ensure the mechanical integrity of the assembly.

An alternate method not used too much at this time is the drum design. The drum construction is somewhat different from the disc, in that the rotor body is of cylindrical construction. By using the hollow drum, conical roots, of bolted construction, can be used for the rotor, again, allowing for stagger adjustments to fine tune the axial compressor to the application if the need arises. The setting is done to a gauge at the factory, as with the stators.

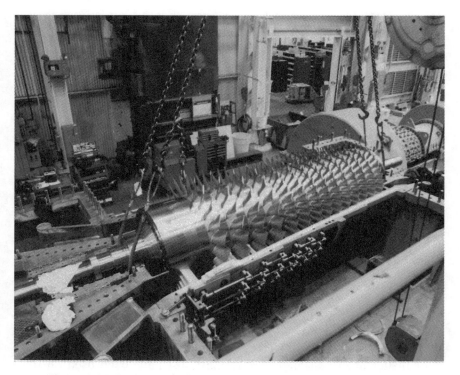

Figure 6-10. A 14-stage axial-flow rotor. (*Courtesy of Elliott Company*)

For smaller compressors, where the speed is relatively high and space is limited, a solid rotor construction is used. This is similar to the disc type of construction, except that the discs are an integral part of the rotor. Blade attachment slots are cut into the rotor, similar to the slots cut into the discs. Rotor blades are rarely, if ever, shrouded.

Rotor material in all cases is low alloy steel with an appropriate heat treatment to match the stresses imposed by the blades and rotor weight. The rotor is generally manufactured from a forging with the material being a chrome-molybdenum alloy such as AISI 4140 or AISI 4340.

Shaft

Shafting takes on several forms to match the various rotor construction methods. Obviously, for the solid rotor, the shaft is a part of the overall rotor. For the shrunk-on discs, the shaft is a continuous member, carrying the discs in the center section. Concentricity of all turns and good control on the roundness of the shaft are critical, if a balanced, smooth running compressor is to result.

The more unique form of shafting is used in the bolted disc and drum designs. Here, the *stub-shaft* design is used. A stub-shaft is fitted to each end of the center body, whether disc or drum. The design must anticipate all possible sources of stress, so the proper shrink can be applied to the interface. An interference fit is used at the interface to ensure concentricity for all operating conditions. It should include a reasonable allowance for momentary overspeed, particularly if a turbine driver is used. The design should consider the potential temperature transients that may be encountered at startup, shutdown, and hot restarts. Some arbitrary allowance should be made for torsional transients, even when not ordinarily anticipated. The shaft material, for the separate shafting, can be made of a different material than that used for the rotor body, although there is not much reason to do it that way. The heat treatment used could be different without compromising the overall rotor.

Blading

Axial compressor blades are usually forged and milled. Precision casting has been used on occasion. The most common material used is a 12 chrome steel, in the AISI 400 series, and is also known as 400 series stainless steel. While the stator blades are occasionally shrouded, the rotor blades are free-standing. Lashing wires have been used on rotor blades, but are generally used to solve a blade vibrational stress problem.

In a new compressor, the wires should not be used because the stress problems should be solved in a fundamental design manner without having to resort to "fixes." When the manufacturer designs the blades, the vibrational characteristics for both rotor and stator blades should be established. The basic bending resonances and the higher orders to which the blades may be excited should be established. Care should be taken to avoid any direct excitation sources, such as splitter vane or stator and guide vane passing frequencies. If possible, any fundamental and lower order resonances should be at least three to four times higher than any of the running speed. The manufacturer should supply a *Campbell diagram* to demonstrate that the compressor is free of direct excitations. When resonances do exist in the operating range, the vendor should demonstrate his understanding of the stress level and provide some assurance to the user that the compressor will not have premature blade failures. One method to convey the information is with a *Goodman diagram.* Some users ask for Goodman diagrams for all stages, regardless of any resonant interferences, to demonstrate that a conservative design concept was used throughout the blading design. While stress levels in the rotor blading are by far the most severe, resonances can occur in the stator blades that have been known to fail when excited by one of the compressor's operating frequencies. This is a rare occurrence, but must still be considered. Most reported failures are caused by rubs or foreign object damage. Regardless of the cause, if a stator blade should break and drop into the gas path where it can be struck by the rotor blades, the wreck is just as traumatic as a direct rotor blade failure.

While much has been said about the axial compressor that would give the impression that the machine is not durable, nothing could be farther from the truth. The compressor has logged hundreds of thousands of hours in trouble-free operation, and will do so if properly designed and operated with reasonable care. It may not be quite as abuse-resistant as the centrifugal. However, as the efficiency and performance of centrifugals is upgraded they tend to become less abuse-resistant. The object of any operating group should be a conscientious effort to properly operate and maintain the equipment so that everyone will benefit.

Bearings

The bearings used in axial compressors are the same journal and thrust type used in the centrifugal compressor. Refer to Chapter 5 for a complete description of these bearings.

For axial compressors, the journal bearings are of the plain sleeve type for the larger, slower speed compressors. They are of the tilting pad type for the smaller, higher speed machines. The sleeve bearing is normally housed in a spherically seated carrier. The bearings require pressure lubrication as do most of the other compressors.

The thrust bearing is generally the tilting-pad type bearing. Most vendors apply the recommendation that the thrust bearing be of the symmetrical design with leveling links. Axial compressors have a high inherent thrust load, so the thrust bearing is quite important in the overall reliability of the compressor. While this is true for the other compressors as well, it deserves an extra emphasis here.

Balance Piston

The axial compressor is inherently always a reaction type of machine. In regards to axial thrust, this means the rotor is subjected to a differential pressure across each rotating blade row. The differential pressures convert to an axial force at each rotor row that totals to a rather high value when taken over the normal number of stages. A thrust bearing would be prohibitive in size to carry the generated thrust. Fortunately, the geometry of the axial provides space for a large balance piston at the discharge end of the compressor. In fact, the construction of the axial compressor rotor is such that the placement of a labyrinth seal on the hub diameter and another labyrinth seal on the shaft forms a balance cavity. The balance piston cavity is normally vented to the suction end of the compressor or to the atmosphere on air machines. The balance piston seal leakage is charged to the compressor as a loss. As in the centrifugal, the return gas represents a head loss due to the heating effect of the return gas, and a direct loss in capacity due to the quantity of gas bypassed.

Seals

Because most axial compressors are in air service, most are equipped with labyrinth type end seals. There are no interstage seals in the machines with unshrouded stator blades. The balance piston seal, a labyrinth type, is the only internal seal. There is no reason that axials cannot use some of the other seals as described in Chapter 5, such as the controlled leakage or the mechanical contact type, if the gas being handled by the compressor needs a more positive seal. If there is any problem at all, it will be that the seal rubbing velocity will be higher due to the larger shaft and relatively high speed. In the past, where axials have been used in process gas service, these problems were overcome.

Capacity Control

Earlier in the chapter, the movement of the stator blades to change the capacity of the axial compressor was discussed. In conjunction with extended shanks and linkage, as mentioned, the axial offers a very good dynamically controlled capacity range. Depending on the range desired, linkage may be installed on the guide vanes only or on the guide vanes and stator vanes starting from the front of the compressor. Control has been used on up to 100% of all the stator vanes; however, the range gained after 50% stator control is generally marginal and not worth the extra expense. The linkage in Figure 6-11 is typical. The linkage is normally connected to one or more power cylinder operators for manual operation from a remote location, or for use as the final control element in an automatic system. Use in an automatic control system is good, particularly if there is a little noise on the control signal. This acts to keep the vanes *live* and the linkage and bearings from fouling and sticking. The sticking of the linkage is probably the largest single problem with the vane control. For manual systems, a regularly scheduled linkage exercise can help keep the movement free. Use of dry clean gas purges are also helpful. The range of the axial is enhanced considerably by the use of movable vanes.

Maintenance

Most of the maintenance ideas presented in Chapter 5 are also applicable to the axial. Concentricity of the rotor clearances in the stator is again stressed. Rotor tip clearances allow a reasonable tolerance. A 10% clearance change is not greatly significant.

When an axial is taken out of service, the blading should be carefully cleaned and inspected. If minor damage is noted, the blade should be repaired or, if the damage is significant, it should be replaced. In most cases, minor, foreign-object "dings" or cracks can be ground out and blended. If several blades in a row show cracking, a note of the accrued operating time of that row should be made and the manufacturer consulted to ensure that a more serious problem isn't beginning to show. With

Figure 6-11. Adjustable stator blade linkage. (*Courtesy of Elliott Company*)

reasonable operation, not too many hours in surge, or no gross upsets, blading should last 50,000 hours or more. If no cracks are detected, blades should be randomly selected and examined by one of the NDT methods. Should any cracks be detected, very strong consideration should be given to replacing the entire blade row in which the cracked blade was located.

If movable stator blades were used, the linkage should be checked for binding and wear. If a clean gas purge was used to keep the linkage clean and dry, the source and supply lines should be inspected to make sure they all work as intended when the compressor is restarted.

Beyond these items, the balance of the maintenance procedure should include the customary bearing and seal checks.

References

1. NASA SP-36, *Aerodynamic Design of Axial-Flow Compressors,* NASA, Washington, D.C., 1965.

2. Alleyne, C. D., Carter, D. R., and Watson, A. P., *Cleaning Turbomachinery Without Disassembly, Online and Offline,* Proceedings of the 24th Turbomachinery Symposium, Texas A&M University, College Station, TX, 1995, pp. 117–127.

3. Horlock, J. H., *Axial Flow Compressors,* Malabar, FL: R. E. Kreiger Publishing Company, 1958.

4. Shepherd, D. G., *Principles of Turbomachinery,* Toronto, Ontario, Canada: The Macmillan Company, 1956.

5. Boyce, M. P., *Gas Turbine Engineering Handbook,* Houston, TX: Gulf Publishing Company, 1982.

6. Lieblein, Seymour, "Analysis of Experimental Low-Speed Loss and Stall Characteristics of Two-Dimensional Compressor Blade Cascades," *RM E57A28, NACA (NASA) 3-19-57,* Declassified 6-24-58.

7. Emery, J. C., Herrig, L. J., Erwin, J. R., and Felix, A. R., *Systematic Two-Dimensional Cascade Tests of NACA 65-Series Compressor Blades at Low Speeds,* Report 1368, NACA (NASA), 1958.

7

Drivers

Introduction

Drivers for compressors must supply torque of a specified value at a certain speed. Whenever the driver speed characteristics are not directly useable, they may be modified by a speed increasing or reducing gear. The only exceptions to the speed-torque criteria are a limited number of the reciprocating compressors. These are the integral engine and the direct steam cylinder driven machines. The balance of the reciprocators align themselves with the other compressors, receiving their input energy by way of shaft torque at a given speed.

The driver is a prime mover capable of developing the required torque at a constant speed or over a range of speeds. The driver's energy source can be either electrical or mechanical. Electrical energy is used by motors, either of the induction or synchronous type, while the mechanical covers a multitude of sources. It may be a fuel, as in internal or external combustion engines, or it may be a gas, such as steam or process gas used in a turbine or expander.

This chapter will describe all common compressor drivers, but as a practical consideration, details on selection or sizing, hazardous area applications, and installations will not be covered. For additional information refer to the National Electrical Code [1] and API publications RP 500 and RP 540 [2, 3] as well as some of the general references given.

Electric Motors

Modern technology has reduced the size of motors, increased their expected life and improved their resistance to dirt and corrosion. Other important developments of the last 30 years are brushless excitation for synchronous motors and new two-speed, single-winding, induction motors.

Long-term cost of ownership is the prime factor in the selection of any equipment. Selection should be based on the least expensive, most efficient motor that will meet the requirements. Improper selection increases operating costs. Oversized motors are commonly purchased either because the actual load requirements are not known or because of anticipated load growth. It is important that the motor be sized for all known design and offdesign operating conditions. Gross oversizing is wasteful because motors perform best and cost less to operate (maximum power factor and efficiency) at the manufacturer's rating. The best checks against improper size are careful review of drive requirements prior to purchase and periodic checks of the individual motors in operation.

Serious consideration should also be given to enclosure selection. Many improvements have been made in recent years in both enclosures and insulation. Therefore, it is important to review purchasing practices to make sure they are based on today's technology.

Review the motor requirements and specifications to make sure that all the unnecessary, nonstandard, special features have been eliminated. Each special requirement such as nonstandard mounting dimensions and nonstandard bearings should be eliminated unless it can be demonstrated the feature is cost effective. In actual practice, many special features are specified because of an isolated case of trouble that occurred years ago. Likewise, some special features may become obsolete through changes in refinery or chemical plant practice or through improved manufacturing techniques.

Synchronous and induction motors cannot always be compared on an equal speed basis. In geared applications such as high-speed centrifugal compressor drives (above 3,600 rpm), the most economical induction motor speed is usually 1,800 rpm. The most economical synchronous motor speed for the same application might be 900 or 1,200 rpm, depending on the horsepower required. For compressor drives above 3,600 rpm, motor prices must be studied within the total evaluation concept.

For 3,600-rpm compressor drives below 5,000 hp, simplicity of installation almost dictates using the two-pole induction motor. No gear is required, and the overall electrical and mechanical installation is the simplest possible.

Electric motors above 5,000 hp are custom-designed for the specific application, taking into consideration compressor characteristics and the power system parameter limitations.

While logistics favor the use of the two-pole motors, there is a break point in mechanical design between the four-pole and two-pole motors. Because opinions vary concerning the desirability of two-pole motors, those contemplating their use should first investigate the "track record" of the motor supplier in the required size range.

A 5,000 hp break point has been used rather arbitrarily, as larger motors are built, ranging in size to 30,000 hp. Generally as the size increases, the synchronous motor becomes more competitive. However, final selection is not only dictated by the driver economics above, but includes the power system as well.

Voltage

The least expensive motors, typically below 200 hp, are low voltage (less than 600 volts). Above 500 hp and up to 5,000 hp, the preferred voltage is 4,000 volts for use on a 4,160 volt system. It should be noted that motor nameplate voltages are commonly about 5% less than the nominal voltage of the system they are used on. However, synchronous motors that may operate at a leading power factor, in general, have a nameplate voltage equal to the nominal system voltage. As the motors become larger, coil capacity becomes more of an issue and increased space is available for insulation, so it becomes economically practical to use higher voltage motors and controls. It is apparent that, as the size of a refinery or chemical plant increases, the distribution system increases, and it is necessary to go to higher system voltages. For distribution system voltages above 5,000 volts, the usual practice is to transform down to 2,400 or 4,160 volts for large motors. Cable and transformer costs frequently make it economical to select motors with higher voltages such as 13.8 kilovolts. Numerous 11 to 14 kV motors are now in service, even outdoors.

Lightning and switching surges that damage motors are related to high-voltage motors. The insulation level of motors is below that of many other types of apparatus such as transformers, switchgear, cables, etc. Because of the low insulation level, the system lightning arresters will not protect the motors adequately. Special surge protection equipment is often needed, particularly for large motors with long cores. Surge protection consists of special low-discharge voltage arresters in parallel with surge capacitors. The capacitor slopes off the wave front to reduce the motor winding turn-to-turn voltage, and the arrester limits the voltage rise to a safe value. Surge capacitors are used without the arrester in some applications To give maximum protection, the arrester should be mounted at the motor terminals.

Enclosures

The enclosure selected affects cost substantially. Corrosive and hazardous atmospheres encountered dictate the protection needed.

Today's standard motor enclosure for indoor applications is the open, drip-proof enclosure for induction and high-speed synchronous motors. For large motors, open, drip-proof construction is available up to about 20,000 hp and is used for squirrel-cage, synchronous, and wound-rotor motors.

For larger motors, the next degree of protection is weather-protected, Type I (WPI), which is an open machine with ventilating passages constructed to minimize the entrance of rain, snow, and airborne particles to the electric parts.

Where a higher degree of protection and longer life is desired, weather-protected, Type II (WP II) motors are recommended. This type of enclosure is used on the motor-driven compressors shown in Chapter 8, Figure 8-15. They are equipped with extensive baffling of the ventilating system so that the air must turn at least three 90° corners before entering the active motor parts. Maximum air-intake velocity is limited to no more than 600 fpm, and blow-through provisions are also provided. In this way, rain, snow, and dirt carried by driving winds will be blown through the motor without entering the active parts. The WP II are the predominant choice for large outdoor motors in petrochemical applications in North America.

Totally Enclosed Motors

The choice for severe applications below 250 hp is the totally enclosed fan-cooled (TEFC) motor (see Figure 7-1). TEFC motors separate the internal and external ventilating air. Breathing is the only way external air ever gets inside. At some size above 250 HP, manufacturers presently switch from a rib-cooled to an exchanger-cooled motor for totally

Figure 7-1. A direct-connected, TEFC two-pole motor driven centrifugal single-stage compressor. (*Courtesy of A-C Compressor Corporation*)

enclosed fan-cooled (TEFC) applications. These exchanger-cooled motors are often referred to as totally enclosed air-to-air cooled (TEAAC) motors. Fin-cooled TEFC motors are becoming available in ever increasing horsepower, and are presently available from some manufacturers at sizes up to 2000 hp. These appear to be the preference of many users for the most severe applications. Above 500 hp (approximately), totally enclosed motors with water-to-air coolers (TEWAC) cost much less than TEFC motors and, in large ratings of synchronous motors, may cost even less than WP II. Large synchronous machines are frequently enclosed and supplied with coolers for mounting in the motor foundation at lower cost than integral-mounted coolers.

Totally enclosed motors offer the highest degree of protection against moisture, corrosive vapors, dust, and dirt. TEWAC motors also have the advantage of reduced noise level.

Division 1 Enclosures

Explosion-proof motors can withstand an internal explosion without igniting a flammable mixture outside the motor. These motors are totally enclosed, fan-cooled, and specially machined to meet Underwriter's Laboratory Standards. Explosion-proof squirrel-cage motors are available up to 3,000 hp at 3,600 rpm, but the larger sizes are usually not practical or economical.

Force-ventilated (F-V) motors are suitable for hazardous locations. Safe air is brought in through a duct system, passed through the motor, and then discharged preferably through another duct system to the limits of the hazardous location. The ventilating ducts should be pressurized to prevent entrance of contaminated air. This indoor construction has been largely replaced by outdoor motors such as weather-protected or totally enclosed types. It might be noted that while the F-V is normally used indoors, there are exceptions where it has been put in outdoor service.

Inert Gas-filled

Inert gas-filled motors can also be used in refineries and chemical plants, but their applications are limited. They have tightly fitted covers and oil seals around the shaft to minimize gas leakage, are continually pressurized with an inert gas or instrument air, and are equipped with an internal air-to-water heat exchanger. Inert gas-filled motors are suitable for any hazardous location but require auxiliaries such as cooling water, gas pressurizing system, and control accessories.

Insulation

Electrical insulation is continually being improved. The motor manufacturers make use of this and other technological developments to put more power into smaller, lighter, more efficient packages. Modern insulating materials can withstand heat, moisture, and corrosive atmospheres, and new metals can withstand more mechanical punishment. Computer design techniques are also helpful.

Insulation systems were first classified according to the material used, and permissible temperatures were established based on the thermal aging characteristics of these materials. For example, Class B insulation was defined as inorganic materials such as mica and glass with organic binders; 130°C was the allowable maximum operating temperature. The present definition of insulation system Class B stipulates that the system be proven ". . . by experience or accepted tests . . . to have adequate life expectancy at its rated temperature, such life expectancy to equal or exceed that of a previously proven and accepted system." The definition is now functional rather than descriptive.

The newest catalogs show standard induction motors designed with Class B insulation for operation in a 40°C ambient with 80°C rise by resistance at 100% load for motors with 100% service factor. Class F insulation, with the capability of operating up to a 105°C rise by resistance, is today frequently offered as standard for machines with a Class B rise, particularly the larger sizes. Many users specify this as a standard. Previously, induction motor ratings were based on temperature rise by thermometer.

These changes require an explanation. National Electrical Manufacturers Association (NEMA) standards previously allowed three methods of temperature determination: (1) thermometer, (2) resistance, and (3) embedded detector. Motor engineers have long recognized that measuring temperature rises by placing a thermometer against the end windings does not give the best indication of insulation temperatures near the conductors in the slot. The average temperature rise of any motor can be measured by resistance. This will give a better indication of the temperature in the hottest part of the winding than will thermometer measurement. On machines equipped with temperature detectors, there will usually be a difference in the readings taken by an embedded detector and by winding resistance, with the detector reading usually slightly higher, because it is embedded near the hottest part of the winding. Adequate placement and selection of the embedded detectors can be a factor in determining the quality of this type of monitoring. For example, longer detectors may provide more of an average temperature, and detectors near the fan end of a long TEFC motor may not see the hottest winding temperatures.

The resistance method gives an average temperature of the whole winding. Some parts will be hotter than others; usually the end turns will be somewhat cooler than parts of the winding in the middle of the iron core. NEMA committee members have been collecting test data on many machines to determine the correlation between temperature measurements by detector and by resistance, and the standards are periodically updated to reflect any of the technology improvements.

NEMA standards do not give any fixed maximum operating temperature by any class of insulation. Briefly, NEMA states that insulation of a given class is a system that can be shown to have suitable thermal endurance when operated at the temperature rise shown in the standards for that type of machine. Standards for synchronous motors and induction motors with a 100% service factor specify 80°C rise by resistance for Class B insulation. Also, the total temperature for any insulation system is dependent upon the equipment to which it is applied. For example, railway motors with Class B insulation have rated standard rises by resistance of 120°C on the armature. Induction motors with service factor have 90°C rise at the service factor load.

Service Factor

For many years it was common practice to give standard open motors a 115% *service factor* rating; that is, the motor would operate at a safe temperature at 15% overload. This has changed for large motors, which are closely tailored to specific applications. Large motors, as used here, include synchronous motors and all induction motors with up to 16 poles (450 rpm at 60 Hz).

New catalogs for large induction motors are based on standard motors with Class B insulation of 80°C rise by resistance, 1.0 service factor. Previously, they were 70°C rise by thermometer, 1.15 service factor.

Service factor is mentioned nowhere in the NEMA standards for large machines. There is no standard for temperature rise or other characteristics at the service factor overload. In fact, the standards explicitly state that the temperature rise tables are for motors with 1.0 service factor. Neither standard synchronous nor enclosed large induction motors have included service factor for several years.

Today, almost all large motors are designed specifically for a particular application and for a specific driven machine. In sizing the motor for the load, the horsepower is usually selected so that additional overload capacity is not required. Therefore, customers should not be required to pay for capability they do not require. With the elimination of the service factor, standard motor base prices have been reduced 4–5% to reflect the savings. Users should specify standard horsepower ratings, without service factor for these reasons:

1. All of the larger standard horsepowers are within or close to 15% steps.
2. As stated in NEMA, using the next larger horsepower avoids exceeding standard temperature rise.
3. The larger horsepower ratings provide increased pull-out torque, starting torque, and pull-up torque.
4. The practice of using 1.0 service factor induction motors would be consistent with that generally followed in selecting horsepower requirements of synchronous motors.

The common practice of using Class F insulated motors with a Class B rise at 1.0 SF in effect provides some obtainable service factor above 1.0 if the user is willing to operate the motor up to the Class F limits in response to some contingency. In many cases this provides at least 15% margin.

In NEMA size ranges, motors with service factors are still available; however, for compressor drives, it would be better if they were not. Experience with operation into the service factor rating has not been satisfactory.

Synchronous Motors

Synchronous motors have definite advantages in some applications. They are the obvious choice to drive large, low-speed reciprocating compressors and similar equipment requiring motor speeds below 600 rpm. They are also useful on many large, high-speed drives. Typical applications of this type are geared, high-speed (above 3,600 rpm), centrifugal compressor drives of several thousand horsepower.

A rule of thumb that was used in the past for constant speed applications was to consider the selection of a synchronous motor where the application horsepower was larger than the speed. This, of course, was only an approximation and tended to favor the selection of a synchronous motor and would be considered too severe by current standards. However, the rule can aid in the selection of the motor type by giving some insight as to when the synchronous might be chosen. For example, applications of several hp per rpm often offer a distinct advantage of the synchronous over the induction motor. In fact, at the lowest speeds, larger sizes and highest hp/rpm ratios may be the only choice.

One interesting characteristic of synchronous motors is their ability to provide power factor correction for the electrical system. Standard synchronous motors are available rated either 100% or 80% leading power factor. At 80% power factor, 60% of the motor-rated kVA is available to be delivered to the system as reactive kVA for improving the system power factor. This leading reactive kVA increases as load decreases. At zero load, with rated field current, the available leading reactive kVA is approximately 80% of the motor rated kVA. The unity power factor machine does not provide any leading current at rated load. However, at reduced loads, with constant field current, the motor will operate at leading power factor. At zero load, the available leading kVA will be about 30% of the motor rated kVA.

Because of their larger size, 80% power factor motors cost 15–20% more than unity power factor motors, but the difference may be less costly than an equivalent bank of capacitors. An advantage of using synchronous motors for power factor correction is that the reactive kVA can be varied at will by field current adjustment. Synchronous motors, furthermore, generate more reactive kVA as voltage decreases (for moderate dips), and therefore tend to stabilize system voltage better than capacitors, since they supply less leading kVA when the voltage is decreased.

When higher than standard pull-out torque is required, 80% leading power factor motors should be considered. The easiest way to design for high pull-out is to provide additional flux, which effectively results in a larger machine and allows a leading power factor. The leading power factor motor may, therefore, be less expensive overall. However, leading power-factor motors are generally "stiffer" electrically, and may need to be evaluated against the higher current pulsations that will result from reciprocating compressors or other pulsating loads. Larger flywheels can normally be applied to compensate for this effect.

In addition to power factor considerations, synchronous motor efficiency is higher than similar induction motors. Efficiencies are shown in Table 7-1 for typical induction and unity power factor synchronous motors. Leading power factor synchronous motors have efficiencies approximately 0.5–1.0% lower.

Table 7-1
Full Load Efficiencies

Hp	600 RPM	1800 RPM	3600 RPM
250	91.0	93.5	94.5
	93.4*	—	—
1,000	93.5	95.4	9.52
	95.5*	—	—
5,000	—	97.2	97.0
	97.2*	97.4*	—
10,000	—	97.5	97.4
—	97.6*	—	
15,000	—	97.8	—
	—	98.1*	—

Synchronous Motors, 1.0 PF
Source: Modified from [7] & [15]

Direct-connected exciters were once common for general purpose and large, high-speed synchronous motors. At low speeds (514 rpm and below), the direct-connected exciter is large and expensive. Motor generator sets and static (rectifier) exciters have been widely used for low-speed synchronous motors and when a number of motors are supplied from a single excitation bus.

Brushless Excitation

One of the most significant developments in recent years is *brushless excitation* for synchronous machines. This development became possible with the availability of reliable, long-life, solid-state control and power devices (diodes, transistors, SCR's, etc.). As the name indicates, there are no brushes, collector rings, or commutators on the motor or exciter. This eliminates brush, collector ring and commutator maintenance and permits the use of synchronous motors in many hazardous (Class 1, Group D, Division 2) and corrosive areas where conventional motors could not be used without extensive additional protection. All the advantages of conventional synchronous motors are retained: constant speed, high efficiency, power factor correction, and varied performance capability. High precision, fast-acting, solid-state field application control is rotor-mounted and provides the same full complement of functions as a conventional synchronizing panel. Brushless excitation is presently almost universally used.

The *exciter* is an AC generator with a stator-mounted field. Direct current for the exciter field is provided from an external source, typically a small variable voltage rectifier mounted at the motor starter. Exciter output is converted to DC through a three-phase, full-wave, silicon-diode bridge rectifier. Thyristors (silicon-controlled rectifiers) switch the current to the motor field and the motor-starting, field-discharge resistors. These semiconductor elements are mounted on heat sinks and assembled on a drum bolted to the rotor or shaft.

Semiconductor control modules gate the thyristors, which switch current to the motor field at the optimum motor speed and precise phase angle. This assures synchronizing with minimum system disturbance. On pull-out, the discharge resistor is reapplied and excitation is removed to provide protection to the rotor winding, shaft, and external electrical system. The control resynchronizes the motor after the cause of pull-out is removed, if sufficient torque is available. The field is automatically applied if the motor synchronizes on reluctance torque. The control is calibrated at the factory and no field adjustment is required. The optimum slip frequency at pull-in is based on total motor and load inertia. All control parts are interchangeable and can be replaced without affecting starting or running operation.

Motor Equations

The following equations are useful in determining the current, voltage, horsepower, torque, and power factors for three phase AC motors:

$$I = .746hp/(1.73 \; E \; \eta \; PF) \tag{7.1}$$

$$kVA = 1.73 \; I \; E \; / \; 1,000 \tag{7.2}$$

$$kW_{input} = kVA \times PF \tag{7.3}$$

$$kW_{shaft} = kW_{input} \times \eta \tag{7.4}$$

$$hp = kW/.746 \tag{7.5}$$

$$T = 5250 \; hp/N \tag{7.6}$$

$$PF = kW/kVA \tag{7.7}$$

where

$$E = \text{volts (line-to-line)}$$
$$I = \text{current (amps)}$$
$$PF = \text{power factor}$$
$$\eta = \text{efficiency}$$
$$hp = \text{horsepower}$$
$$kW = \text{kilowatts}$$
$$kVA = \text{kilovoltamperes}$$
$$N = \text{speed, rpm}$$
$$T = \text{torque, ft-lb}$$

A typical medium-size, squirrel-cage motor is designed to operate at 2–3% slip (97–98% of synchronous speed). Synchronous speed is determined by the power system frequency and the stator winding configuration. If the stator is wound to produce one north and one south magnetic pole, it is a two-pole motor. There is always an even number of poles (two, four, six, eight, etc.). The synchronous speed is

$$N = 2 \times 60f/P \tag{7.8}$$

where

  N = speed, rpm
  f = frequency, Hz
  P = number of poles

The actual operating speed will be slightly less by the amount of slip. Slip varies with motor size, load, and application. Typically, the larger and more efficient the motor, the less full-load slip. A standard 10-hp motor may have 2½% slip; whereas, motors over 1,000 hp may have less than ½% slip. Operating slip can be approximated by multiplying % load by full-load slip.

Compressor and Motor

Coordinating a motor with driven equipment involves selecting the proper motor horsepower, comparing motor and driven equipment speed-torque curves, determining required starting time and effect of momentary voltage losses, providing for thrust conditions by using proper coupling and end-play limits, and coordinating bearing lubrication requirements.

Motor horsepower required to operate the compressor is dictated by process plant operating conditions. Usually there are minimum conditions, normal (design) conditions, and maximum conditions. Economic sizing of the motor depends on exercising good judgment when determining capacity to be provided in the motor over the usual maximum operating conditions. Two other factors should be considered: capability of the compressor casing and likelihood that plant requirements will dictate a future change in the rotor to take advantage of casing capability. For a guide to the actual size, API generally recommends that the motor be sized at 110% of the greatest power required by the compressor. This allows some margin for compressor deterioration.

Selecting Compressor Motors

The first step in selecting motors for large compressors (1,500 hp and over) is to determine the motor voltage, speed, and enclosure type.

Motor voltage selection is determined by economics and by the availability of adequate system capacity to permit motor starting without excessive voltage drop. Restrictions by the utility or by the size of the plant's generating capacity may limit the maximum drop to less than 20%. In some cases, system dips as small as 5% or less may be the maximum tolerable. For example, many utilities insist on limiting dips to this magnitude at the customer interface, and some motors will be installed at a system location that can influence large amounts of critical lighting or other similar loads. Applications of this type will often require some type of soft start. However, in the usual refinery or chemical plant process unit, large blowers and compressors are very seldom started once the plant has been on stream for some time. The undesirable effects normally associated with local drops as high as 20% or more have been tolerable when occurring infrequently.

The motor cost is but one facet of any cost study for selecting voltage level. The study must compare installed cost of motor, starting equipment, transformers, and power and control cables at the various levels under consideration, as well as plant standards.

Higher voltage levels such as 13,200 volts are sometimes more economical overall, although the cost of the motor itself is higher than it would be at 2,300 or 4,000 volts. For example, if the plant has a 13.8 kV distribution system, it can be more economical to install a 13,200-volt motor than to provide the primary switchgear required for lower voltage motors. Operating experience at levels as high as 13,200 volts has been good, but it is not extensive when compared to lower voltages. The number of manufacturers with experience at these levels is more limited. Also, there are motor design considerations that limit the minimum size at which these high voltage levels can be applied. The motor manufacturer probably would not recommend a 3,000-hp, 13,200-volt motor as a first choice.

When the plant distribution voltage is above utilization levels, for instance 23 kV or higher, economics will usually favor the 2,300- or 4,000-volt motor. However, each application has its own peculiarities. An examination of the relative cost of alternate schemes sometimes favors 4,160 volts. This is true particularly with motors in the 4,000 hp and larger sizes, and short circuit levels are above 150 MVA.

Where speed-increasing gears are a consideration, the 3,600-rpm motor is eliminated for all practical purposes because of higher cost, less favorable torque characteristics, and mechanical design considerations.

The motor and gear combination must provide the proper input speed to the compressor. Therefore, speed selection should consider whether the 1,800 rpm, or the 1,200 rpm motor and the corresponding gear provides the more economical combination. Before selecting motor speed, motor characteristics should be obtained for both speeds. The speed-torque and speed-current curves, power factor and efficiency values should be compared and evaluated. The most important considerations in matching the motor to a compressor are:

1. Motor speed-torque characteristics
2. Load accelerating torque requirements
3. Motor supply voltage during acceleration

For the slower speed compressors, an evaluation must be made concerning the use of direct-connected motors that are more expensive, or the use of a less expensive 1,800 rpm or a 1,200 rpm motor with a speed-reducing gear. As before, the evaluation must include all factors, such as installation and the extra space, as well as some factor for additional maintenance.

Starting Characteristics

The closer the motor and compressor speed-load curves are to each other, the longer the driver will take to reach full speed, and the hotter the motor will get (see Figure 7-2).

A motor speed-torque curve for a compressor (250 to 1,000 hp or more) does not look the same as a smaller machine (10 or 20 hp) (see Figure 7-3). NEMA Standard MG-1 gives minimum locked rotor-torque values as a function of motor size at 1,800 rpm (see Table 7-2).

Before final motor selection can be completed, a comparison of the selected compressor's required speed-torque must be made with the pro-

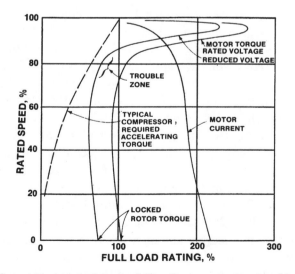

Figure 7-2. Speed-load curve for a centrifugal compressor and motor. Startup difficulty could occur at the "trouble zone" where the curves are very close [7].

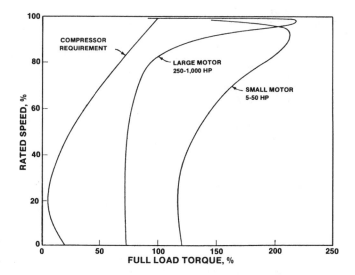

Figure 7-3. Typical speed-torque curves for large and small motors [7].

Table 7-2
Locked Rotor Torque vs. Motor Size

Rated hp	Minimum locked rotor torque percent of full load torque
5	185
10	165
100	125
250	80
2000	60

Source: [7]

posed motor curves. Each kind of compressor has its own unique curve, as illustrated by Figure 7-4. Whenever feasible, either from the compressor design or process considerations, unloading should be considered. Unfortunately, some compressors can not be readily unloaded. For the higher inertia loads, the torsional inertia must be taken into account to determine the starting time. For lower inertia loads, such as with some of the rotary compressors, the motor represents the principal inertia. Probably the worst assumption to make is that compressors follow the centrifugal square law curve, or that there is a typical compressor curve. While a little unusual, Figure 7-5 illustrates a standard motor for a typical compressor.

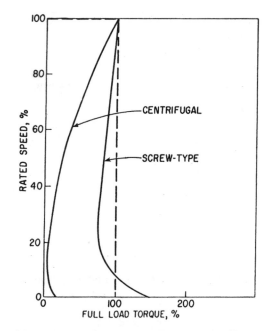

Figure 7-4. Speed-torque curve for two compressor types. Each type of compressor has its own curve [7].

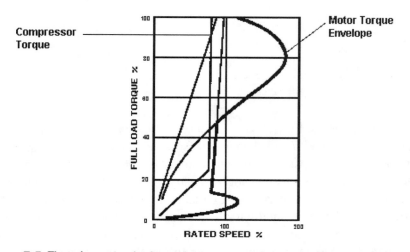

Figure 7-5. There is no standard speed-torque curve for accelerating compressors. Attempting to make a standard motor to meet all needs (heavy line) would impose unnecessary demands on motor design [7].

Some engineers specify motors for across-the-line starting. Motor output torque varies approximately as the square of applied voltage. A 10% voltage drop means a 20% drop in torque, which is enough to keep some drives from ever reaching full speed. This relationship should be borne in mind for operation as well as starting. A lower than required voltage while operating may appear as an overloaded compressor.

Starting Time

The voltage drop during starting should be calculated using the speed current data, locked rotor power factor, and the distribution system constants. Speed-torque values at this reduced voltage can be calculated, assuming that the torque varies as the voltage squared. These values are compared to driven equipment curves.

In comparing speed-torque characteristics, driven equipment torque requirement at any speed must not exceed 95% of motor torque at that speed. Also, the driven equipment speed-torque curve for the loaded condition is used. If less than 5% accelerating torque is available, it is considered doubtful that the motor can successfully accelerate within the allowable time. Theoretically, 1% difference should permit acceleration; however, 5% provides the margin for data and calculation inaccuracies. Because the time for the motor to accelerate through a given speed interval is inversely proportional to the accelerating torque, close approaches of the load-to-motor starting torque may result in excessively long starting times. Even if the motor does not stall at some sub-operating speed, as would be expected if the accelerating torque margin went to zero, the time to accelerate may result in the motor exceeding its allowable starting thermal limit and, at best, experiencing a protection relay trip prior to reaching the operating speed. The loaded curve is suggested for the basic evaluation because in many process applications, unloading may not be feasible. Also, if available, it presents extra margin. It should be pointed out that higher starting torque motors are more expensive. There is a possibility that they will require higher starting currents. To add insult to injury, induction motors with the higher starting torque design will probably be somewhat less efficient. Synchronous motors designed for higher starting torque may be difficult to find.

Starting times for large motors driving high-inertia loads, such as centrifugal compressors, can be 20 seconds or longer. The motor draws locked-rotor current for most of this period. These high currents maintained for such long periods cause winding and rotor temperature to rise rapidly.

If a motor has sufficient torque available at all points along the speed-torque curve, the starting time should be calculated. An approximation of motor starting time can be obtained by summation of starting time increments calculated for several speed intervals. Five or six speed intervals should be used. Use small intervals when accelerating torque changes are large, large intervals when torque changes are small.

$$\Delta t = \frac{2\pi\Delta N}{\Delta T \times g} \times WR^2 \qquad\qquad (7.9)$$

where

Δt = incremental starting time, sec

ΔN = speed interval, rpm

ΔT = average accelerating torque over the speed interval (difference between motor and load torque)

g = gravitational constant

WR^2 = torsional moment of inertia

Before leaving the starting time subject, it should be mentioned that once a reasonable starting time has been established, there is no merit in penalizing the motor by doubling the time. As mentioned earlier, gross oversizing as opposed to a conservative approach leads to an inefficient overall operation.

An early appraisal of motor and driven equipment speed-torque characteristics, particularly at the reduced voltage occurring during starting, is necessary. The following data should be obtained:

Motor

1. Speed-torque curve
2. Speed-current curve
3. Moment of inertia (WR^2)
4. Locked rotor power factor
5. Time constant for open circuit voltage (when motor control will use delayed transfer to alternate sources on voltage loss). This value must include the effect of any capacitors applied on the load side of the motor controller.

Driven Equipment

1. Unloaded speed-torque curve (zero flow through compressor)
2. Loaded speed-torque curve (compressor design point)

3. Speed-torque curve, intermediate discharge pressure (refrigeration service for restart evaluation)
4. Moment of inertia including gear, at the motor shaft
5. Gear ratio

Enclosure Selection

The selection of the motor enclosure type involves not only economics but also assessment of two factors: area hazard classification and other area operation conditions.

The determination of area hazard classifications has been aided greatly by publication of API Standard RP 500. This standard provides an engineering guide for assessing the degree of hazard. The standard should be used as a guide by anyone responsible for determining area classification or applying equipment in a hazardous area. It is not a substitute for sound engineering judgment.

Operating conditions that also must be considered in selecting the enclosure type are exposure to airborne dust, dirt and moisture, the possibility of corrosive dust or vapors in the area, and the expected maximum ambient temperature.

The compressor or blower installation in a typical refinery or chemical process unit is not out-of-doors completely. Some form of shelter often is provided, ranging from only a roof to a completely closed building. When process equipment such as a centrifugal gas compressor, which is not hazardous in normal operation, is present in the shelter, the hazard classification depends on the extent to that which the shelter restricts ventilation. The extent of the shelter provided determines the area classification and the type of motor enclosure that should be applied.

It is generally agreed that a shelter with a roof having ridge ventilation and with curtain walls not extending lower than 8 feet above the operating platform would be freely ventilated. Because a gas compressor would not be a source of hazard, except under abnormal conditions such as an equipment failure, this type of compressor shelter is usually classified as a Division 2 area.

In accordance with the RP 500 definition, any shelter having more obstruction to air passage than a "roof and one wall closed" limit is considered to have restricted ventilation and is classified Division 1.

Enclosure Applications

Motor enclosure types for compressor service under shelters are applied as follows:

Division 1 Areas
   a. Force ventilated
   b. Totally enclosed inert gas or instrument air-filled

Division 2 and Nonhazardous Areas
   a. Drip-proof
   b. Modified drip-proof (air intake at bottom)

The force-ventilated type is preferred when a Division 1 classification is necessary. The standard drip-proof enclosure is preferred in Division 2 or in safe locations. For an outdoor installation, serious consideration should be given to using NEMA Type 2 weather-protected enclosures rather than the completely enclosed types.

In the force-ventilated or drip-proof types, the cooling air passes directly over the motor winding insulation. As a rule of thumb, a motor requires about 4,000 cfm of cooling air per thousand horsepower. Therefore, the possibilities of airborne dust and dirt collecting on the winding must be considered carefully. Filters in the air intake will lessen the hazards of this condition. Not much can be done about moisture drawn in with the cooling air, but careful selection of motor location inside the shelter can minimize the amount of wind-blown rain striking the motor.

On first examination, the totally enclosed inert gas or air-filled enclosure with a closed ventilating circuit using a gas-to-water heat exchanger would seem the best answer for all locations. However, the cost of this enclosure runs at least 75% more than the drip-proof enclosure. Also, the auxiliary requirements of this type enclosure must be considered. It requires an external cooling water supply of good quality and high reliability. Where salt water or fresh water with corrosive impurities is used, double-tube heat exchangers are necessary. A cooling water failure alarm and subsequent automatic shutdown feature and enclosure moisture detector and alarm must also be provided. In addition, an external supply of inert gas or instrument air for leakage makeup is required. All factors considered, this type enclosure requires a fairly extensive auxiliary installation. The service factor for this enclosure is normally 1.0.

The force-ventilated enclosure, in addition to the disadvantage of contact between the cooling air and the winding, requires external ducting and a pressuring blower. Standby blower arrangements are necessary because the motor is not self-ventilating on loss of the pressurizing blower. The service factor for this enclosure is normally 1.0.

The drip-proof enclosure offers a ventilating system that is not dependent on external auxiliaries. Its installed cost is the lowest of all the enclosure types. Standard drip-proof ratings above 500 rpm may have a 1.15

service factor. The winding insulation is exposed to all the hazards associated with direct contact with cooling air. Also, the higher noise levels of this type may be objectionable, if noise level is a design consideration.

Recognizing all the factors mentioned, the drip-proof type has been the favored enclosure for service in shelters classified Division 2 or in safe areas because it offers adequate protection at the lowest installed cost (see Figure 7-6). In most applications, air-intake filters and screens have been provided. In one instance, a modified drip-proof enclosure was installed with provision for future filter installation.

Variable Frequency Drives

A relatively new innovation for use in electric motor compressor drives is the *variable frequency* power source. Fundamentally, the power source converts an existing three-phase source into DC then uses an inverter to convert back to a variable frequency supply. Thyristors or transistors are used to switch the output at the required frequency.

Figure 7-6. An integral geared compressor driven by a motor with an open drip-proof enclosure. The enclosure also includes 85dba sound attenuation. (*Courtesy of Elliott Company*)

Motor

The electric motor is basically a standard, single-speed, single-voltage motor. When receiving a variable frequency, the motor will operate at a variable speed. The motor may be either an induction or synchronous motor. The decision as to which type to apply is essentially similar to that for a single-speed application.

While the motor is basically standard, there are several items that may be considered when the variable frequency system and motor are purchased:

1. The insulation may need to be upgraded to protect from surge voltages.
2. The winding reactance may need to be decreased to improve thyristor commutation.
3. The base temperature rise may need to be decreased to improve motor performance.

The merits of a variable-speed motor would appear to be obvious, as many compressors in the past have benefited from the variable speed available in a steam turbine. A compressor may be adjusted as required to meet the process needs. The advent of the variable-frequency drive returns some of the benefits to the process operator that were lost when the more favorable electric energy caused motors to replace steam turbines.

The inverters are either voltage source or current source (see Figure 7-7a and b). There are other variations, but they apply to drivers smaller than the ones used with compressors. However, pulse-width-modulated (PWM) (see Figure 7-7c), transistorized units are less complicated and are relatively maintenance-free with reliable units available to at least 500 hp. For all but the smaller compressors, the current source inverter is the one typically used. With a six-step voltage source, a rule of thumb has been to size the motor at two-thirds of its rating so as not to exceed the insulation temperature rise. For current source motors, the output torque is not constant with decreased speed, which fortunately is compatible with most compressors, as torque tends to follow speed. For current source drives, one needs to upsize the motor captive transformer by approximately 15% to account for harmonic heating effects.

Inverters do not output a pure sine wave but synthesize the output wave with pulses. Because of the pulses, harmonics are presented to the motor and, hence, the somewhat higher losses. Common systems are either 6 pulse or 12 pulse. This definition comes from the number of

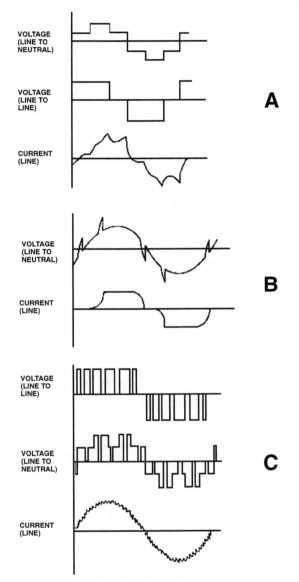

VOLTAGE
(LINE TO
NEUTRAL)

VOLTAGE
(LINE TO
LINE)

CURRENT
(LINE)

A

VOLTAGE
(LINE TO
NEUTRAL)

CURRENT
(LINE)

B

VOLTAGE
(LINE TO
LINE)

VOLTAGE
(LINE TO
NEUTRAL)

CURRENT
(LINE)

C

Figure 7-7. Output from three inverters: (A.) Voltage source, (B.) Current source, (C.) Pulse width-modulated source.

pulses used to simulate the output wave form. The more pulses, the less severe the harmonics; however, the cost also goes up. The same issue also applies to harmonic reduction at the drive's rectifier input with the external power supply receiving the benefit of the higher pulse count. Figures 7-8, 7-9, and 7-10 are schematics of the rectifier and inverter cir-

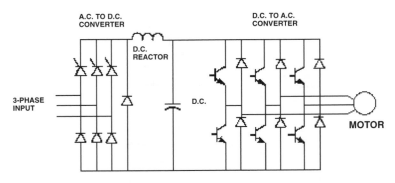

Figure 7-8. Schematic of the rectifier and inverter circuit for a voltage source inverter.

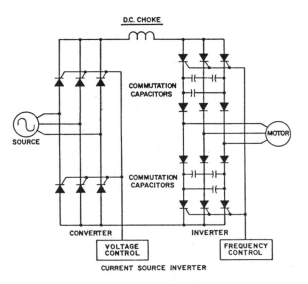

Figure 7-9. Schematic of the rectifier and inverter circuit for a current source inverter.

cuit for a basic six-pulse system. For a 12-pulse system, phase shift transformers are added together with an additional six thyristors. The phase shift transformers shift the output of the second set of thyristors by 30°. If more pulses are needed, additional transformers and thyristors can be added in groups of six.

While cost will probably decrease with increased usage, this is a factor to be evaluated before a decision to use the variable frequency is made. Because of the harmonics, the output torque contains some unsteady torsional components. These can be handled by evaluating the compressor train torsional response. This will be further covered in a later chapter.

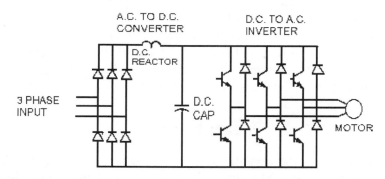

Figure 7-10. Schematic of the rectifier and inverter circuit for a pulse width-modulated inverter.

Application, relative to size, is not too much of a problem as the variable frequency drive has been used to 40,000 hp. High horsepower drives have been in service in Europe for a significant period of time [4]. In the U.S., the applications have been somewhat more modest in size but, as the popularity grows, the size will doubtless follow.

Probably the last hurdle is the cabinet size, which houses the electronics. It is large enough to be a factor in location. If located in an air-conditioned space, the air conditioning must be sized to accommodate the additional heat load, which is significant on the larger drives.

Variable-frequency drive technology is constantly improving in step with the advances in power electronic device technology and with the associated microprocessor controls. The following list of desirable features is offered:

• Minimum input harmonics
• Maximum input power-factor throughout the speed range
• Minimum output harmonics and torsional excitations
• Minimal tuning and setup required
• Minimum maintenance, with maximum reliability
• Permits the use of a standard or, at least, a more-standard motor
• Higher drive efficiency, lower cooling requirement
• Lower component count with a smaller footprint
• Lower cost

At present, PWM current-source drives are available in sizes ranging upward into the thousands of horsepower range, as are stepped-PWM voltage source drives. Both of these newer type drives offer minimal extra voltage stress and do not require the derating of standard motors. Because of the rapid growth of the power electronic technology, improved drives will continue to join the marketplace. This should continue until the above list is thoroughly satisfied.

Steam Turbines

One of the first questions the designer must answer concerns which type steam turbine should be used. The *back pressure turbine* is selected when process steam demands are greater than the steam required for process drivers such as large compressors. This type turbine is also selected when various steam levels are required by the process.

The *condensing turbine* is selected when steam demand for process drivers is greater than the low-pressure process steam requirements. It is also selected when no high pressure steam is available.

The *induction-type turbine* is selected when excess steam is available at an intermediate pressure.

The *extraction turbine* is selected when there is a demand for intermediate-pressure steam and, in particular, when there is a variation in the amount of steam required. The extraction turbine generally falls into two classes:

1. The controlled type, which is selected when there are large intermediate pressure process requirements with fluctuations in demand.
2. The uncontrolled type, in which there are small intermediate pressure process requirements with little change in demand.

The guidelines are quite general, but will at least act as introduction to the types of turbines available to the process designer. Sometimes the decision on which type turbine to select is not obvious. The back pressure turbine is most frequently selected. It has lower capital cost, simple construction, is the most suitable turbine for high speeds, and is generally more reliable.

The condensing turbine has several advantages and disadvantages over the back pressure turbine. The advantages are that it requires less change in the live steam for various turbine loads and is therefore easier to control. It also requires less steam because the enthalpy drop is larger. Finally, only one steam level is affected for a change in power requirements.

The disadvantages are that the condensing turbine has a high capital cost because it is larger than a back-pressure type. It develops high specific volumes of steam in the exhaust end, and there is the additional cost for a condenser and other auxiliary equipment. The condensing turbine has a lower overall reliability because the condenser, ejectors, extraction pumps, and other auxiliary equipment add to the complexity of the operation. Because high specific volumes are developed in the condensing turbine, longer exhaust blades must be used. In comparison, the back pressure turbine uses shorter blades with decreased risk of blade excitation.

The condensing steam turbine has a relatively low thermal efficiency because about two-thirds of the steam enthalpy is lost to cooling water in the condenser. Expensive boiler feedwater treatment is required to remove chlorides, salts, and silicates, which can be deposited on the blades causing premature failure. The blades are already under erosion conditions because of water drops present in the condensing steam. Even with these disadvantages, the condensing turbine is still selected, especially in a process that requires very large compressor drivers and relatively low amounts of process steam.

A rule of thumb: Select an extraction steam turbine when 15 to 20% of the driver power requirements can be supplied by the extracted steam.

Both extraction and induction turbines provide some significant advantages to process plant designers. Some of these are as follows:

1. Process steam can be supplied at two or more pressures without having to purchase boilers operating at different pressures or having to throttle steam, which is a waste of useful energy.
2. Process steam requirements can be controlled at a suitable pressure and volume required by the process and maintained at these conditions by extraction or induction turbines.
3. It is easier to make a plant steam balance using extraction or induction turbines.
4. Process steam requirements and driver steam requirements can be optimized.
5. Process plants usually operate under fluctuating flow and pressure conditions. Extraction and induction turbines provide the flexibility for these fluctuating operating conditions.

Extraction or induction turbines are not without their problems. Some of these are as follows:

1. Turbine blades can be excited by steam flowing through the intermediate nozzles, which can cause premature blade failure.

2. Excess steam extraction can cause starving in parts of the turbine, which will result in overheating.

3. To control the intermediate pressures, extra valves are required.

4. Extra nozzles require a longer turbine shaft, which increases the span between bearings which can produce rotor dynamics problems.

5. Extraction and induction turbines may be up to 5% less efficient than back-pressure turbines.

Steam Temperature

As steam temperatures increase, more expensive materials must be used to manufacture the turbine. Above 750°F, the price increase will be about 5% to about 850°F. Another 5% will be added for temperatures between 850°F and 900°F, and another 5% for temperatures above 900°F.

Steam turbine designers have selected as a standard for single-stage turbines a limit of 600 psia and 750°F maximum inlet steam conditions.

The upper limit in exhaust pressure for back pressure multistage turbines varies between 350 psia and 500 psia. For small, single-stage, back-pressure turbines, the standard is somewhere between atmospheric and 65 psia.

Speed

Turbine speeds are limited by the centrifugal stress that can be applied to the blades and blade roots. For a 3,000-hp turbine, a speed of about 14,000 rpm can be expected, but a speed of only 8,000 rpm can be expected from a 10,000-hp turbine. Higher horsepowers and speeds can be obtained in special turbines, and are frequently specified for large centrifugal compressors. These turbines need very careful checking for lateral critical speeds, torsional critical speeds, radial bearing stabilities, thrust bearings, balance, and permissible flange loadings.

A Campbell diagram is frequently used to determine the effect of multiple excitation frequencies in high-speed steam turbines. Figure 7-11 shows a Campbell diagram for a condensing steam turbine. If this particular turbine operates at a speed of 8,750 rpm, the turbine blades would not be excited. But, if the turbine speed is reduced to 7,500 rpm, the turbine blades would be excited at four times running speed. If the turbine were operated at 10,000 rpm, a three-times running speed excitation would be encountered. What this means is that any vibration in the

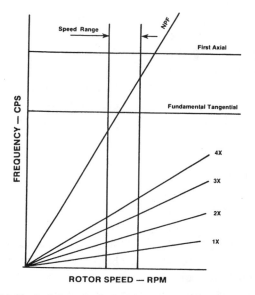

Figure 7-11. Typical Campbell diagram. (*Courtesy of Elliott Company*)

twelfth-stage turbine blades would be reinforced every third or fourth vibration of the blade. In other words, severe blade vibration could be expected at 7,500 rpm and 10,000 rpm, in this particular turbine.

Sometimes, because of process requirements, it is impossible to avoid some excitation frequencies. If the Campbell diagram shows this will occur, then the blade in question must be carefully designed to keep stresses low. When properly addressed in design, operation can take place in an area of excitation. The major variables affecting turbine selection may be listed as follows:

1. Horsepower and speed of the driven machine.
2. Steam pressure and temperature available or to be decided.
3. Steam needed for process, so that a back-pressure turbine can be considered.
4. Steam cost, and the value of turbine efficiency. Should it be single-stage or multistage? Should it be single-valve or multivalve? Is the steam an inexpensive process by-product, or is the entire cost of generating the steam chargeable to the driver?
5. Should extraction for feedwater heating be considered?
6. Should a condensing turbine with extraction for process be considered?

7. Control system, speed control, pressure control, process control. What speed or pressure variation can be tolerated and how fast must the system respond?
8. Safety features such as overspeed trip, low-oil trip, remote-solenoid trip, vibration monitor, or other special monitoring of temperature, temperature changes, and casing and rotor expansion.

The price range from the minimum single-stage turbine to the most efficient multistage is quite wide.

Operation Principles

The steam turbine is basically a steam expander, expanding the steam over a given pressure range along an isentropic path. Space does not permit a complete development of all the aspects of the turbine. However, a few fundamentals will be reviewed to help give an understanding of how turbines fall into different efficiency ranges.

Refer to Figure 7-12 for a diagram showing the various turbine stage losses. Efficiency, ηs, is defined as the actual enthalpy change divided by the isentropic enthalpy change or

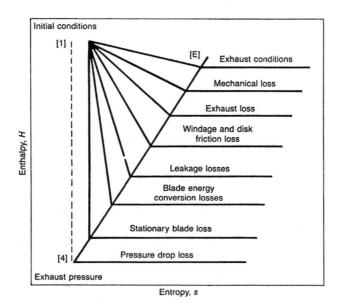

Figure 7-12. Various turbine stage losses. (*Courtesy of Elliott Company*)

$$\eta_s = \frac{H_1 - H_E}{H_1 - H_4} \tag{7.10}$$

where

H = enthalpy (Figure 7-12)

As was discussed in the centrifugal compressor chapter, there is a geometric relationship that can be used to determine performance.

$$C = \sqrt{\Delta h_a \times 2g} \tag{7.11}$$

where

C = steam velocity

Δh_a = isentropic enthalpy drop

$$u = \pi d N \tag{7.12}$$

where

u = blade velocity
d = pitch diameter
N = rotational speed

Stage efficiency is calculated by Equation 7.13 (see Figure 7-13)

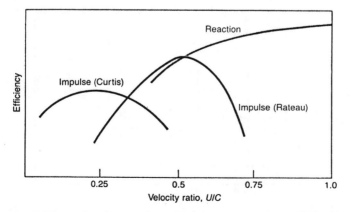

Figure 7-13. Stage efficiency vs. velocity ratio. (*Courtesy of Elliott Company*)

$$\eta = u/C \qquad\qquad\qquad\qquad (7. 13)$$

Without stretching the imagination too much, a high enthalpy drop, Δh, across a stage will raise the steam velocity, C, lowering the velocity ratio, and drop the efficiency. Of course, the wheel speed, u, could be increased but this is also limited by stress considerations. Two conclusions can be drawn from the relationships. One is that single-stage turbines tend not to be efficient unless steam conditions are quite correct, and the other is that, if efficiency is to be maintained, a multistage is required.

A few other general comments can be made. A partial-admission turbine is not as efficient as a full-admission turbine, as seen in Figure 7-14. In actual practice, full admission is not widely used in the small turbines. In many cases, the steam flow is small relative to the size of the turbine and only a partial arc of admission is required to get the area to support the needed steam velocity, as was just shown with the equations. Whenever there is a choice, another consideration must be weighed. The steam has to cross the horizontal joint, which may cause design and operational problems. If the casing is cored for internal passages, the cost is high compared to the alternative of taking the steam out of the casing above the joint and piping it in below the joint. This may not seem all that bad. However, due to thermal expansion problems, the crossover pipe has a tendency to crack during operation. Presumably, the larger turbines use the 360° admission more for the necessity of physical space for steam admission than for the efficiency.

For wide ranges of operation, multiple valves help keep the efficiencies high (see Figure 7-15). As power output is increased, the lower curve flow

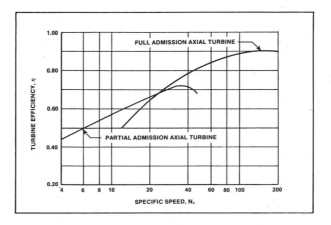

Figure 7-14. Efficiencies of full-admission and partial-admission turbines.

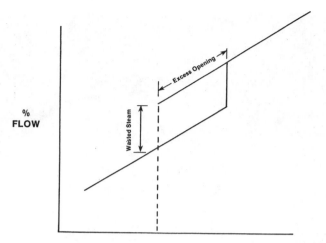

Figure 7-15. In a multivalve turbine, power is increased by increasing the admission area to provide more flow. This results in a step effect as illustrated.

limit is encountered. To increase flow, more area must be provided, represented by the solid vertical line to the upper curve. More power is now available because more steam can be admitted. This stair-step approach may be done several times, until the full arc of admission (360°) is achieved. This action takes place in a multivalve turbine (see Figures 7-16 and 7-17). If a wide range of operation is anticipated, this type of turbine should be considered and its higher cost evaluated. With a multivalve turbine, when load drops, the upper curve dictates the flow until the solid vertical line, at which the lift automatically closes a valve dropping the turbine to the lower characteristic, and all is well. However, if cost was such that a single-valve turbine was chosen, for a nominal additional cost, a hand valve can be furnished, which can be manually opened and closed at the load point presented by the solid vertical line. When the operator becomes complacent and leaves the valve open, as indicated by the excess opening, the area indicated as wasted steam is in effect. Therefore, though the addition of hand valves is an economical method of widening the efficient operating range, it does present a problem.

The same curve may be used to represent a turbine specified for a greater power than needed, which would be the upper curve, when all that was needed was the lower curve. Again, the same effect, except with no recourse other than poor operation.

Figure 7-16. Multivalve steam turbine. (*Courtesy of Demag Delaval Turbomachinery Corp.*)

Steam Turbine Rating

As with the motor driver, the steam turbine must be matched to the compressor. Also, a turbine rating of 110% of the maximum power required by the compressor should be specified. This should be at the compressor's normal speed point. The turbine speed should include a maximum continuous speed 105% of the normal compressor speed. API Standards 611 and 612 cover general purpose and special purpose steam turbines [7, 8].

A few notes on the application of steam turbines, which are frequently overlooked, should be mentioned. One which is misunderstood more than any other is the exhaust pressure, particularly on a condensing turbine. If the condenser cooling medium fluctuates, say summer to winter, careful consideration must be given to the normal value of the condenser pressure. For most efficient operation, the value realized most of the time should be specified. If a lower value can be achieved with unusually cold coolant, then this value should also be given. The latter should be stressed as not being a long-term operating point, but given primarily to

Figure 7-17. Multivalve steam turbine with the upper half removed. (*Courtesy of G.E. Industrial & Power Systems*)

use in evaluating the stress on the latter stages blading. A change from 4 in Hg to 2 in Hg doubles the specific volume, which doubles the exhaust volume raising velocities considerably. If the turbine vendor understands the situation, steps can be taken at the time of design to keep a few cold days from causing the operator to have to take corrective action to prevent blade damage.

Conversely, letting the vendor design his exhaust end to the lower value will cause him to grossly oversize, at extra cost and ultimately at reduced efficiency during the normal period.

Concerning condensing turbines, be sure to obtain an adequately sized vacuum breaker for the condenser. There are no words to describe the agony of watching a turbine or the driven compressor tear itself up, because the train was tripped and it continued to rotate with condenser vacuum that couldn't be broken.

Gas Engines

When considering the possibility of using internal combustion drivers, evaluate process requirements and costs. If a low-cost, gaseous fuel is available, gas engines and gas turbines may surpass other drivers in economical installation and operation. In the initial process design stage, a method of establishing the cost of purchase, installation, and operation for drivers is needed.

After the type of driven equipment is selected, the horsepower requirements must be estimated. Unless the horsepower requirements happen to fit one of the gas turbine sizes available, a gas engine will be needed.

The large, heavy-duty, integral engine-compressor has long been the workhorse of the gas compression industry. The in-line or V-shaped power cylinder, horizontal compressor cylinder configuration is familiar.

Other variations in compression using the reciprocating engine driver are the reciprocating compressor frame and the centrifugal compressor.

The separate engine-driven reciprocating compressor frame arrangement will have approximately the same first cost as the integral type but slightly higher installation costs. The principal advantage of the separate engine-driven frame is that more compressor cylinders may be used by adding throws to the frame. Integral types are limited in this respect, and many throws are necessary in high-pressure or multiple-service applications.

While not common, engine-driven centrifugal compressors are used. This combination is desirable for low-ratio applications and in fuel-cost situations where the high engine efficiency is attractive. The difference in rotating speeds (engine 300–600 rpm, compressor 3,000–5,000 rpm +) requires the use of a speed increaser.

The primary advantage of the gas engine is the fuel rate (efficiency) (see Figure 7-18). Again, this is a plot of an average of the manufacturers' published data. It can be seen that rates vary from 6,500–7,000 Btu/ bhp-hour for the large units to 8,500–9,000 Btu/bhp-hour for the smaller ones. On this basis, large gas engines have thermal efficiencies of 35–39%.

Gas Turbines

Most gas turbine manufacturers manufacture a mechanical drive turbine. This unit generally includes the auxiliary and control equipment

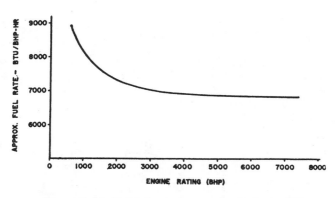

Figure 7-18. Approximate gas engine fuel rates [7].

necessary to start, to sequence to idle speed, ramp to full speed, monitor basic functions, and automatically shut down the unit in event of serious trouble. Start-up control may be performed manually by an operator, with acceleration to idle controlled automatically. More often, gas turbines are started in fully automatic mode. All speed control setting for process control is done by the user.

Gas Turbine Types

Gas turbines are classified as single-shaft or two-shaft and simple or regenerative cycle. A two-shaft configuration is distinguished by the lack of a mechanical connection between the gas generator (power-producing) system and the power turbine (power-utilizing) system. This arrangement allows greater speed-horsepower flexibility than the single-shaft in which all compressors and turbines are mechanically connected. Also, the starting requirements and controls are simplified, because the load compressor does not require acceleration during the start cycle.

A variation of the two-shaft gas turbine is the aircraft derivative machine. The gas generator is an aircraft jet engine or an adaptation thereof. The power turbine may be of a custom design. Alternately, aircraft based designs may be used. They are covered in API Standard 616 [12].

Simple and regenerative cycle turbines differ in configuration, efficiency, and equipment cost (see Figures 7-19, 7-20, and 7-21). The regenerative turbine costs more because of the heat exchanger and extra ducting required to preheat the combustion air using the turbine exhaust gas. An increase in efficiency of about 6–7% results. Simple-cycle turbines normally have thermal efficiencies of 20–26% and regenerative

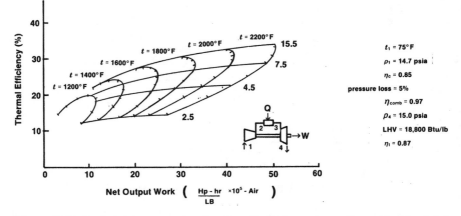

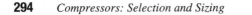

Figure 7-19. Performance map showing the effect of pressure ratio and turbine inlet temperature on a simple cycle [13].

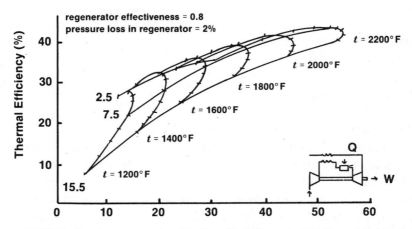

Figure 7-20. Performance map showing the effect of pressure ratio and turbine inlet temperature on a regenerative cycle [13].

types 28–32%. However, because most process applications of gas turbines use the exhaust gas heat in steam-producing boilers or other process uses, simple-cycle units are more common. With exhaust heat recovery, simple-cycle gas turbine efficiency meets or exceeds regenerative cycle efficiency.

Another reason for the increased use of gas turbines as prime movers in the process industry is the high thermodynamic cycle efficiencies and subsequent low operating cost.

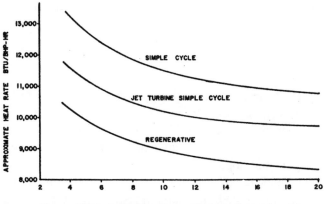

Figure 7-21. Approximate gas turbine fuel rates [7].

Simple-cycle efficiency does not usually mean as much to process users as total-cycle efficiency, because the gas turbine is not usually economic in process applications without some type of heat recovery. Total-cycle efficiency is most important in any economic evaluation. In a cycle with heat recovery, the only major loss that is charged to the cycle is the heat exhausting from the boiler stack. With the good comes the bad. Gas turbine maintenance is generally somewhat higher in cost and should be included in the total evaluation.

Gas Turbine Economics

The costs associated with the use of gas turbines, or any item of capital equipment, are broken down into three categories:

1. Purchase price of equipment: turbine, compressor, auxiliary equipment, and heat recovery equipment, if desired.
2. Installation of equipment: hauling, foundations, assembly, debugging, and buildings, if required.
3. Operating costs: fuel, lubricating oil and cooling water, and normal, preventive and emergency maintenance.

Of the gas turbine-compressor combinations previously mentioned, gas turbines are by far the most frequently used as drivers for centrifugal compressors. Next in use would be axial compressors and, on occasion, the gas turbine is used as a driver for the larger screw compressors. There are also a few installations where the gas turbine is used with the reciprocating compressor.

Sizing and Application

The matching of the gas turbine to the driven compressor in both speed and horsepower is somewhat of a guessing game. While the use of helper steam turbines permits the use of 110% of compressor maximum power criterion, the speed is not as easy to match. This makes a gear necessary in most of the single-shaft applications. The gear option may also be required for some of the "land-based" design two-shaft gas turbines. For the aircraft derivative gas turbine, speed may sometimes match the centrifugal speed, only because the output turbine may be a more custom design. A single-shaft turbine should have approximately a 15% speed turndown, which is normally used as a minimum of 90% and maximum of 105%.

Gas turbine frame sizes are limited, and custom sizing is not done as it is with steam turbines; therefore, the increment matching can only be done with a helper turbine. There is some benefit besides the power matching when used with a single-shaft turbine. The steam turbine can be used as the starter. For two-shaft turbines, a dedicated starter is needed on the gasifier turbine compressor package.

Expansion Turbines

An *expansion turbine* (also called *turboexpander*) converts gas or vapor energy into mechanical work as the gas or vapor expands through the turbine. The internal energy of the gas decreases as work is done. The exit temperature of the gas may be very low. Therefore, the expander has the ability to act as a refrigerator in the separation and liquefaction of gases.

Types

There are two main types of expansion turbines: axial flow and radial flow. *Axial flow expansion turbines* are like conventional steam turbines. They may be single-stage or multistage with impulse or reaction blading, or some combination of the two. Turbines of this type are used as power recovery turbines. They are used where flow rates, inlet temperatures, or total energy drops are very high.

Radial-flow expansion turbines are normally single-stage, with combination impulse reaction blades and a rotor resembling a centrifugal semi-open compressor impeller. Radial-flow expansion turbines are used primarily for low temperature service, but they may also be used to recover power.

Operation Limits

Common operating limits for turboexpanders are an enthalpy drop of 40–50 Btu/lb/stage of expansion and a rotor tip speed of 1,000 ft/s or higher. Turboexpanders are available with inlet pressures up to 2,500 psig and inlet temperatures over 1,000°F. Allowable liquid production can be as high as 20% of the discharge weight.

Power Recovery

When a large flow of gas is reduced from a high pressure to some lower pressure, or a high-temperature process stream (waste heat) is available at moderate pressures, turboexpanders should be considered to recover the energy (see Figure 7-22). The turbine can drive a compressor

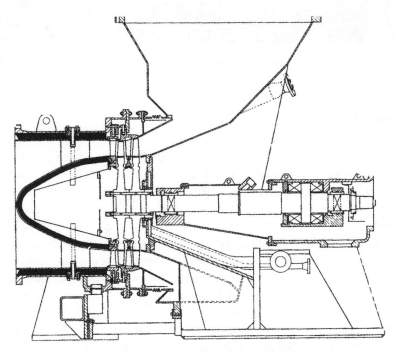

Figure 7-22. FCC power recovery turbine. (*Courtesy of Elliott Company*)

recovering otherwise wasted energy (see Figure 7-23). There will be a temperature drop in the expander so the inlet gas may require heating or drying to avoid low exhaust temperatures or condensation.

Figure 7-23. Power recovery turbine and compressor on the test stand. (*Courtesy of Elliott Company*)

Refrigeration

Turboexpanders may be used in closed cycles with a pure gas such as nitrogen, which is alternately compressed and expanded to provide the required refrigeration through a heat exchanger. Open cycles are also used in which the process stream to be cooled passes through the expander. Liquefaction of natural gas is an example. Open cycles eliminate the need for a low-temperature heat exchanger. Open cycles can also be used to recover condensable products from process streams. Water vapor and CO_2 must be reduced to very low levels in open cycle systems to prevent solidification and fouling of equipment. A compressor expander is shown in Figure 7-24.

The cryogenic expanders are ideal candidates for magnetic bearings (covered in Chapter 5). In process service, elaborate seals are required to

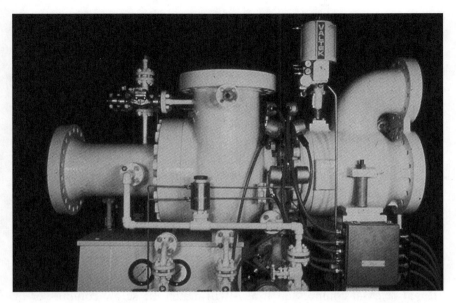

Figure 7-24. Cryogenic expander-compressor. (*Courtesy of Mafi-Trench Corp.*)

keep lubricating oil from entering the gas stream. Also the hydrodynamic bearing design is critical because of the high speeds normally required. By use of magnetic bearings, not only can the high speeds be achieved without regard to bearing speed limitations, but the seals can be reduced to the use of simple labyrinth seals in most applications.

Condensation

Radial-inflow turboexpanders can be designed to handle relatively large amounts of condensation with very little loss of efficiency. Axial turboexpanders can also tolerate some condensation, but usually at a loss of efficiency.

Several arbitrary rules of thumb have been suggested to account for the reduction in efficiency when moisture particles are present. One is to multiply the efficiency by the mean vapor content, by weight. For example, if expansion is started in the superheated region and ends with 6% by weight of liquid, the mean vapor content is 97%. If the design efficiency is 78%, the adjusted value of efficiency is 0.78 (0.97) = 0.757 or 75.7%. Some designers prefer to assume a loss equal to twice the value used in the example.

Expander Applications

One application for hot gas expanders is for use in refineries to recover the energy in the tail gas of a catalytic cracking unit. In the recent past, other uses included certain process applications that used air for an oxygen source. If the residual pressure after the process reaction was sufficiently high, an expander was used to recover the remaining energy. A variation might have included auxiliary firing if the residual oxygen content was high enough to support the combustion. Unfortunately, the economics were diminished because of environmental concerns with stack emissions and the advent of direct oxygen use. A potential for the hot gas expander is in binary bottoming cycles, to convert low level heat to usable shaft work.

While many of the applications use the expander to drive a generator, a compressor is a good alternative candidate for the load. Expanders are generally custom-sized and can, therefore, be readily matched to the centrifugal or axial compressor. It also will match the screw compressor of the dry type, at least in the larger frames. Basic sizing to the compressor should follow the same guidelines as for steam turbines.

The expander is somewhat difficult to govern, and is sometimes used with a helper steam turbine for use in starting and speed control.

A variation of the application of expanders to a compressor train is to include an induction motor-generator. This arrangement does not provide precise speed control, but works well in other aspects. The motor-generator acts as a starter to bring the compressor train to speed and permits the process to start. As the expander begins to recover energy, it first takes load from the compressor and when excess torque becomes available, the induction machine acts as a generator. Because an induction generator is a varless machine, it must be tied to the grid to produce power. This system is inherently speed regulated: the faster the generator turns the more load it takes from the expander. Regulation is within a few percent, generally in a range acceptable to a centrifugal or axial compressor.

References

1. *National Electrical Code 1984,* National Fire Protection Association, Quincy, MA, 1983.
2. API RP 500, *Recommended Practice for Classification of Locations for Electrical Installations in Petroleum Facilities,* First Edition, Washington, D.C.: American Petroleum Institute, 1991.

3. API RP 540, *Electrical Installations in Petroleum Processing Plants,* Third Edition, Washington, DC: American Petroleum Institute, 1991.

4. Jeumont-Schneider, R. Champrade, *High Power Adjustable Speed Drive with Synchronous Motor,* Power Conversion International, Sept./Oct. 1979, pp. 83–89.

5. Scholey, D., "How Adjustable Frequency Drivers Affect Induction Motor Operation," *Plant Engineering,* July 26, 1984, pp. 44–47.

6. Lazor, D. A. and Bryson, E. J., "Efficiency Improvements for Steam Turbine Drivers," *Hydrocarbon Processing,* January 1984, pp. 61–63.

7. Evans, Frank L., Jr., *Equipment Design Handbook for Refineries and Chemical Plants,* Vol. 1, Houston, TX: Gulf Publishing Company, 1979.

8. Scholey, D., "Induction Motors for Variable Frequency Power Supplies," *IEEE Transactions,* July 1982, pp. 368–372.

9. LeMone, C. P., *Large Adjustable Speed Drives and Their Application to a High Speed Centrifugal Compressor,* Proceedings of the 10th Annual Turbomachinery Symposium, Texas A&M University, College Station, TX, 1981, pp. 81–85.

10. API Standard 611, *General-Purpose Steam Turbines for Refinery Services,* Third Edition, Washington, DC: American Petroleum Institute, 1988, Reaffirmed 1991.

11. API Standard 612, *Special-Purpose Steam Turbines for Petroleum, Chemical, and Gas Industry Services,* Fourth Edition, Washington, DC: American Petroleum Institute, 1995.

12. API Standard 616, *Combustion Gas Turbines for Refinery Service,* Third Edition, Washington, DC: American Petroleum Institute, 1992.

13. Boyce, Meherwan P., *Gas Turbine Engineering Handbook,* Houston, TX: Gulf Publishing Company, 1982.

14. Siemens, *High-Voltage Three-Phase Motors,* Catalogue M-2 Supplement, 1975, Siemens West Germany.

8

Accessories

Introduction

As with many consumer goods, compressors cannot be purchased without accessories. Some of the accessories are essential to the basic operation of the compressor, such as lubrication systems and couplings. Other accessories, such as the anti-surge control equipment, are optional in order to enhance operation but are not essential. The accessories may be purchased from the equipment vendor or they may be purchased from another party—a decision open to the purchaser. Generally, the accessories essential to the operation are purchased from the equipment vendor as part of the original purchase. The lubrication system and couplings are classic examples of this option. Intercoolers, while essential, are not normally purchased from the equipment vendor if the application is a process compressor.

This chapter will cover some of the more common accessory items for compressors such as the lubrication system, gears, coupling, instrumentation, vibration monitoring, and process control. The subject is broad and far-reaching. It is hoped that, for the first-time user, this discussion will be a good introduction and, for the veteran, it may offer another perspective on the subject.

Lubrication Systems

Lubrication is a fundamental requirement for all compressors with the exception of those equipped with an alternative form of bearing such as the magnetic bearing. If it is a tiny unit, the lubricant may be sealed into the rolling element bearings by the bearing manufacturer. In process service, lubrication of bearings takes on a more elaborate form. Some of the smaller units will probably use an attached oiler or an oil mist system. Because this affects only the smaller units, this section will deal primarily with those compressors using force-feed lubrication.

The ring-oiled bearing might be considered the most fundamental and basic of the lube systems (see Figure 8-1). The ring rides on top of the shaft and is dragged at part-shaft speed by friction. The lower portion of the ring resides in a reservoir of oil. In its most primitive state, the reservoir is the lubricant source and heat sink. The rotating ring moves oil from the reservoir to the upper portion of the bearing. Here the ring and shaft interface causing some of the oil to be removed. The oil enters through grooves cut into the surface of the bearing, where it is carried to the minimum clearance area by the journal pumping action. The next level of sophistication is to add circulation and cooling to the reservoir. An alternative to the separate

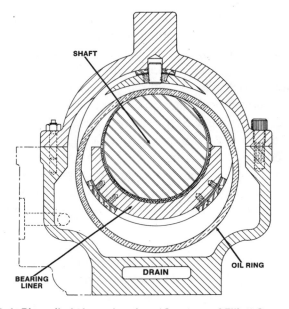

Figure 8-1. Ring-oiled sleeve bearing. (*Courtesy of Elliott Company*)

circulation system is to connect the reservoir to a pressure-fed external lube system in use with the balance of the compressor train bearings.

Chapter 3, which discussed various reciprocating compressors, stated that many reciprocators use a pressure-fed lubrication system for the frame bearings. This system is built into the crankcase in many applications. The basics of these systems follow the fundamental criteria which will be discussed with the fully separate system. The larger reciprocating compressors may use a separate frame lubrication system.

For the rotary, centrifugal, and axial compressors, a separate lubrication system is used and in some cases, seal oil and control oil are also supplied from this system. Chapter 4 mentioned that the oil used for flooding is taken from the lubrication system. Because the secondary duties of cooling the process stream and timing the rotors overshadow the primary job of lubing the bearings, the lube oil system may be misnamed for this compressor. As a model to guide the discussion, the basic system as covered by API Standard 614, "Lubrication, Shaft-Sealing, and Control Oil Systems for Special-Purpose Applications," [1] will be used. In the opinion of some vendors, this system is an overkill, but it can easily be tailored to fit any system by scaling down as required. The standard then can be fully or partially invoked or in some smaller system, the standard can be used as an outline and guide.

A basic pressurized lube system consists of a reservoir, pump, cooler, filter, control valves, relief valves, pressure and temperature switches, gauges, and piping. The oil is pumped from the reservoir, cooled and filtered, pressure controlled, and directed to the bearings by way of a supply header. A drain header collects the oil exiting from the bearings, and gravity flows it back into the reservoir (see Figure 8-2). If control oil is required for a power positioner on a steam turbine governor valve, additional control valves are used to establish the two levels of pressure needed because the control oil is normally at a significantly higher pressure than that needed by the bearings (see Figure 8-3). Note that an accumulator has been added to improve transient response for the turbine governor. Bearings normally operate in a 15 to 18 psig range with some variation from vendor to vendor. Control oil is generally in the 100 to 150 psig range.

If oil film or mechanical contact seals are used, another pressure level must be established. This pressure level is difficult to generalize as the seal pressure is a differential above the process gas pressure. For a mechanical contact seal, it is in the range of 35 to 50 psid to the gas and must follow the gas pressure from startup to shutdown. This generates an

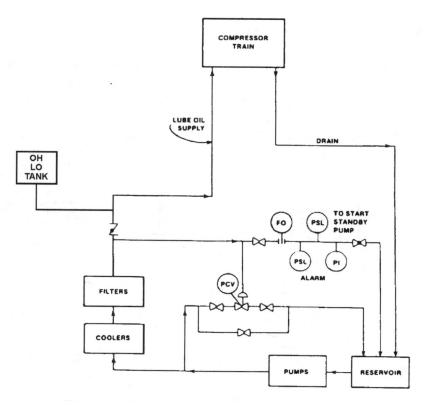

Figure 8-2. Block diagram of a pressurized lube oil system.

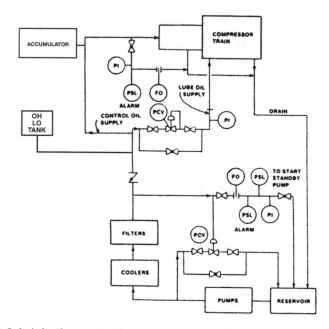

Figure 8-3. Lubrication system for a compressor that requires two levels of pressure.

additional design consideration, which will be discussed in more detail later. For the oil film and pumping bushing seals, the pressure is only a few psi above the gas pressure; however, an elevated tank is required. This tank forms the basis for the manometric differential pressure control for the seals as well as a backup supply reservoir to the seals in the case of seal oil supply failure. Figure 8-4 is a block diagram of a lube oil sys-

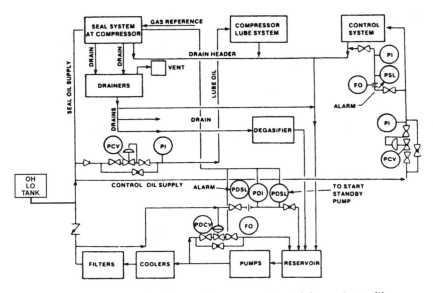

Figure 8-4. Lube oil block diagram used for a compressor lube system with a control system and a seal oil system.

tem similar to the one shown in Figure 8-3 with the seal system added. Figure 8-5 is a schematic drawing of a combined lube and seal system as would be furnished for a compressor with mechanical contact seals.

In the following paragraphs, various available options will be discussed. It is hoped that by using the options best suited for a given application even an inexperienced user might be able to specify a lube system for a compressor.

Reservoir

The reservoir is the lube oil storage tank. In some of the packaged compressors, it is built into the package base, and, in some standardized compressors, it is built into the compressor frame. In a reciprocating

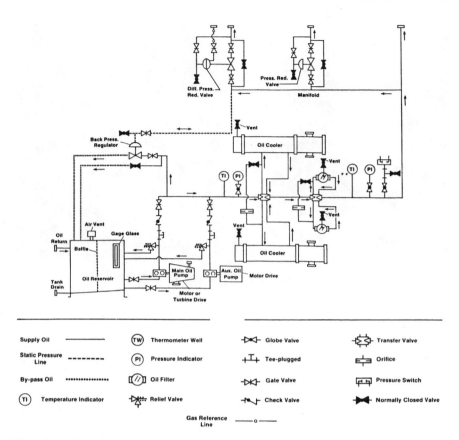

Supply Oil	————	(TW)	Thermometer Well	▷◁	Globe Valve	⟨⟩	Transfer Valve
Static Pressure Line	‐ ‐ ‐ ‐ ‐ ‐	(PI)	Pressure Indicator	⊥	Tee-plugged	⬡	Orifice
By-pass Oil	••••••••••••••	⟨//⟩	Oil Filter	▷◁	Gate Valve	⬡	Pressure Switch
(TI)	Temperature Indicator	▷◁	Relief Valve	◁	Check Valve	◀▶	Normally Closed Valve

Gas Reference Line ——o——

Figure 8-5. Schematic of a combined lube and seal system for a compressor with mechanical contact seals. (*Modified courtesy of Elliott Company*)

compressor, it is in the crankcase. When there is a choice, which there is for the larger compressors, it is recommended the reservoir be separate from the base (see Figure 8-6). As various reservoir requirements are covered, the reasoning for this will become clear.

The reservoir should be designed to prevent the entrance of dirt and water. This means sealing the top and raising top side openings above the surface. API 614 uses a one-inch dimension. The bottom should slope so the reservoir may be completely drained. The pump suctions should be located at the high side of the sloped bottom. Some form of openings should be provided to permit internal inspection and facilitate cleaning. The gravity return lines should enter the reservoir above the maximum oil level on the side away from the pump suction. The pressurized lines

Figure 8-6. Lube oil console. (*Courtesy of Elliott Company*)

should be individually returned, including the relief valves, and piped into stilling tubes that discharge below the suction loss oil level. An automatically closing fill opening should be installed in the top of the reservoir and should include a strainer to prevent entrance of foreign material with the oil during fill operations. A breather-filter cap should be used to cover the fill opening. A flanged opening should be placed on the top and blind flanged for an optional, user-furnished vent stack. Some form of level-indicating device should be provided, mounted on the side of the reservoir. A top-mounted dipstick is also required, with the dipstick marked in liquid units of the type in use at the plant location.

The reservoir should be sized for five minutes of normal flow, with a retention time of eight minutes. The retention time should be calculated using normal flow and total volume below the minimum operating level. Provision must be made for the oil rundown from the field located piping. It should be checked on all systems, but particularly on the larger sizes. It is quite embarrassing to take a new compressor through commissioning, have a shutdown and overflow the reservoir on rundown, especially if all the company executives are there to witness the event. Additional features for the reservoirs and the defined operating levels are shown in Figure 8-7.

Heaters should be considered for the reservoir. While they are normally thought of as cold-weather features, they aid in keeping the oil dry if the compressor is shutdown long enough for the oil to cool. The heater

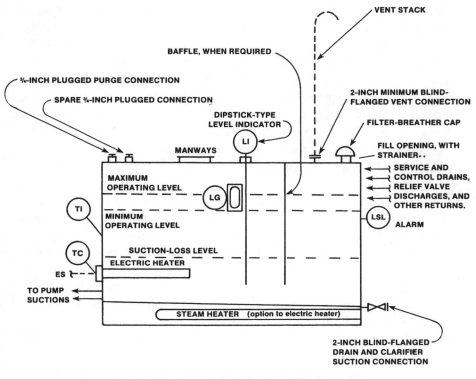

Figure 8-7. Reservoir features. (*Modified from [1]*)

can be either steam or electric. A 15-watts-per-square-inch maximum limit on the watt density should be used in sizing the electric heater.

The internals of the reservoir should be coated unless the reservoir is constructed of 300 series stainless steel. API 614 mandates the use of 304L, 321, or 347 stainless steel processed to ASTM A 240. For the critical equipment units, the stainless steel reservoir is a good idea. A decision must be made by the user relative to the general purpose units. A good coating can keep the internals clean and free of corrosion if applied properly. When the idea of stainless steel reservoirs was introduced, it met with immediate resistance, but as the alternatives were considered, it began to gain acceptance.

Pumps and Drivers

In many ways the pump is the heart of the lubrication system as it is the only active element. It must furnish sufficient capacity at a high enough pressure to satisfy the entire train connected to its lube system. The pump driver must be sized to be able to start with cold oil, generally 50°F or colder. It must have enough power to operate under all conditions, including the highest seal pressures expected, which, on a refrigeration compressor, is the shutdown stagnation condition. This pressure is equal to the saturation pressure of the refrigerant at ambient temperatures and can be much higher than normal operating pressure. Also the pump should be checked for minimum viscosity operation, particularly at the higher pressures. Pump failures have been traced to violation of the minimum allowable viscosity limit on higher pressure operation resulting in rotor contact.

The pumps can either be rotary positive displacement or centrifugal. The rotary positive displacement pump of the helical-lobe type is recommended. Centrifugal pumps should only be used for large systems, over 500 gallons per minute, if a suitable rotary screw pump is not available. The centrifugal should have as steep a pressure flow characteristic curve as can be found. The normal low head rise centrifugal pump will lose capacity as the filters get dirty and cause the system to starve for oil and go into an unscheduled shutdown. Oversizing the centrifugal and running it normally in the overload region is not an adequate solution to the problem.

On some of the smaller, standardized compressors, the main lube oil pump may be shaft driven. Others use rotary gear pumps. These have been used for quite some time, and the desire for change or modification usually falls on deaf ears. A good quality standby pump should be connected in parallel to the rotary gear pump in case the first one fails. As a general rule, on even the smaller systems, there should be a main and a full-sized standby pump. On the larger systems, as the equipment is put into unspared service, the two pumps are a must. In some plants a third, smaller pump is used, which is referred to as a coastdown or emergency pump. This pump is used as a contingency in the event of a full power failure when both the main and standby pumps fail due to loss of power or for any other reason. Of course, the main compressor should go into the shutdown mode using the emergency pump to supply oil to the bearings during the coastdown cycle. The whole premise for this course of action is that a source of energy, not affected by a power outage, is available. A better alternative using an overhead tank for compressor coastdown will be discussed later.

The pumps may be driven by any of several drivers. For many years the favorite arrangement has been a steam turbine for the main pump and an electric motor for the standby. It is a nice combination in that if the steam supply is disrupted by power failure, there is some residual pressure as the system decays, in many cases, allowing the pump to deliver some capacity. This may be the only argument for the shaft-driven pump which delivers oil while the compressor shaft is turning. Unfortunately, even though the pump will turn, it doesn't necessarily deliver oil because of long suction lines, air leaks, or the capacity is just not available at the lower speeds. In some cases there is no steam available and some smaller compressors do not have the steam turbine option, primarily because of cost. In these cases, the two pumps should be motor driven. If possible, the motors should receive power feed from two independent power sources. If an emergency or coastdown pump option is selected, it should be run from a totally independent power system. Some large plants use an uninterruptible power supply (UPS) to operate devices during a power outage. Sufficient additional capacity might be added to the batteries to run a small motor. The motor can be direct current and will not tax the inverter on the UPS. If a separate battery system is installed only for the coastdown pump, the odds are it will not be operational after a period of time due to lack of battery maintenance. That is why it should be combined with other plant functions that require the backup system to be operational. Other power sources, such as air motors, have been used but without great success.

Pump casings should be made of steel if possible, which, on the smaller compressor systems, will not be practical. Most petrochemical plants live in the fear of fire, and the use of cast iron casings in a hydrocarbon plant is not a good idea.

The pumps should be piped with a flooded suction to avoid having priming problems. This is difficult to do on a shaft-driven pump. This procedure precludes mounting the pumping equipment on the top of the reservoir. If top-mounted equipment is desirable to keep the system compact, then vertical pumps, operating below the oil level in the reservoir, should be considered. When this arrangement is used, the need for steel in the pump casing is eliminated. The oil reservoir must be strengthened to carry the extra weight without excessive deflection. While this is one way to maintain a compact system, it is only practical on the smaller systems where component maintenance is not as difficult.

A strainer should be used in the pump suction line temporarily for a centrifugal pump and permanently for the rotary positive displacement pump. For the permanent installations, a Y-type strainer with an austenitic stainless strainer basket should be used. The cross-sectional area of any strainer should be at least 150% of the normal flow area. In all strainer installations, a compound pressure gauge should be installed between the strainer and pump suction.

Booster Pumps

In some applications (usually high pressure compressors using oil film seals) alternative pump schemes should be considered. It may be that the desired seal pressure is not achievable by one set of pumps or the quantity required by the seal is small relative to the main pump capacity. There are times when booster pumps are needed; however, if the reason is energy, it would be worth reviewing the economics very carefully, because reliability tends to suffer with the booster. The booster pumps are paired into a main and standby and are configured to take suction from the lower pressure system. Sufficient interlocks have to be supplied to the drivers so that if the main pumps shutdown, the boosters come down. Other problems may arise when the controls are set or trimmed because the system is usually quite sensitive to the high system gain caused by the high pressure at the valves.

Pump Sizing

Pumps should be sized for 1.2 times the system's normal flow requirement, with a minimum of 10 gpm above the normal flow. If the booster arrangement is used, an additional capacity is needed equal to the flow of both booster pumps running simultaneously. A centrifugal pump should be within the range of 50 to 110% of the best efficiency point when running at normal capacity. The pump, as was previously mentioned, should have a steep curve. API 614 mandates a minimum of 5% pressure rise to shutoff. This minimum seems to be on the low side. A 15% rise to shutoff would come closer to maintaining minimum oil flow with dirty filters. See Chapter 7 for various driver's sizing guidelines.

Pump Couplings

For pumps above 25 hp, flexible disk, spacer-type couplings should be used. The flexible elements must be selected for compatibility with the plant atmosphere. For the smaller systems, a non-spacer coupling may be adequate, but the coupling should be of good quality. This is not the place to save money. Coupling guards should be furnished as a part of the lube system.

Relief Valves

Whenever possible, external relief valves should be furnished. On the larger systems, this is mandatory. Internal relief valves will lead to premature pump failure if allowed to operate open for more than a few minutes, since the hot oil is returned directly to the suction. An upset period, the time when a pump may open the relief, is probably not the time when an operator would detect the open valve and shut the pump down to reset the valve.

External relief valves should be of the full-flow, non-chattering (modulating) style. In oil systems with relatively low pressure levels, the hydraulic-type relief valve becomes attractive because the plug lift is smooth, and instability during lifting is eliminated. The valves should be located as close to the pump as possible to provide fast reaction time.

Pressure Control Valves

For positive displacement pumps, a bypass-type control valve should be furnished to set the primary lube system pressure. The valve should be able to maintain system pressure during pump startup and pump transfers, which includes relieving the capacity of one pump, while both are running. The valve should provide stable, constant pressure during these transients. Flow turndown of 8 to 1 is not unusual. Multiple valves in parallel should be used if a single valve is not suitable. The valve should be sized to operate between 10 and 90% of the flow coefficient (C_v). Additional pressure control valves should be furnished as required to provide any of the intermediate pressure levels.

On the smaller, non-API type systems, the relief valve is also the pressure control valve. This definitely must be an external valve. While a compromise for the smaller system, the requirements are also not so severe.

Startup Control

A pressure switch located so as to sense falling pressure at the earliest moment should be used to activate the standby pump. The switch should be so connected to the system as to permit testing the startup circuit without shutting down the compressor. Figure 8-8 shows four piping arrangements. The figure at b would be the recommended method.

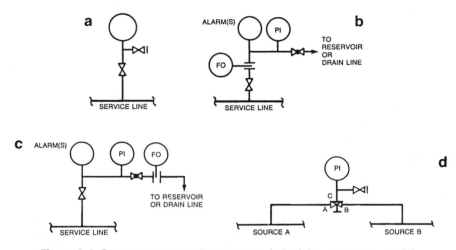

Figure 8-8. Pressure gauge and pressure switch piping arrangements [1].

It should be emphasized at this point that the speed of response is critical. The pressure transient pressure should not fall to less than 50% of the difference in pressure between the standby pump start pressure and the low oil pressure trip pressure. This is normally achievable with good design practice and the use of a switch and direct wiring. There is some tendency to use a transmitter and control through a remote computer. The latter arrangement is difficult to check on a shop test and normally is too slow to meet the requirement. An accumulator can be added and must be used if the requirement cannot be met. This additional hardware contributes to higher initial cost and possible reliability problems in the future. The direct switch method is therefore highly recommended.

Provision must be made in the wiring to maintain power to the standby pump driver. This will prevent power interruption due to the pressure switch contacts opening when lubricant pressure is restored. Cycling of the standby pump will occur if the circuit is not maintained. A reset will have to be furnished to permit shutting down the standby pump when the main pump is back on stream.

Check Valves

Check valves must be used on each pump discharge to prevent back-flow through the idle pump. On those systems where a flooded suction is not furnished, an orificed bypass around the appropriate check valve should be furnished to keep the suction line of the non-flooded suction pump charged. Sometimes the check valve is drilled to provide this feature at a minimum cost, but this is not the best practice.

Coolers

API 614 mandates twin coolers. For a compressor in critical service or, even in some cases, an unspared compressor, it is important to be able to switch coolers without requiring a main unit shutdown in the case of a problem. For shell and tube coolers using water, a good design outlet temperature is 120°F. The normal expected input from the compressor is approximately 140°F. A removable bundle design is generally recommended above 5 square feet, though this may vary from plant to plant. The classic API tube size has been ⅝ inch outside diameter with a minimum of 18 BWG tube wall thickness. Again, plant standards may require a larger and heavier tube, which is no problem as the cooler size increases, but causes problems with the smaller sizes. In fact, the API minimum requirement may cause procurement problems and is not generally used on the smaller standard compressor lube systems. Most coolers are of a size which brings them under the ASME pressure vessel code rules. If so, the cooler should bear the ASME pressure vessel code stamp. For all the noise made about it, the extra cost to the vendor is not that severe. Cooler shells, channels, and covers should be steel. Tubes should be inhibited admiralty and the tube sheets should be naval brass, unless the plant standards call for different material, in which case, the vendor must be informed by way of the specification for the equipment. On some packaged standard compressors, the materials of construction are not an option. The materials used should be reviewed to ensure the compatibility with plant operations. If an incompatibility is noted and cannot be changed, the cooler may require changing once the unit is received. The vendor should be fully aware of the plan so as not to void his warranty on the balance of the equipment. If any of this becomes a problem, another vendor should be considered or, if the balance of the equipment is so attractive, it should be owner warranted.

If cooling water is not available, air-cooled exchangers can be furnished, even in relatively small sizes. The cost is higher than a shell and tube-water-cooled exchanger. Also the outlet oil temperature will be higher than that from the water-cooled exchanger. This is no particular problem if the compressor designer is aware of the higher temperature. More oil will have to be circulated to make up for the loss of the temperature differential.

Twin coolers should include a pre-piped vent and orificed fill line to permit filling the idle cooler prior to being put into service. A drain valve, located at the low point of the cooler, should be furnished for both the oil and the water side (if water is used).

Filters

While there may be reasons to only use one cooler, filters should be dual on all but the smallest standard compressors. The filter should remove particles to a nominal size of 10 microns. A filtration to a smaller particle size should be considered; however, getting the system clean to that level will take more patience. The transfer between filters will occur frequently during the initial operation period and the dual filters will be found quite handy.

The filter elements should be replaceable and should be corrosion resistant. The filter should not contain any type of internal relief valve that would permit the bypassing of the dirty oil.

The filters should be located downstream of the coolers and should be equipped with a vent and orificed fill line to permit air removal prior to being put into service to prevent shocking the system. External lifts for the filter covers should be furnished if the covers are too heavy for an operator to safely handle. A suggested weight is 35 pounds, deferring however to operator safety. An adequate valved drain should be furnished for each filter body to permit easy removal of dirty oil and sludge.

API mandates that the pressure drop across a clean filter element be no more than 15% of the allowable pressure drop when dirty. An upper limit of 5 psi drop is set for clean filters. This is a reasonable criterion. If a little arithmetic is performed, the head rise for a centrifugal pump may be calculated. The specifications on both of these items must be coordinated and made compatible.

As with the coolers, the larger filter bodies come under the jurisdiction of the ASME unfired pressure vessel code. If they do, the filter should receive the code stamp. The filter bodies and the heads should be constructed of steel.

Transfer Valves

For the twin arrangements, a two-way, six-ported, continuous-flow valve will be required. If it is a tapered plug type, a means of lifting the plug will be required. The purpose of the tapered plug valve is to provide a tight metal-to-metal seat to prevent or at least minimize leakage on reseating after a transfer. Cylindrical bore valves of the two-way, six-port type valves are available. These valves use O-rings to act as seals to prevent leakage. Another arrangement, particularly well suited for use with coolers, is the dual three-way valve arrangement. The valves are oriented so that the stems face each other. The stems are connected by an extension shaft. For smaller sizes, a simple handle will serve as an operator. On the larger sizes, a geared operator may be required. Regardless of construction, under no circumstances should valve failure block oil flow. Valve body materials should be steel with stainless steel trim.

The most versatile arrangement, from an operations point of view, is individual transfer valves, one for the twin coolers and one for the twin filters. The use of one valve for both the filter and cooler results in a loss in flexibility, as the cooler maintenance interval is usually somewhat longer term than a filter changeout. Should one cooler be out of service, and the filter that is paired with the operating cooler be fouled, the compressor will have to shutdown. Each user will have to review the extra cost of a second valve against the operation limit.

Some transfer valves block flow better than others, making it necessary to evaluate the design chosen. Valve design and manufacturing technology has improved to the extent that leakage should not be the problem it once was.

Accumulators

Accumulators can be used to help stabilize the lube system against pressure transients such as that from the turbine power operator during a large correction. For a sizing rule of thumb, the system pressure should not vary by more than 10%, while the turbine servo travels full stroke in a one second interval. The role of accumulators for pump switching was covered earlier in the section on Startup Control.

The preferred accumulator is the bladder type as shown in Figure 8-9. Body material should be 300 series stainless steel, in accordance with ASTM A 240. Either a manual precharge valve or automatic charging system can be used, based primarily on the user's preference. The alternative choice is a *direct contact* accumulator as seen in Figure 8-10. This accumulator has several problems. The gas used to pressure the top of the oil will eventually be absorbed by the oil, which can lead to drain line foam-

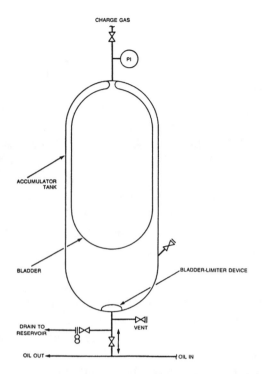

Figure 8-9. Manual precharge bladder type accumulator [1].

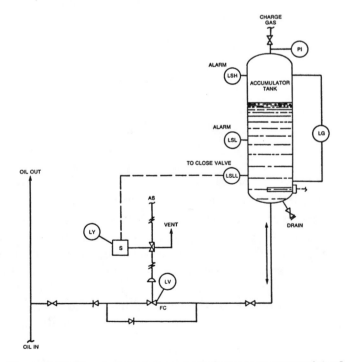

Figure 8-10. Direct contact manual precharge type accumulator [1].

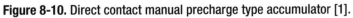

ing. The other problem is that if the oil level isn't carefully controlled, it may dump gas into the oil system. The direct contact accumulator is used primarily when the requirement exceeds the available capacity of the commercial bladder type. With either type of accumulator, additional check valves may be required in the oil system to keep the accumulator from delaying the standby pump start or other similar functions.

Seal Oil Overhead Tank

For liquid film seals, an overhead tank is normally used. The tank functions as part of the differential pressure control. The process gas is referenced to the top of the tank, and the tank's physical height becomes a manometric leg. The oil level is controlled in the tank. Figure 8-11 shows an alternate arrangement for the overhead seal tank, including the recommended operating levels and volumes. The material of the seal oil tank should be stainless, the same as that recommended for the accumulator.

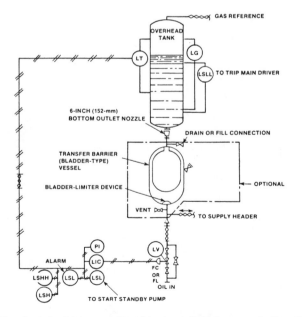

Figure 8-11. Direct contact type overhead tank module for a seal oil system, including an optional transfer barrier (bladder type) vessel [1].

Lube Oil Overhead Tank

An alternative to the coastdown pump approach is the overhead lube oil tank. It has the valuable feature of being motivated by gravity. If reasonably maintained, it should be one of the more reliable methods available. It will safely permit the coastdown of a compressor without losing the bearings due to lack of lubrication during power failure. Figure 8-12 shows two schemes for the tank arrangement. While both systems are in successful operation, the method at (b) requires less instrumentation and would be considered somewhat less prone to problems.

The tank should be placed at an elevation that provides a static head less than the low oil pressure trip switch setting so as to not interfere with the normal trip function. Normally this means that the static head should not be more than the equivalent of 5 psi. The minimum sizing should be based on a three-minute rundown cycle. Adding a bit more to the time is probably not all that costly.

For this tank, stainless steel should be specified considering its importance, and because it is elevated, it will not receive the inspection for main-

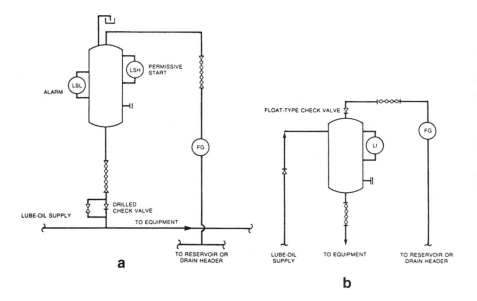

Figure 8-12. Two arrangements for a lube oil coastdown tank. Arrangement "b" is recommended [1].

tenance as the grade elevation equipment. Provision should be made for the cleaning of the tank. If the alternate arrangement in the figure at (b) is selected, the tank will be a code vessel and will require an ASME stamp.

Seal Oil Drainers

When oil buffered seals are used, oil will move past an inner seal toward the process side of the compressor. The oil is prevented from moving into the compressor by a set of labyrinths and is captured in an inner drain cavity. From the cavity, it is piped to the outside where it is collected in either a pot or trap. Figure 8-13 shows several alternative arrangements and equipment. The user must choose between automatic or manual drainers. If the gas from the top of the drainers is to be directly returned to the compressor, it is important that mist eliminators be used. The oil collected in the drainers is reclaimed or disposed of, based on the level of contamination and the user's disposal practices.

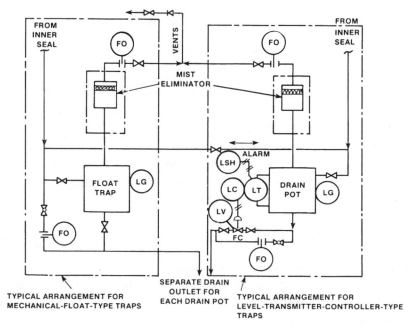

Figure 8-13. Arrangements for inner seal drain traps [1].

Degassing Drum

If the oil from the drainers is not contaminated, it can be returned to the reservoir for immediate reuse. If this be the case, a *degassing drum* is recommended to remove the entrained gases prior to returning to the reservoir. For flammable gases, this practice helps eliminate the explosion hazard in the reservoir. If a suitable low pressure recovery system is available, the gases removed can also be recovered. The alternative is to go to the flare with hydrocarbons. Figure 8-14 is a diagram of a typical

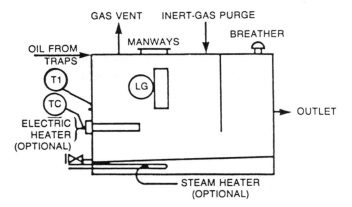

Figure 8-14. Typical degassing drum arrangement [1].

degassing drum. Materials of construction should be consistent with the balance of the lube system. Sizing of the drum should be three times the guaranteed total inner seal leakage from all seals over a 24-hour period. If required to operate the drum at a positive pressure to match a gas recovery system, the suggested design may eliminate the breather and operate as a pressure vessel. The vessel will, of course, need to follow the appropriate pressure vessel design practices.

Piping

Interconnecting piping should be stainless, consistent with the balance of the system. In critical equipment applications, this is definitely recommended. The user must decide for the smaller system to what extent stainless is feasible. For all systems, the piping should follow the recommendations and mandates in API 614. Fundamentally, this means minimize fittings, bend when possible, and fabricate by welding. It really makes no difference how small a lube system is; screwed pipe lube systems leak. On the small systems where welding or seal welding of the screwed fittings is not feasible, fabrication using the thickest walled tubing that the vendor will furnish will make a better system. With a little persuasion, the vendor may consider stainless tubing.

System Review

In the course of obtaining a new lube system, or when revising an existing one, there is a point in the design when the overall system should be reviewed. With a lube system, in particular, it is very easy to get totally engrossed in the individual parts and forget the functioning of the parts together. The possibility of an accumulator overriding the stand-by pump start function was previously discussed as an example of this problem. The system should be reviewed keeping the basic compressor operating parameters in mind at all times. The startup function should be reviewed in a systematic manner, from the time the pump is initially started to the time the overhead tanks automatically fill. From startup, the review should look at the running operations and the steady state conditions of each item. Some questions that should be answered include: Is the accumulator charged with oil? Did it charge properly to the desired gas pressure? Is the standby pump interlocked to prevent it from cycling if the main pump drops off the line? After looking at the possible upsets that could occur during normal operation, shutdown should be reviewed. If there are seals, do the pressures appear to track the process pressure, particularly if this pressure rises on shutdown? Is there enough power in the driver to take the additional head? Do the valves sequence properly? Are the valves properly sized for each step? This is a sample of the questions that might be reviewed, and not intended to be a comprehensive list. The review should be tailored to the system being considered and the length of the list should reflect the complexity. The review of existing systems for operational improvement can also be considered.

Dry Gas Seal Systems

System Design Considerations

One of the main functions of the seal gas system is to supply clean (filtered) dry gas to the primary seal. Figure 8-15 shows a typical seal gas

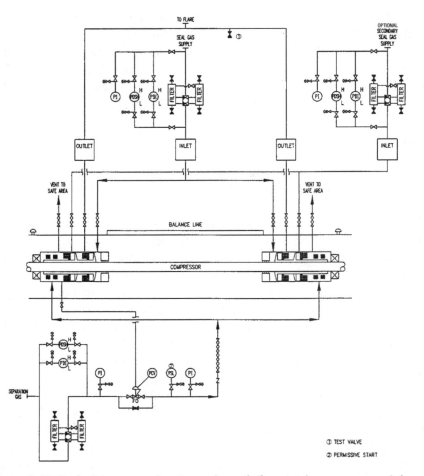

Figure 8-15. Typical dry gas seal system schematic for a tandem arrangement dry gas seal.

system for a tandem seal arrangement. Note the optional secondary seal gas supply to be used with tandem seal having an internal baffle. Figure 8-16 shows a typical seal gas system for a double-opposed seal arrangement. The source of the seal supply (tandem) and buffer gas (double) is normally compressor discharge gas, but is not limited to that source. An alternate supply of gas may be required for startup, if the suction side of the compressor can go sub atmospheric. The system design pressure must be able to accommodate the higher of either settle out pressure or system relief pressure. Care must be taken to exclude all liquids from any of the

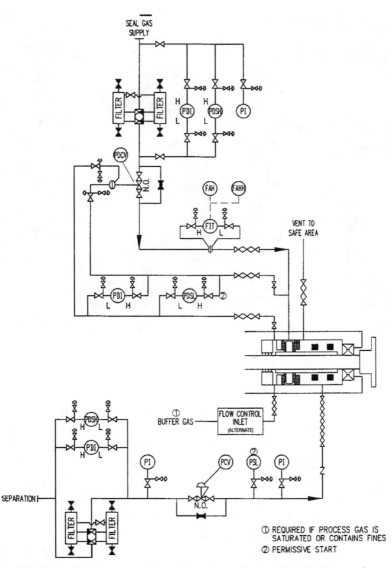

Figure 8-16. Typical dry gas seal system schematic for a double-opposed arrangement dry gas seal.

buffer streams. This may require heat tracing or liquid knock-out provisions and sometimes both. The material of construction is normally austinetic stainless steel. For many of the systems, stainless tubing will be adequate (see Figure 8-17). When pressures are high or the system is very large, rigid stainless pipe is required.

Figure 8-17. Dry gas seal console. (*Courtesy of A-C Compressor Corporation*)

Dry Gas Seal System Control

The control system should be kept as simple as possible to maintain reliability. For tandem seals, the system control for the seal gas supply has traditionally been a differential pressure control, using a direct-operated control. The differential pressure usually was set at 10 to 25 psid. The flow using this control is generally much higher than needed. The differential pressure control can be modified to measure the pressure drop across an orifice and the control converted to a volume control. A velocity across the inner labyrinth of 5 fps would be considered sufficient to prevent back diffusion. This arrangement does have the advantage of reducing the seal gas requirement. For a practical design, the velocity should be increased to at least 10 to 15 fps to allow for wear. Many of the compressor manufacturers will recommend a higher value for more margin.

For the double-opposed seal, normally an inert gas is injected between the two opposed seals. This gas is pressure controlled to maintain a differential pressure higher than the process side pressure. The supply to this seal is critical because a failure will permit the differential pressure across the outer seal to reverse, which will result in a seal failure. This seal arrangement usually incorporates a buffer to the process side of the seal to keep any dirty process gas away from the dry gas seal. The leakage to process from the gas seal is too low to keep any dirt from reaching the seal. However, there is somewhat of a problem. If the buffer is differential pressure controlled, this differential pressure must be added to the process pressure which in turn sets the injection pressure. For this reason a volume control is a better choice because the differential pressure involved is not as great.

The dry gas seal has one of two types of barrier seal (seal between the bearing and dry gas seal) either a labyrinth or single or double carbon rings. Normally the seal system includes provision to supply buffer gas to the barrier seal, also known as a separation seal. The gas to this seal is referred to as separation gas. One reason for choosing the carbon ring style barrier seal is to keep the separation gas usage to a minimum. The gas is normally nitrogen. The basic control is by a direct-operated pressure control valve.

For the tandem arrangement gas seal, a primary seal vent must be provided to vent the leakage across the process side seal. This vent may be to flare or other suitable gas disposal point. The back pressure under normal conditions should be kept to a low value. A small amount of back pressure is recommended to keep a positive differential across the secondary seal. Leakage measurement may be provided in the vent line to provide health monitoring of the primary seal. Unfortunately, the rotameter, which would be the obvious choice, should not be used because of its lack of reliability. If an orifice or needle valve is used to set the back pressure to the seal vent, pressure upstream of the restriction can be measured for a relative flow measurement. This type of reading does provide trend data that may be used to judge the seal's performance.

The balance of the controls consists of the required pressure switches and/or transmitters to provide monitoring of the system and alarm and shutdown functions for the critical buffers. Filter differential normally is also monitored and alarmed on high filter differential pressure.

Dry Gas Seal System Filters

Filters must be provided for all gas buffers in places where the gas will pass through the gas seal. The gas should be filtered to a nominal value of 2 microns. On most critical systems, dual filters will be required to permit servicing the filters without having to shut down the compressor. This requires the use of a transfer valve that can be switched without causing a flow interruption. If liquids are anticipated, heat tracing or an alternate gas source should be the first consideration. Should this not prove to be practical, a liquid separation system must be included with the filters. On systems with heat tracing, liquid separation may be desirable as a backup.

Gears

Whenever there is an inherent speed mismatch between a compressor and driver, several solutions are available. For small sizes, V-belts offer advantages in flexibility. For compressors where the power levels approach 100 hp, a more positive drive should be considered. Depending on the application, this value may move up or down by 25 hp.

Most of the discussion has been about compressors over 75 hp and has leaned toward critical or semi-critical equipment. This type equipment requires a gear unit external to the compressor with the gears arranged for increasing or reducing speed as dictated by the application (see Figure 8-18). Alternately, the gear may be integral to the compressor as shown in Figures 8-19 and 8-20. API has two gear standards, API 613 [9] for special purpose gears and API 677 for general purpose gears.

Figure 8-18. Compressor driven by an electric motor through a speed increasing gear. The motor enclosure is a WP2. (*Courtesy of A-C Compressor Corporation*)

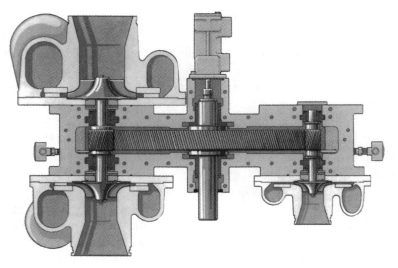

Figure 8-19. Section illustration of the gearing of an integral geared compressor. (*Courtesy of Elliott Company*)

Figure 8-20. Cutaway of a gear-mounted multistage centrifugal compressor. (*Courtesy of Nuovo Pignone*)

Gear Design and Application

The special purpose gear unit should have a specified API 613 service factor of 1.4 as a minimum. The service factor used in the gear design should be selected to meet the requirements of API for the application.

The gear unit must be tuned to the system in which it operates. The vendor with the train responsibility must analyze the system to be certain the gear unit will not transmit torsional or lateral vibrations to the rest of the system or be damaged by system excitations.

If the gear unit is retrofit, for example when replacing a turbine with a motor and gear unit, the system, in like manner, must be thoroughly analyzed. In either case, the gear supplier must furnish the user with adequate data on the mass elastics of the unit, inertias, bearing characteristics, and the like, to permit the user's own or a consultant's analysis.

Resonant responses must not coincide with excitation frequencies of rotational shaft speed, especially gear meshing frequency (the speed of a shaft times the number of teeth of the gear on that shaft), or other identified system frequencies; otherwise, a self-excited system will exist. Lateral response criteria should conform to API 613.

An important consideration of a gear unit is the *pitch line velocity* (PLV), which is the product of the gear or pinion pitch diameter and the shaft speed of the gear or pinion.

A general classification of pitch line velocity and corresponding American Gear Manufacturers Association (AGMA) quality values are:

- 20,000 fpm or less—moderate service—AGMA 12
- 20,000 to 30,000 fpm—medium service—AGMA 13
- 30,000 fpm or higher—severe service—AGMA 14

To promote even wear, a hunting tooth design is desirable. In this design, a pinion tooth does not contact a given gear tooth more than once until it has meshed with all the other gear teeth.

For maximum reliability, a gear-unit ratio must be within the capability of a single-step application. Practically, ratios may approach 6:1, but the preferred ratios are in the 3:1 to 5:1 range.

A double-helical design should be considered, particularly for critical service. Rotation of the helices should be such that they will be apex leading, that is, teeth engage at the centers with mesh progressing to the tooth outer ends. Additionally, rotation should permit the force from the gear mesh to cause the gear to load the lower half of its bearings. The double-helical gear generates no thrust load of its own and can be operated with no thrust bearing if flexible disc or limited end-float type couplings are used on each shaft. The advantage of this approach is the elimination of the horsepower loss associated with a thrust bearing.

The low speed gear shaft and the housing must be designed to permit installation of a stub shaft for a torsiograph unit if an operational problem occurs. API 613 gives the details of the shaft end requirements for attaching a torsiograph. This should be done on all synchronous motor compressors and on multiple driver or multiple compressor case trains.

Rotors and Shafts

The shaft and gear teeth of a pinion should be of an integral design. The gear should be a forging with the gear shrunk on a keyless shaft for critical service with a pitch line velocity over 20,000 feet per minute. Under 20,000 feet per minute, the gear may be "rimmed," that is, design-welded in place.

The gear teeth should be heat-treated for proper strength and through hardened. Alternatively, surface hardening by carburizing or nitriding is used. Flame and induction surface hardening are also alternative hardening methods but normally are used less than carburizing or nitriding. For new gears, through hardening is preferred, using the surface hardening for later upgrades. Requirements for strength and hardness must result in an adequate durability and tooth stress as determined by API 613. The surface finish of the teeth must be 20 microinches Arithmetic Average Roughness Height (Ra) or better.

Gear elements with adequate hardness and surface finish have a greater resistance to initial scoring, a destructive phenomena found in meshing teeth.

For adequate operation, there are many features that should be incorporated in the rotors. They include: (1) good surface finish, (2) modified tooth forms, (3) high AGMA quality number, (4) hunting tooth design, (5) no resonant responses, (6) high damping journal bearings, (7) good contact pattern between teeth, and (8) good rotor balance. Other factors to be considered for quiet gear operation are proper diametral pitch, pressure angle, helix angle, overlap ratio, backlash, and minimum apex runout. (For the meaning of these terms, refer to API and AGMA standards.) API 613 has a requirement for maximum apex run-out under subject of "Axial stability of the meshing pair."

Shafts must be stiff enough to prevent deflection that would adversely affect the tooth contact in operation. Internal alignment should be carefully performed. This is another area that can directly affect tooth contact.

Each shaft should have an integral coupling flange rather than a removable coupling hub. If removable hubs must be used for some reason, they should be of the keyless, hydraulically dilated design.

Bearings and Seals

The journal bearings must have stiffness and damping properties sufficient to prevent bearing contributed vibrations and to result in proper gear contact. The stiffness and damping properties of the journal bearings affect the rotor system dynamics. Normally, stabilizing bearings, such as tilting pad and three-lobe, are needed to prevent shaft oil whirl. The tilting pad bearing can be seen in Figure 5-38. Figure 8-21 shows a typical three-lobe journal bearing.

If a thrust bearing is used, it should be of a tilting pad, self-leveling design. With a double-helical gear unit, thrust bearings, if used at all, should be on the low speed shaft only to accommodate loads in both axial directions. Journal and thrust bearings must be split for removal and installation without having to remove the coupling hubs.

Seals should be non-contacting, multiple point labyrinths. The housing and seals should be drilled and the housing tapped to accommodate a dry

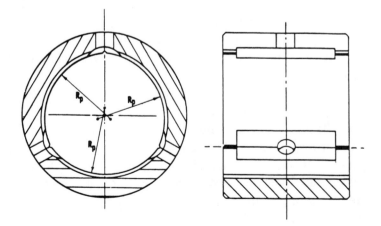

Figure 8-21. Three-lobe journal bearing. (*Courtesy of Turbocare, A Division of Demag Delaval Turbomachinery Corp., Houston facility*)

gas purge connection. Appropriate shoulders or shaft fingers should be immediately inside the seal to aid in the prevention of oil leaks.

Housing

Either cast from cast iron or carbon steel or fabricated from carbon steel, the housing (gear box) must be rigid to prevent deflection under loads. The housing must be thoroughly stress-relieved to maintain accuracy. The housing should be horizontally split at the shaft centerlines. The horizontal joint should be finished to prevent oil leaks with a metal-to-metal fit. Ample side clearance between housing and rotors will prevent oil pumping and will permit free fall of oil to the sump. Sump depth, adequate windage baffles, and generously sized return lines prevent oil level buildup or foaming in the housing. Uneven impingement of hot oil on the housing causes differential thermal growth of the unit and must be avoided.

All piping and instrument connections should be made on the lower half of the housing to aid in removal of the top half. Other connections, if they must be made in the top half, should have easily accessible joints.

Lubrication

A *dry sump* design should be employed. The gear unit for a train with a central lube oil system should be designed for the turbine grade oils of the system. Typically, 150 Saybolt Seconds Universal (SSU) oils at 100°F (ISO 32) with an inlet temperature of 110°F to 120°F are adequate.

If the gear unit application must have its own oil supply, such as with some retrofits, the intent of the API 614 should be employed.

All lube piping should be stainless steel. The stainless steel should be one of the 300 series, preferably "L" or low carbon grade, such as type 304L.

Couplings

Introduction

The purpose of a flexible coupling is to transmit torque from one piece of rotating equipment to another, while accepting at the same time a small amount of misalignment. Flexible coupling misalignment is expressed, as an order of magnitude, in thousands of an inch. Actual misalignment, expressed in coupling terms, is angular in nature and expressed in angular units, that is, degrees. How many is a function of the coupling type and installation. An installation variable is the equipment movement due to the temperature changes taking place in the machines as they go from the non-operating state to operation. Some angular values will be used in the discussion of the various types, but again, these are for reference only. Each application must be reviewed using the type coupling selected and the specific design proposed by the vendor.

Ratings

It is recommended that couplings and coupling-to-shaft juncture should be specified for a capability of 150% of normal torque for flexible element couplings and 175% for gear couplings. The coupling rating should be based on the maximum continuous misalignment capability of the coupling. Further, the coupling and coupling-to-shaft juncture should be rated for a capability of 115% of any specified maximum transient torque. Special purpose couplings are covered in API Standard 671 [6].

Spacers

There are a number of features that all couplings have in common. One is the need for a spacer. API 671 calls for an 18-inch spacer minimum. This is reasonable for smaller units, say to 5,000 hp; however, as the size of train increases to 15,000 to 20,000 hp, a 24-inch spacer should be considered. Above that size, longer spacers, 30 to 36 inches, are in order. The spacer first of all provides for unit separation and maintenance space. Secondly, the longer the spacer, the less the angular deflection of the coupling at its flexure point for a given offset. This makes absolute equipment alignment less critical.

For the smaller units, under 200 hp, this requirement may be relaxed. Space for service is not as critical, but a minimum of 5 inches should be used.

Hubs

The attachment of the coupling to the shaft has been the source of many problems in the industry. The following practices are recommended for critical equipment.

1. If a coupling hub does not have to be removed for maintenance, seal removal, or impeller removal, it should be integral with the shaft.

2. On slow-speed equipment, where the hub does not have to be removed for maintenance and integral hubs are not available from the supplier, a straight fit can be used with a reasonably heavy shrink—1.25 mil/ in. of diameter. Tolerances may allow this value to vary from 1.0 to 1.5 mil/in. of diameter; however, 1.0 mil/in. may be too loose for some applications, while 1.5 mil/in. over-stresses hubs at maximum bore sizes. In light of this, each application should be carefully evaluated if there is no experience at hand to use as a guide. Of course, field removal at the higher shrink fit is more difficult. Keys should not be used.

   Where removal is necessary, keyed hubs on straight shafts should use 0.5 mil/in. to 0.75 mil/in. shrink fit. Caution should be exercised in these applications.

3. If the hub removal is necessary, such as required on compressor with non-split seals, a tapered hub fit on the shaft should be used. The removable hubs should have tapped puller holes. The shaft should be keyless with the preferred method of installation and removal by use of hydraulic dilation. Two injection ports 180° apart should be used whether injection is through the shaft or through the hub. Shrink fits should be 2 to 2.5 mil/in. of diameter. API 671 recommends 1.5 mil/in. minimum, but experience indicates the heavier shrink may be required. For the juncture rating calculation, a friction value of .12 is recommended.

   In the NEMA standard, ¾ in./ft taper has been used for keyed tapered hubs. However, API 671 recommends a ½ in./ft taper for hydraulic release, keyless hubs (see API 671, Appendix E). The ½ in./ft seems a reasonable compromise, but there is still a strong following for the NEMA taper.

   Contact area for hydraulically removable, keyless hubs should be a minimum 85% when using bluing and a taper gauge. It is recommended that a set of gauges, ring and plug, be purchased with a new unit for maintenance and replacement purposes. A male and female cast iron lapping block set should also be made at the same time to help achieve the required fits by lapping.

   For installation, care should be used to install the hub at the proper location on the shaft taper to provide the interference needed. A jig should be used and allowance made for hub shrinkage on cooling.

4. If keyed tapered shafts are furnished, as may be true with some installations, particularly for machines under 1,000 hp with small shafts, two keys are recommended. A shrink fit of 1 to 1.5 mil/in.

should be used. Unfortunately, heavier shrinks are impractical because of hub stress.

For maintenance and stocking, it is sometimes desirable for plants with a large number of pieces of equipment to minimize the number of coupling sizes. Also, plant-to-plant coupling exchanges may be made if the size permits. However, caution should be exercised because what appears interchangeable may not be. Because of the tuning of torsional criticals, the original equipment vendor may have had a coupling custom-made for a system. In the fine tuning of torsional criticals, couplings are the logical focus as the other components must be established before a system torsional study can even be made. A coupling tuned for one system may not be a logical choice for another. Therefore, before arbitrary trades are made, this aspect should be reviewed. The same comments apply to lateral criticals; however, a new critical speed analysis may not be necessary if the replacement coupling weight is within 10% of the original. If heavier, a new lateral analysis should be performed.

Gear Couplings

Gear couplings are no longer the primary coupling of choice, having been superseded by the flexible element coupling. When used for critical equipment, the gear coupling should be specified as the high speed type regardless of speed for most applications. This class of coupling can be balanced and has the highest grade material. Plain hub style gear couplings are usually called a marine type. When the application permits, this is the preferred type. Since first preference is integral hubs, it follows that the marine type must then be used. Another advantage to this type is that all wearing parts are removable with the center section. The disadvantage of the marine type coupling is that it has a higher overhung moment because the teeth are on the spacer and the weight is further removed from the bearing centerline. This may cause a problem with lateral criticals. When this is true, teeth on the hub must be considered. This requires a removable hub design because the hub itself now becomes a wearing part. Teeth on the hub can be furnished as a standard arrangement with teeth near the outboard end or a low moment arrangement with the teeth on the inboard end of the hub. The latter, which is attractive in reducing overhung weight, tends to have problems with clearance for sleeve withdrawal and, therefore, should only be used if rotor dynamics problems cannot be solved in another manner.

It should be noted that a mix and match has been used with a marine type plain hub on one end of the coupling and a toothed hub on the other. This, however, is most extraordinary and is mentioned only to indicate a possibility for problem solving.

Gear teeth hardness should be a minimum of 45 on the Rockwell C. scale. Hardness of the teeth having the greater face width (generally the sleeve teeth) must be equal to, or preferably greater than, the hardness of the mating teeth.

Lubrication may be continuous from the lube oil system or grease packed. For longer continuous operation and where the coupling speeds are high, continuous lubrication is preferred. Where maintenance intervals permit and the separation forces on grease are not too high, grease is a good solution. There are newer coupling greases on the market whose oil and thickener are the same density and, therefore, are not subject to centrifugal separation. Run times on conventional grease lubricated couplings are 8–18 months.

When continuous lubrication is called for, the couplings should be supplied with oil filtered to a minimum of 2 microns. Experience shows that filtering to ½ micron is possible and desirable. Usually this is done with a separate filter; however, the whole system may be filtered to this level. The oil quantity furnished, per gear mesh, should be 3 gpm minimum.

Alignment

The absolute angular misalignment capability of a gear coupling is a function of the tooth form and backlash. The total angular misalignment at each mesh will be the total of the angular component and the angular result of parallel offset, both of which are the vector sums of horizontal and vertical misalignments. Values from ½°/mesh to 6°/mesh may be found in the catalogs. However, this higher number is the value at which the teeth exceed the clearance and the coupling truly locks up.

Equation 8.1 is based on a maximum sliding velocity during misalignment of 5 ips and Equation 8.2 is based on 8 ips. Research by the Naval Boiler and Turbine Laboratory [7] developed these values. Experience indicates that the more conservative range of 1–3½ ips is most desirable.

$$\alpha = 5{,}500/dN \tag{8.1}$$

$$\alpha = 8{,}800/dN \tag{8.2}$$

where

α = angular misalignment per mesh, deg
d = pitch diameter of the teeth, in.
N = shaft speed, rpm

Flexible Element Couplings

Flexible element couplings transmit torque and accommodate misalignment by use of flexible diaphragms or discs. These elements are of metal construction and may be used singularly or in packs. Refer to Figure 8-22 for a multiple element coupling of the diaphragm type. A single element, diaphragm type is shown in Figures 8-23 and 8-24. A flexible disk is illustrated by Figure 8-25.

The flexible element coupling requires no lubrication, which is a distinct advantage. The need for lubrication is always a problem with gear couplings. This advantage is partially offset in the flexible element by the need for corrosion-resistant materials for the normal chemical plant atmosphere. Inconel 718 and 15-5 PH have been used and have been rea-

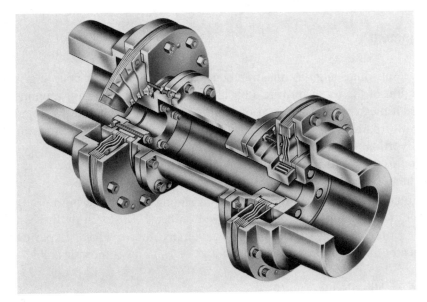

Figure 8-22. Cutaway of a multiple diaphragm flexible element coupling. (*Courtesy of Zurn Industries, Inc., Mechanical Drives Division*)

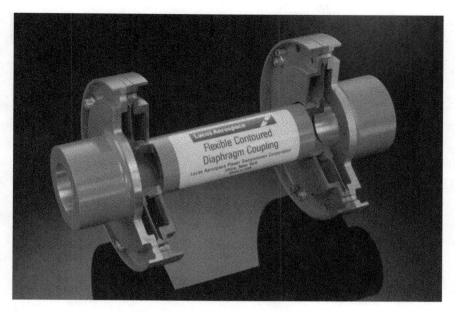

Figure 8-23. Cutaway of a single diaphragm flexible element coupling. (*Courtesy of Lucas Aerospace Power Transmission*)

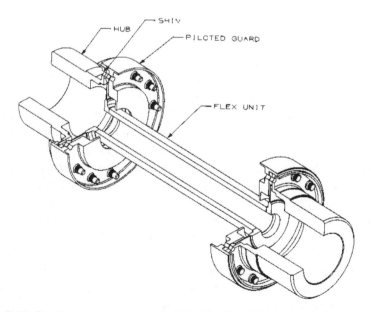

Figure 8-24. Section drawing with parts identification of a single diaphragm flexible element coupling. (*Courtesy of Lucas Aerospace Power Transmission*)

Figure 8-25. Multiple disk flexible element type coupling. (*Courtesy of Kop-Flex Power Transmission Products*)

sonably successful in chloride atmospheres, and 300 series stainless steel has been used successfully in hydrogen sulfide environments.

Another advantage of the flexible element coupling is the lower bending moment imposed on shaft ends. This can be a problem with the gear coupling, particularly if the loading is high or lubrication poor. The flexible element, because of the elastic member, has a predictable bending moment which is normally much lower than the comparable gear coupling [8]. Axial loads transmitted are much less than for a gear type coupling. This greatly reduces thrust bearing loads. Absence of radial clearances through the major components makes it possible to obtain a precision repeatable balance. Flexible element couplings are available in both single and multiple element form.

Flexible element couplings tend to be somewhat heavier than the comparably rated gear coupling. On a retrofit, heed the earlier warning about lateral criticals. The coupling can handle axial misalignment but is more restrictive than the gear type.

Limited End-Float Couplings

The flexible element coupling is by the nature of its design a limited end-float coupling. In gear applications where flexible element couplings are used on both the low and high speed sides, the gear, if double-helical, may be operated without need of a thrust bearing. The gear elements are allowed to float between the compressor thrust bearing and the motor magnetic center.

For gear couplings, to guard against transmitting thrust continuously from the motor to driven equipment bearings, use gear-type flexible couplings with the limited end-float provision. The following requirements are specified for all large motors:

1. The coupling is required to have a no more than ¼-inch end-float (if a gear type).
2. The motor end-float must exceed the coupling end-float by ½ inch, making the minimum permissible motor end-float ½ inch.
3. The motor magnetic center must be within ³⁄₃₂ inch of the motor's geometric center.
4. The motor centering force must be sufficient to return the rotor to magnetic center against friction of a new, properly lubricated gear coupling.
5. The momentary end thrust transmitted by the motor to the coupling and thus to the driven equipment must not exceed 100 pounds per thousand horsepower. (This provision covers thrust transmitted when breaking away the engaged coupling gear teeth from their rest position or when the coupling separators touch when starting.)

The gear coupling must be modified to provide a limited end-float feature. This feature is desirable when sleeve bearing machines without thrust carrying capability are used. On shutdown, the units can float to their limit stops, for example, on a motor. On shutdown, the motor will float to a bearing face. On restart, the friction in the teeth keeps the rotor on a thrust face, which is not capable of carrying load and causes a bearing failure. To prevent this, a feature is added to a gear coupling limiting this end-float. Happily, the flexible disc coupling is an inherently limited end-float design. On a motor gear compressor arrangement, the parasitic power used by the gear thrust bearings can be eliminated by omitting the thrust bearing entirely and using both a high-speed and low-speed limited end-float coupling.

Instrumentation

Overview

Compressor instrumentation has moved from some simple pressure and temperature gauges to a field all of its own. Systems now in the marketplace appear to rival the telemetry used by NASA on the space flights. Thirty plus years ago, vibration monitoring was non-existent for all but some hand-held velocity type of pickup equipment. The name Bently was known to only a few friends of Don Bently, who were aware of his experiments with proximity probes. Today the name is associated with a large vibration monitoring equipment company with several competitors. The advent of the vibration probe and monitor opened up a new world to the compressor engineer. It was wonderful and frightening all at the same time. This little probe saw things that had been present in compressor operations for a long time but were unknown to both the design engineer and the operator. The measurement data came in faster than the technologists could analyze the information. An awful lot of time was spent worrying about the insignificant data, only because it was there. With the maturing of the measurement, part of the vibration story can be the rotor dynamics field. The ability to observe the shaft dynamics in real time by way of orbits was truly fascinating.

The subject of vibration instrumentation can fill many pages, but is only one phase of the world of compressor instrumentation. Therefore, the coverage here can only be brief.

Pressure

The most fundamental aspect of basic instrumentation for monitoring compressor performance is pressure. That is what the compressor does, it delivers gas at a pressure. In most cases, the measurement instrument is a common gauge. In covering these items, the basics of how the instrument works or why will be left to the reader. As was mentioned, space permits only a superficial treatment. In the era of computer monitoring, the pressure transducer or transmitter is becoming more prominent. This instrument converts the input signal into another form proportional to the input. For example, an electrical transducer or transmitter would convert a pressure signal to a proportional electrical output signal, a logical form for input to a computer or data logger. Chapter 10 discusses the basic requirements of pressure instrumentation in more detail.

It should be pointed out that it is probably not necessary to make as many redundant taps as called for in a code test. However, as long as taps are made, they can be made to provide information as accurately as possible. There is no point in putting a calibrated gauge or transducer into a pressure connection that is incapable of delivering a reliable reading.

Differential pressure is included in the pressure instrument class. Good differential readout gauges are still not all that common. Fortunately, in the transducer category, they are more readily available. Figure 8-26 covers some installation details for pressure-oriented instrument piping, supplementing the information presented earlier in Figure 8-8.

Temperature

Verifying temperature is the second most important aspect of any compressor operation. As with pressure, the basic form of measurement is a simple temperature gauge. The construction of the gauges is quite varied, ranging from a bimetallic device to the filled systems. When transmission is involved, the sensor becomes quite simple, taking the form of a thermocouple or a resistance temperature detector (RTD). The monitor does the translation from the native signal to a temperature readout or signal proportional to temperature.

Temperature switches are also part of the temperature instrument family. Although not sophisticated, entire compressors are influenced by the temperature switch. It is now becoming more popular to take a thermocouple signal to the computer and generate a switch closure as part of the computer signal.

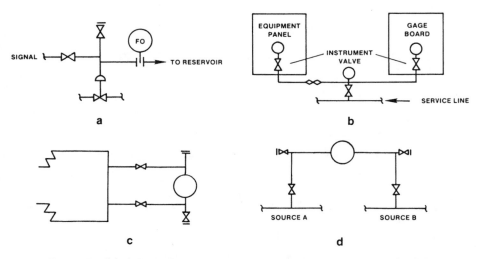

Figure 8-26. Installation details for pressure oriented instrument piping [1].

API 670, "Vibration, Axial Position, and Bearing Temperature Monitoring System" includes temperature in its scope. For several years, radial and thrust bearings have been instrumented using either thermocouples or RTDs. Each user specifying the instrumentation had the bearings fitted in his own way. While this gave good data in some instances, it was not consistent. Furthermore, the data from one user could not be compared with the data from another. In fact, because of different influences by the various compressor vendors, one large user could not correlate his own experience. While the addition of bearing temperature monitoring to the API standard has numerous benefits, the more immediate one is the establishment of a standard method of installation. Figures 8-27 and 8-28 show the recommended installation position on the radial and thrust bearing respectively.

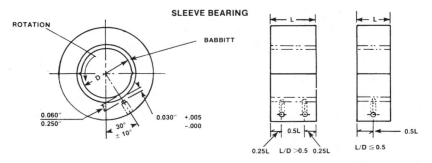

Figure 8-27. Typical radial bearing temperature sensor installation.

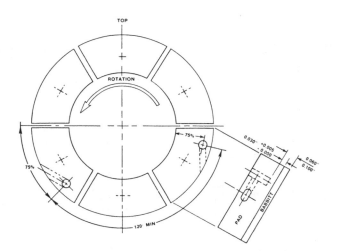

Figure 8-28. Typical thrust bearing temperature sensor installation.

The two recognized standard sensors are the ISA type J thermocouple (iron-constantan) and the 100 ohm at 0°C platinum 3-wire RTD. Additional attributes such as TFE insulation and stainless overbraid are specified. The sensors are installed in a drilled hole at the location shown, with the objective being to place the sensor approximately .030 inches to the rear of the base of the babbitt. Surprisingly, steel conducts at approximately the same coefficient as the babbitt, so there is no significant temperature drop at the metal interface. The sensor is potted in place, with some of the overbraid included, to provide strain relief. An alternate to potting is to use a spring and clip arrangement, which has the advantage of an easy sensor replacement. Figure 8-29 shows a radial bearing with a temperature sensor installed. Figure 8-30 depicts an instrumented thrust bearing.

Flow

Flow is another one of the basic compressor parameters. It can be deferred back to pressure, since most of the flow involving compressors is measured flow by a primary device such as an orifice and a differential pressure sensor as discussed in the section on pressure. For plant use,

Figure 8-29. Radial bearing with a temperature sensor installed. (*Courtesy of A-C Compressor Corporation*)

Figure 8-30. Thrust bearing pad with a temperature sensor installed. (*Courtesy of Turbocare, A Division of Demag Delaval Turbomachinery Corp., Houston facility*)

where relative flow is generally of more interest, the flow is usually recorded on a flow chart. The chart is calibrated in square root units. Unfortunately, unless all the constants can be located, the relative flow is not of much value. If a differential transducer is connected across a primary element with a known bore and pipe size, and if the element was calibrated, the calibration chart will provide the raw data that will permit the generation of meaningful flow information.

Torque

Torque meter instrumented couplings are available using the *strain gauge* for the measuring element. A standard coupling is sent to the torque meter supplier for application of the electronics and the strain gauges. The torsional strain, as measured by the gauges, is telemetered to a stationary cylindrical receiver placed concentric to the coupling. The received data are processed and converted to a signal proportional to the transmitted torque. It may be displayed locally and/or transmitted to a remote location such as a control room. Figure 8-31 shows a coupling instrumented for torque measurement.

Speed

With the introduction of the new instruments, speed is basically taken for granted. It is a very important parameter for reciprocating compressors, however, because speed is one of the factors in generating displaced volume. For the axial and the centrifugal compressor, speed offers a multiple influence. In the fan laws stated in Chapter 5, speed was the common parameter in both capacity and head. In fact, since head is proportional to speed squared, it becomes quite important that the speed be accurate.

The electronic counter circuit contributed to the development of modern tachometers. By using a toothed wheel and a magnetic pickup with the counter, a direct reading digital speed output may be derived. While not too common, the signal can be put through a digital-to-analog converter and an analog meter reading made available. The digital readout is useful for performance testing because it requires no interpolation. The analog meter reading is good for startups or gross adjustments where the rate of response is part of the information.

Figure 8-31. Multiple diaphragm flexible element coupling with a strain gauge type torque meter. (*Courtesy of Zurn Industries, Inc., Mechanical Drives Division*)

Rod Drop Monitor

Rod drop monitoring is a monitoring system used on horizontal reciprocating compressors. Most horizontal reciprocating compressors use wear bands also known as rider rings to support the piston in its cylinder. The wear bands are generally made of a plastic material which in time abrades and requires replacement. Should the wear go unchecked, contact with the cylinder wall by the piston will occur with the potential for serious damage to the cylinder and the piston. Also the packing may be damaged which could lead to a gas release.

Compressors should be stopped periodically to check for wear. The inspection may occur every few months. To accomplish this the compressor must be stopped, blocked in, and purged if the gas is hazardous. A valve is removed from the cylinder and a feeler gauge inserted through the valve opening. A measurement is taken between the lower side of the piston and the cylinder wall.

An alternate method of measurement is the use of a rod drop monitor. The monitor consists of a proximity probe located in a vertical position in the packing case. Another probe is used to develop a once-per-revolution timing pulse. The probe gap voltage is read with a remote monitor. The timing pulse is used to gate the probe reading to allow for an instantaneous rod position reading. The advantage of taking the reading at a discreet point in the stroke is that the effect of scratches or coatings on the rod can be minimized. Also, the reading can be timed to occur at or near bottom dead center (BDC). Here the dynamic forces have a minimal effect. Also the location of the center of the piston is known at that instant. With this information, the monitor can make a geometric correction to rod position reading providing a display that indicates the amount of wear band wear. Figure 8-32 shows a diagram of the rod drop sensors.

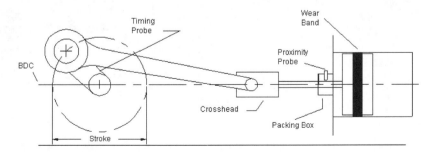

Figure 8-32. Line drawing showing instrumentation locations for a rod drop monitor.

The rod drop monitor is recommended for all non-lubricated compressor applications, because there is no lubricant to act as a buffer to prevent piston-to-cylinder contact on the loss of the wear band. Lubricated compressors handling gases with traces of water or gas components that can degrade the local lubricant are candidates. Hydrogen compressors should be considered for monitors because hydrogen is a difficult gas in itself and may contain trace quantities of water. While sweet gas compressors, as are found in pipeline service, would normally not be considered a problem, the rod drop monitor may be used to signal a loss of lubricant and the compressor can be shutdown before damage can occur.

Molecular Weight

This is another tough one, as there are no direct molecular weight meters. In many plants molecular weight can be obtained indirectly and in most instances does not change all that rapidly. Actually, gas analysis is a more fundamental piece of data. There are "on stream" gas chromatographs but they are quite expensive. With an equation of state, which is geared to handle a broad line of gases by having a large stored base of gas constants, the gas analysis is the input of choice. Getting all this together into a compressor analysis program is somewhat of a chore, but is in the realm of reason.

A control room readout of molecular weight or, more correctly, a control room computer readout of molecular weight coming in at real time is currently available for the newer computer controlled plants.

Vibration

The vibration equipment available is probably best approached by discussing the basic sensors and what they do or don't do. Because monitors now run from A to Z, the reader can obtain vendor literature or dig into the multitude of papers and articles on the monitoring, hardware, software, and, in some cases, philosophy of instrumentation.

Vibration Sensors

Sensors are divided into two general classes:

1. Seismic transducers
   (a) Accelerometers
   (b) Velocity Transducers
2. Proximity Transducers

Sensors respond either to amplitude or to displacement of the vibration. Seismic sensors are also frequency sensitive. Figure 8-33 shows a comparison of various methods of sensing vibration. Velocity sensors have a sensitivity directly proportional to frequency and amplitude;

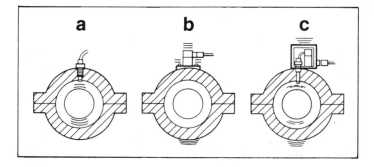

Figure 8-33. Three methods for measuring shaft vibration. (*Courtesy of Bently Nevada*)

whereas, accelerometers show a sensitivity directly proportional to the amplitude as the square of the frequency. It is appropriate to use seismic sensors when the frequency of vibration is high, since high acceleration forces are involved at high frequencies and the sensitivity of seismic sensors increases with the frequency. Conversely, if the frequency is low, the proximity sensor would be favored (see Figure 8-34).

Often, two or more types of sensors may be used in conjunction with each other to give more "visibility" to what is going on. Monitoring of a turbine generator is a good example. Information may be collected about shaft thrust, eccentricity, rpm, bearing wear, gearbox wear, and others for maintenance and protection analysis.

Sensors may be used with a number of monitoring devices. Most sensors give a voltage output signal proportional to the vibration level. The output signal is interpreted as either peak-to-peak voltage, peak voltage, or rms voltage signals. Figure 8-35 illustrates these values on a sine wave.

Seismic Sensors

Accelerometers

Characterized by high frequency response, accelerometers are compact and rugged, ideal for mounting on machinery cases, foundations, piping, etc. Applications to gear trains and rolling element bearings are typical.

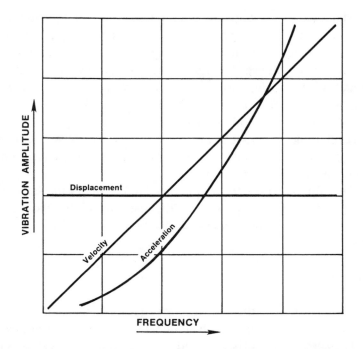

Figure 8-34. The relationship of displacement, velocity, and acceleration to vibration amplitude and frequency.

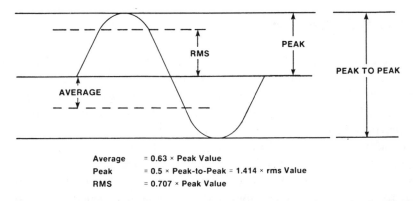

Figure 8-35. The relationship between peak, peak-to-peak, and average amplitudes for a sine wave.

A crystal material is excited by the force imposed on it by an internally mounted mass. A voltage is produced by the crystal proportional to acceleration. This voltage is then amplified by a charge amplifier type signal conditioner from whence the signal can be transmitted long distances (1,000 feet is not uncommon) to the monitor/readout unit. It is calibrated in terms of gravitational units (g), which are proportional to force. Force is one of the most reliable indicators of equipment distress.

The accelerometer output is measured in terms of pico coulombs per gravitational unit. The nominal signal level output of the charge amplifier/signal conditioner is 100 mv/g.

The size of mass within the accelerometer determines the self-resonant frequency of the sensors. The smaller the mass, the higher the frequency. Accelerometers are usually operated in a range below this self-resonant frequency.

Velocity Sensors

Since acceleration is the second derivative of displacement, a piezoelectric accelerometer sensor with an integrator becomes a velocity transducer. This arrangement is gradually superseding the self-generating moving-coil velocity sensor (where a coil of wire moves relative to a magnetic field).

The signal levels from these various velocity sensors are comparable. An advantage of the self-generating moving coil type is that no excitation voltage is required to drive it. However, it could be affected by high magnetic fields generated around heavy electrical equipment, a problem that the accelerometer type is immune to.

Proximity Sensors

Proximity sensors are non-contacting devices that measure the relative distance between the probe tip and the conductive surface that it is observing. In operation, a modulator/demodulator generates a high frequency RF signal that is then applied to a coil on the tip of the probe. The signal is radiated into the area surrounding the probe tip. If there is no conductive material within the range of the RF field, the entire signal is returned to the modulator/demodulator. As a conductive material begins to intersect the RF field, eddy currents are set up in the conductive material, resulting in an energy loss and, consequently, a decreasing signal return to the modulator/demodulator. At some point approaching zero gap, the modulator/demodulator drops to zero.

For most applications, the modulator/demodulator is calibrated to a scale of 200 millivolts per mil. This means that for each one mil of gap change, there will be a corresponding 200 millivolt change. Five mils of gap change will produce a one volt change in output of the modulator/demodulator.

If the observed surface is moving, the modulator/demodulator output varies in direct proportion to the peak-to-peak movement of the observed surface. Having a flat frequency response from DC to 10,000 Hz, the transducer is able to accurately follow motion at frequencies in excess of those typically encountered.

There are several types of information available from each probe and modulator/demodulator. Average gap, or position, data are available from the DC output. Thus, the system can be used for measurement of average position, eccentricity, concentricity, thickness, etc.

The AC component of the modulator/demodulator output is an indication of dynamic motion, or vibration. This signal provides data relating to the peak-to-peak amplitude, frequency, and form of the dynamic action of the observed surface.

Radial Shaft Vibration

Two vibration sensors, mounted at 90° to the axis of a metallic shaft and 90° apart around the shaft, can provide orbital information and display vibration eccentricity (see Figure 8-36).

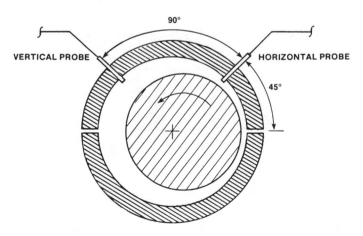

Figure 8-36. Radial shaft vibration probes.

Axial Shaft Motion

By using the positional information available from a proximity probe, another valuable parameter can be measured. The axial position or axial shift detection can be added to the monitored information. Probes are generally installed at the shaft end in pairs to provide redundancy. The probes, preferably, should be sensing the axial position from two different surfaces. A typical arrangement can be seen in Figure 8-37.

Continuous vs. Periodic Monitoring

There are two types of condition monitoring: continuous and periodic. Continuous monitoring, as its name implies, examines measurements taken on a continuous basis. Periodic monitoring is based on measurements taken at regular time intervals.

With the advent of computerized real-time systems, the distinction between continuous and periodic condition monitoring must be modified. Though technically periodic, a scanning system operating fast enough to protect against catastrophic failure is considered continuous. Most people consider one second or faster scan rates as continuous. A scan rate of one second is defined as monitoring each point once each second.

The argument that an analog system dedicated to each measurement parameter provides better protection than a scanning system simply is not true. An analog system has inherent time delays that result in finite

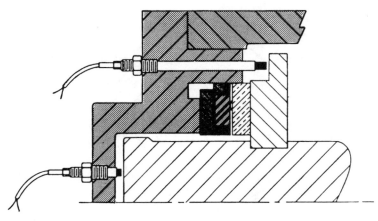

Figure 8-37. Typical axial proximity probe installation. (*Courtesy of Turbocare, A Division of Demag Delaval Turbomachinery Corp. Houston facility*)

response time to a step increase. Some users have demanded computer systems with exceptionally high scan rates to duplicate the response they thought they had with analog systems. In actuality, a relatively low scan rate of 1–2 seconds is normally sufficient to duplicate the performance of analog condition monitoring systems. Periodic monitoring once meant taking measurements manually at intervals that varied from one week to three or six months. Periodic monitoring can now be redefined as measurements taken at intervals that are too long to provide protection against a sudden failure. In general, measurements taken at intervals longer than 5 seconds can be considered periodic.

What are the advantages and limitations of continuous and periodic monitoring? Continuous monitoring requires a relatively large initial expenditure. But once installed, cost of operation is quite low. Periodic monitoring has a low initial cost, but is manpower intensive and therefore has a relatively high continuing cost.

Continuous Monitoring

Continuous monitoring is necessary on critical machines where problems can develop rapidly and have severe financial consequences. Typical machines in this category are unspared process compressors. Remotely located machinery such as pipeline gas compressors also require continuous monitoring. Also, continuous monitoring may be dictated by safety considerations. Even though the cost of a failure is small, machines should be continuously monitored if a failure will result in hazards to personnel. Figure 8-38 depicts a typical continuous monitoring system.

Periodic Monitoring

Periodic monitoring is typically applied to less critical machinery where advance warning of deteriorating conditions will show a positive return on investment. Another form of periodic monitoring is the detailed analysis of dynamic data from critical machines. Signals from sensors installed for continuous monitoring of overall vibration level, together with additional temporary sensors, are spectrum analyzed and compared against previously accumulated data. This is sometimes referred to as signature analysis. In some cases, changes in vibration signatures will provide earlier warning of deterioration than changes in overall vibration level.

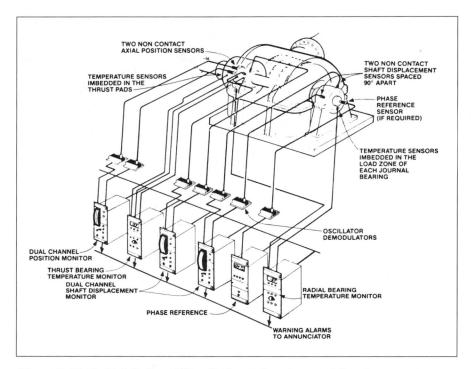

Figure 8-38. Typical continuous monitoring system for a centrifugal compressor. (*Reprinted by permission of* Sound and Vibration)

Control

Compressors are normally one part of an overall process train. When a compressor is selected for use in a process, consideration must be given to the required operating range. It will become quickly apparent when a compressor with a fixed capacity is installed in a process with a variable capacity requirement. Modifying capacity of the individual compressors was discussed in the earlier chapters.

A few basics on manual and automatic control will be covered to familiarize the user with some of the considerations necessary to acclimate a compressor to the process environment.

Control systems for process compressors become an absolute necessity whenever deviations from a single operating point occur in the process or system. These systems consist of a mechanism that can sense performance and, by making a calculation, adjust the process controls with satisfactory speed, accuracy, and stability.

Almost all compressor operations will be more satisfactory if a control system is included. Hence, it should always be considered during the original planning phase. Manual operation will result in an operating pattern that differs considerably from a controlled one. Proper evaluation of patterns is required before the need for a control system can be established.

Figure 8-39 illustrates the operating characteristics for a simulated manual and automatic control. In any specific case, the process engineer and the compressor design engineer must cooperate in establishing some of the needs for and capabilities of the compressor.

The design of a control system is a job of considerable magnitude and a general understanding of the engineering approach is necessary to prevent underestimation of the work to be done.

Analysis of the Controlled System

An automatic control system is made up of four blocks: (1) process, (2) transducer, (3) controller, and (4) final element.

The process block produces the required control parameter. The transducer block converts this signal and transmits it to the controller block, which sends the signal to the final element where it is translated into a usable parameter of the proper type to modify the process. Hence, the process block can be described as being derived from the complete physical system.

In the same manner, the transducer block can be defined as a signal selector-converter. The main function is that of interpreting a signal to the controller. The controller block compares this signal to a reference signal called the set point, and issues a corrected signal to the final element. The final element physically changes the process as directed by the controller block. The determination of the final element depends upon the control parameter selected. The elements may be suction throttle valves, discharge bleed valves, governor speed changers, inlet guide vanes, slide valves, variable volume pockets, movable stator vanes, or other compressor elements.

These basic blocks may be arranged to control various parameters as shown in Figure 8-40. Actual applications of the various block arrangements should help in understanding them.

Pressure Control at Variable Speed

In a specific application, a large, mixed hydrocarbon gas system may be set up with a turbine-driven compressor as shown in Figure 8-40(a).

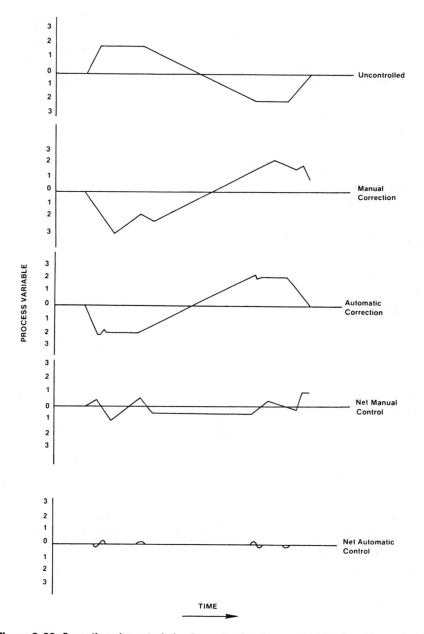

Figure 8-39. Operating characteristics for a simulated manual and automatic control.

The process demands that a constant pressure be held on the header at various loads (flow). Specifically, the control system operates as follows:

1. The pressure transmitter (transducer block) senses the process pressure. It converts this signal to a signal proportional to the process pressure and sends it to the pressure controller.
2. The pressure controller (controller block) amplifies the transmitter signal and sends a modified signal to the final element. Depending on the system requirements, the controller block may include additional correction factors, integral and derivative (reset and rate). This is called a three-mode controller.
3. The final element in this case is speed control. This mechanism varies the turbine-governor speed setting over a predetermined range.

A pressure rise in the reactor would cause an increase in process pressure above the set value, send a signal to the governor, and thus, reduce the speed, lower the flow, and thereby, maintain the desired system pressure.

Volume Control at Variable Speed

A turbine-driven screw compressor might be applied to a catalyst regeneration process. The nature of the process then will require constant volume control to maintain a required output temperature in the regenerator. The arrangement is shown in Figure 8-40(b) and would occur as follows:

1. The flow transmitter (transducer block) senses the flow element differential pressure, converts this signal to a signal proportional to the process flow, and sends it to the flow controller.
2. The flow controller (controller block) amplifies the transmitter signal and sends a modified signal to the final element. The three-mode controller will probably be used.
3. The final element is speed control, which is accomplished by a mechanism that varies the turbine-governor speed setting.

An increase in flow over set point would cause a signal to reach the governor and reduce the speed to maintain the desired system flow.

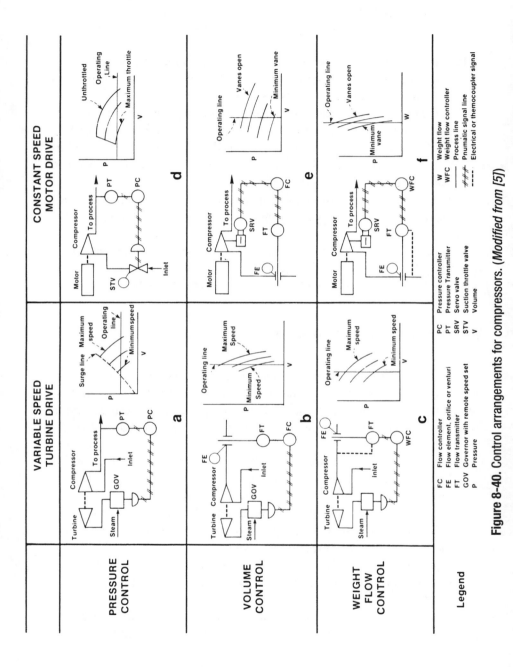

Figure 8-40. Control arrangements for compressors. *(Modified from [5])*

Weight Flow Control with Variable Stator Vanes

A turbine-driven axial compressor is used to supply air to a blast furnace. The furnace requires a certain constant weight flow, depending upon the size of furnace and the size of the charge. An example of a possible arrangement is shown in Figure 8-40(c) and can be described as follows:

1. The compensated flow transmitter determines the process flow. It converts this quantity to a signal that is proportional to the process flow and sends it to the flow controller. The transmitter could be a pneumatic device using a venturi primary element, with compensation for pressure by a pressure element and compensation for temperature by a thermocouple. The output would be a pneumatic signal that is proportional to weight flow.
2. The flow controller amplifies the transmitter signal and sends a modified signal to the final element. Rest and rate correction factors may also be required.
3. The final element is speed control. This is accomplished with a mechanism that varies the turbine-governor speed setting.

An increase in weight flow over the set point would cause a signal to reach the governor, which would reduce the turbine speed to maintain the desired flow.

Pressure Control at Constant Speed

A plant air system could resemble Figure 8-40(d) where a motor-driven centrifugal compressor is used.

1. The pressure transmitter senses the process pressure and converts it to a signal that is proportional to it.
2. The three-mode pressure controller amplifies the transmitter signal and sends a modified signal to the final element.
3. The final element is a suction throttle valve that reduces the flow of air into the compressor.

A process pressure increase over a set value would cause a signal to reach the suction throttle valve and would close the valve in order to reduce the inlet pressure.

Volume Control at Constant Speed

A motor-driven compressor is used as an oxidation system for a chemical unit. The control scheme for this arrangement is shown in Figure 8-40(e). Description of the control blocks follows.

1. The flow transmitter senses the process flow, converts this to a signal that is proportional, and sends this signal to the flow controller.
2. The flow controller amplifies the transmitter signal and sends a modified signal to the final element.
3. The final element is the compressor guide-vane mechanism. The guide vanes are adjusted by means of a positioning cylinder. This cylinder is operated by a servo valve that receives a signal from the flow controller.

Here, an increase in flow above the set point causes a signal to reach the final element, which will result in the closing of the guide vanes to decrease flow.

Weight Flow Control at Constant Speed

Motor-driven reciprocating compressors are sometimes used in tonnage oxygen plants. To maintain a uniform output, the plant must be supplied with a constant weight flow of air. As ambient conditions change weight flow, a control system as shown in Figure 8-40(f) can be used to keep the plant supplied with the proper quantity of air. The necessary steps in this system are:

1. The compensated flow transmitter senses the process weight flow. It converts this signal to one that is proportional to the process flow and sends it to the weight flow controller.
2. The weight flow controller amplifies the transmitter signal and sends a modified signal to the final element.
3. The final element is the compressor guide-vane mechanism. The variable clearance points are adjusted by means of a positioning cylinder that is operated by a servo valve in response to a signal from the flow controller.

An increase in flow over the set point causes a signal to reach the final element and will result in the closing of the guide vanes to decrease flow.

Anti-Surge Control

Surge is part of the inherent operating characteristics of centrifugal and axial compressors. It is probably one of the most misunderstood of all the characteristics of these compressors. Regard for surge seems to vary from complacency to terror based sometimes on fact and experience and other times on rumor. While surge was described in Chapters 5 and 6 for the centrifugal and axial compressor, a quick review to introduce anti-surge control is included here.

As flow is reduced in an axial or centrifugal compressor, there is a minimum limit when the geometric form of the internal blading can no longer move gas forward through the machine in a stable fashion. Unfortunately, the flow cannot remain static, so if it no longer moves forward, the residual volume in the machine moves in the reverse direction. The audible sound, normally associated with surge, comes from the reverse-flowing gas meeting forward-moving gas in the inlet. The noise is similar to thunder being caused by pressure waves of air colliding due to the local heating from lightning.

Is the surge phenomenon harmful? This can be compared to a wasp or bee sting to a human. If the human is healthy and not allergic, there is only temporary discomfort but no permanent damage. If the human is allergic, has other health problems, or is hit by a whole nest of wasps or bees at one time, we have another story. Similarly, a low-stressed, conservatively designed machine with no large amount of process temperature sensitivity can withstand surge occasionally with the only problem being the disturbance in the process from unsteady flow. If, however, the machine has highly stressed parts or exhibits other marginal design parameters, then we have a problem. Likewise, if the gas is sensitive to large temperature rises, where either process gas decomposition or temperature reactions can occur, we have an allergy. Finally, if any machine is surged long enough (wasp nest), problems are likely to occur. To understand the mechanics, the following occur in surge.

1. Temperature—during the back flow, gas at discharge temperature is introduced to the inlet. If more than one cycle occurs, the same gas is reheated and returned, getting hotter with each cycle.

2. Load—during the back flow the shaft torque is reduced, then restored with feed forward giving a torsional pulse with each cycle.

3. Component stress—during the back flow and forward reversals, all the components involved with the gas propulsion are loaded and unloaded, placing blading in a cyclic loading mode.

By evaluating the three physical parameters above as they apply to any compressor, one may anticipate the possibilities for problems.

Whenever normal operating limits or startup require the compressor flow to be less than the minimum flow for surge, an automatic anti-surge control should be considered. Process variables that affect the surge flow limit are flow, differential pressure, inlet temperature, molecular weight, other gas properties, and speed. To evaluate the possibility of surge, the variation of the above limits must be considered together with a compressor performance curve, normally polytropic head plotted against inlet flow in actual operating conditions. This curve can also be plotted in terms of pressure and flow for different inlet temperatures and flows if the variations are not too great. The head capacity curve, while not as straight forward, is somewhat more useful if many variations must be considered. Figure 8-41 is a centrifugal compressor curve. Included is the unstable region normally not shown.

To control or prevent surge, gas must be bypassed around the compressor in order to increase inlet flow. On a gas turbine or in a multistage compressor, the bypass may have to come from more than one point in the compressor. In air service with atmospheric inlets, bypass means dumping back to atmosphere. In gas application or closed systems, bypassing must be done by returning discharge gas to the inlet with provision made to remove the heat of compression either with a shell and tube exchanger or by direct contact such as flashing liquid to vapor.

There seems to be one problem with anti-surge control, which is its name. Both surge control and anti-surge are used. Anti seems to be what the control is supposed to do, so it will be used here.

The best anti-surge control is the simplest and most basic that will do the job. The most obvious parameter is minimum-flow measurement, or if there is a relatively steep pressure-flow characteristic, the differential pressure may be used. The latter parameter allows for a much faster response system, as flow measurement response is generally slow; however, the speed of response need only be fast enough to accept expected transients. One major problem with the conventional methods of measurement and control is the need to move the set point for initiation of the control signal away from the exact surge point to allow some safety factor for control response time and other parameters not directly included

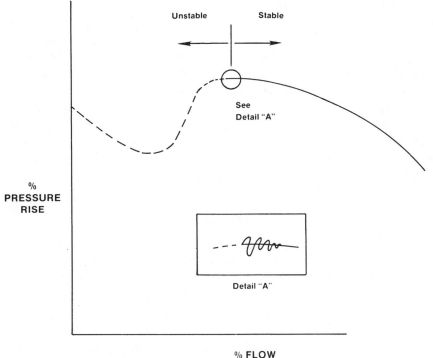

Figure 8-41. Unstable flow region of a centrifugal compressor curve.

in the surge control. A proper speed of response and other more exact variable compensations are ways of narrowing this margin, which represents wasted power and wasted capacity.

Variables such as speed, temperature, and molecular weight can be included as part of the compensation. These may be linearized and put into the system as analog weighting functions or, with computer control, be put into the equation in a more exact fashion. Inlet temperature and speed present little problem as signals proportional to these parameters are commonly measured. A well-designed, properly compensated anti-surge control with a good response can probably function to a 5% flow margin on a conventional centrifugal compressor. On an axial compressor, differential pressure is normally used and a limit can only be defined for a given application. Needless to say, close control can be held. The computer controller also adds intelligence to permit the system to selectively narrow the margin based on plant conditions. With the high cost of power there has been a desire to operate without surge margin or to the level of incipient surge. Incipient surge is a narrow margin between full stability but before full complete flow reversal. This is represented at detail A of Figure 8-41. Various methods have been devised, such as measuring stage differential pressure and looking at the unsteady component of the signal. The unsteady component does increase with the onset of surge; however, on multi-impeller compressors, either all impellers must be instrumented or a guess must be made, based on probability, as to which impeller is the one most likely to go into surge first. Other methods have been proposed that require internal instrumentation which, however, have the same limits just presented, as well as the problem of simply existing in a hostile environment.

References

1. API Standard 614, *Lubrication, Shaft-Sealing, and Control Oil Systems for Special-Purpose Applications,* Third Edition, Washington, D.C.: American Petroleum Institute, 1992.

2. API Standard 670, *Vibration, Axial Position and Bearing Temperature Monitoring System,* Third Edition, Washington, D.C.: American Petroleum Institute, 1993.

3. Gilstrap, Mark, "Transducer Selection for Vibration Monitoring of Rotating Machinery," *Sound and Vibration,* Vol. 18, No. 2, February 1984, pp. 22–24.

4. Mitchell, John S., "How to Develop a Machinery Monitoring Program," *Sound and Vibration,* Vol. 18, No. 2, February 1984, pp. 14–20.

5. Brown, Royce N., "Control System for Centrifugal Gas Compressors," *Chemical Engineering,* February 17, 1964, pp. 135–138.

6. API Standard 671, *Special-Purpose Couplings for Refinery Services,* Second Edition, Washington, D.C.: American Petroleum Institute, 1990, Reaffirmed 1993.

7. Boylan, William, *Marine Application of Dental Couplings,* Paper 26—1966, Society of Naval Architects & Marine Engineers, May 1966.

8. Bloch, Heinz P., *Less Costly Turboequipment Uprates Through Optimized Coupling Selection,* Proceedings of the 4th Annual Turbomachinery Symposium, Texas A&M University, College Station, TX, 1975, pp. 149–152.

9. API Standard 613, *Special-Purpose Gear Units for Refinery Services,* Fourth Edition, Washington, D.C.: American Petroleum Institute, 1995.

10. Brown, Royce N., *An Experimental Investigation of a Pneumatic Closed Loop Anti-Surge Control for Centrifugal and Axial Flow Compressors,* Master's Thesis, University of Wisconsin, Madison, Wisconsin, 1966.

11. Bloch, Heinz P., "Use Keyless Couplings for Large Compressor Shafts," *Hydrocarbon Processing,* April 1976, pp. 181–186.

12. Feltman, P. L., Southcott, J. F., and Sweeney, J. M., *Dry Gas Seal Retrofit,* Proceedings of the 24th Turbomachinery Symposium, Texas A&M University, College Station, TX, 1995, pp. 221– 229.

13. Schultheis, S. M. "Rider Band Wear Measurement in Reciprocating Compressors," *Orbit,* Vol. 16 No. 4, Bently, Nevada, December 1995, pp. 12–14.

14. API Standard 677, *General-Purpose Gear Units for Petroleum, Chemical, and Gas Service Industries,* Second Edition, Washington, D.C.: American Petroleum Institute, 1997.

9

Dynamics

Introduction

One of the basic problems with any machine is unwanted vibration. Because the machine has dynamic rotation, or rotation plus reciprocating action, vibration will be present primarily because of unbalance. In a reciprocating compressor, the unbalance can be controlled by design, if the number of cylinders can be selected and properly arranged relative to the process requirement. In the rotary compressors, including the dynamic types, the designer has more control, as these compressors have no inherent shaking forces. While unbalance is a major source of vibration in compressors, careful attention to the rotating element in the entire design and manufacturing cycle and proper field maintenance can solve many of the problems.

While unbalance may be the forcing function, resonances within the compressor are another factor in vibration. The resonances take almost insignificant exciting forces and magnify them to a level to cause problems. These resonances must be identified and somehow controlled. Tuning them out of the operating range is preferred. If this proves impractical, as a last resort, damping can be used to minimize the effect of resonance.

Other sources, such as compression temperature rise, driver-induced vibration, or component problems (bows) can contribute to the machine shaking. These must be treated as they occur. As a minimum, care must be used to understand the nature of the sources to keep them from inter-acting with the resonant frequencies of the compressor. The best remedy is to stop the excitation at the source. If this is not possible, selective tun-ing and proper application of damping must be used.

If the vibrations are left unchecked, damage to the compressor will occur, such as premature failure of bearings and seals, or packing rubs, and, in extreme cases, major component fatigue and failure. Secondary effects may result from the shaking forces being transmitted through the frame to the foundation. Damage may result to the foundation, particular-ly with reciprocating compressors. Interaction with other equipment in the area is also a concern. If there are people working in the area of a shaking compressor, there may be motion sickness or other physiological effects. Because of the potential for damage or problems, a basic under-standing of the nature of, and possible remedies to, the more common vibrations is necessary.

Balance

Basics

Before the subject of balancing can be properly discussed, unbalance should be described. An unbalance exists when the mass center is dis-placed from the rotating center. Figure 9-1a shows a massless disk with a finite mass, m, located at radius, r, from the center of rotation. If the disk rotates at an angular velocity of ω, the force, F, exerted by the finite mass, m, is

$$F = m\,r\,\omega^2 \qquad\qquad (9.1)$$

where

$\omega = 2\,\pi\,N$ (shaft speed), rad/unit time

Equation 9.1 is the basic equation for unbalance. For such a simple arrangement, balancing (referred to as *static balance*) could be done by placing the shaft on knife edges. Initially, the location of the mass would rotate the disk gravitationally until the mass was on the bottom. If

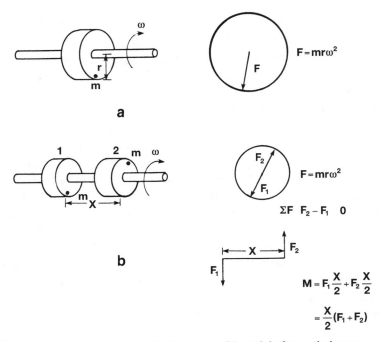

Figure 9-1. A one mass (a) and a two mass (b) model of an unbalance.

weights were added opposite the location of m, until the mass of the weights equaled the mass, m, and were located at the same radius, r, the disk would be motionless, regardless of the position of the angular placement of the disk on the edges.

In Figure 9-1b, two masses are located exactly 180° apart, but on two different weightless disks. The two disks are separated by a distance x. If this rotor were placed on the knife edges, it could be placed in any angular position desired, and it would not rotate. The rotor is now in static balance. The diagram also shows that the forces, when viewed from the end of the rotors, including the summation of forces, appear to be zero. However, if the figure is viewed longitudinally, the vectors, F, are separated by the distance, x. If a summation of moments is made, a moment unbalance is revealed. This type of balance is referred to as a *dynamic unbalance*. Correcting this type of unbalance requires spinning the rotor as in a balance machine. This method takes the name of *dynamic balancing*.

The balancing of a rotor certainly has changed from a "black art" with the introduction of more sophisticated equipment. While it is now much more scientific, it certainly is not a job for the novice. For several obvious reasons, the various balance methods will not be detailed. Rather, the effects of balance results are of more interest to the average person working with the compressor.

Unbalance

Figure 9-2 is a chart showing the residual unbalance and the resulting force in pounds. It can be generated from Equation 9.1 using the unbalance in inch-ounces at several different speeds. The real problem is that there has not been any practical method of relating unbalance to rotor

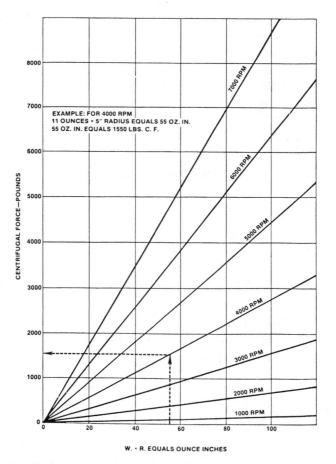

Figure 9-2. Residual unbalance and the resulting force in pounds. (*Courtesy of A-C Compressor Corporation*)

vibration. The more sophisticated damped unbalance response computer programs tend to close the gap, but to someone at a balance machine, these programs are somewhat remote. Also, the accuracy is dependent on the program and the skill of the user. Like the balance machine, it is not simply a matter of reading an answer on a printout.

Since compressors had to be balanced, and though acceptable vibration levels had been determined, the two still elude each other. Another method was devised by the field personnel in charge of keeping the compressors running. By trial and error, the 0.1 g criterion evolved. This simply says that if the force due to the residual unbalance is less than 10% of the rotor weight, the rotor will no longer respond to the unbalance. If Figure 9-2 is used, the value of 55 inch-ounce yields 1,550 pounds of force at 4,000 rpm. Such a rotor might weigh 15,000 to 20,000 lbs. This would be a very large compressor. To exactly meet the 10% criteria, the rotor would have to weigh 15,500 lbs. To the end that a balancer did not want to use a curve and back-calculate, a more straightforward method was developed. Equation 9.2 was used in API Standard 617, fourth edition and earlier standards

$$U = 56,347 W/N^2 \tag{9.2}$$

For a comparison, the example can be evaluated

$$U = 56,347 \times 15,500/(4000)^2$$

$$U = 54.6 \text{ in-oz}$$

This value falls near the .1g limit, which just happens to be the basis for the equation.

Balance machines rate their sensitivity in eccentricity, e, where e is the apparent center of gravity shift due to the unbalance. To obtain e, simply divide the inch-ounce of unbalance by the rotor weight in ounces, in this case,

$$e = 54.6/(15,500 \times 16)$$

$$e = 220 \ \mu in.$$

This balance level can be readily achieved.

In the API Standard 617, fifth edition, the allowable value for residual unbalance was changed to

$$U = 4 \text{ W/N} \tag{9.3}$$

where

U = allowable unbalance, in.-oz.

W = rotor overall weight or journal static weight (if unbalance taken for each end of rotor)

N = operating speed, alternatively maximum continuous speed, rpm

The equation is based on Mil-Std-167 (SHIPS) dated 1954, a Navy specification [2]. The basis for the change was that the allowable residual unbalance for rotors operating under 14,000 rpm was lowered, while it permits higher residual unbalance values for rotors above 14,000 rpm. The balance is more realistically achievable on the balance machine.

If the previous example is applied in Equation 9.3,

$$U = 4 \times 15,500/4000$$

$$U = 15.5 \text{ in-oz}$$

The value is lower than the .1g would call for, but most users would prefer to install a compressor rotor balanced with some residual unbalance margin.

For comparison, the eccentricity is calculated for the value from Equation 9.3.

$$e = 15.5/(15,500 \times 16)$$

$$e = 62.5$$

Balance machines are limited to the 25 to 50 μin sensitivity level, indicating that Equation 9.3 for the example will take the balance close to the limit.

The international standard ISO 1940 is based on a linear equation similar to Equation 9.3 [3]. The specification calls for balance quality level by Grade Numbers with the lower the number the lower the permissible unbalance value. Equation 9.3 when restated in the ISO terms would require a Grade level of 0.66 or G 0.66 in ISO terms.

Figure 9-3 is a plot of the two equations for residual unbalance and includes two common ISO grades for comparison.

When the sensitivity in microinches in considered, it can be seen why, when a component is mandrel balanced, the required runouts will be very small. If the runout is larger than the balance machine sensitivity, the component could actually be unbalanced. Since mandrels do have runout, the alignment of the balanced component with shaft runout is desirable; however, even here, the differential between the mandrel runout and shaft may be high enough to miss the desired level—a reason for trim balance. It is also probably the best argument for the progressive stacking and balance as recommended by API.

Balance Methods

There are basically three scenarios used in balancing. Within these three are various methods.

Total Unbalance 1kp Rotor

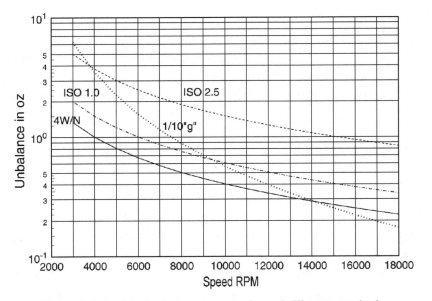

Figure 9-3. Residual unbalance, comparison of different standards.

Shop Balance Machine

This is where most balancing of rotors takes place. The machines range from heavy, direct-driven to the light, belt-driven type. There are numerous shop-made versions to balance a particular type of rotor, especially for the high speed, very light rotors. The machines spin the rotors at a speed considerably less than operating speed. The force generated by the unbalanced rotor, recalling the simple force equation is measured. Because of the range of rotor weights, it is not difficult to see why several sizes of balance machines are needed. Also, the engineer's worry of balancing a light rotor in a too-large machine is well-founded because the sensitivity changes with the construction. The forces generated by the small rotor on the big machine do not cause much reaction and, therefore, lose sensitivity.

The reaction of the bearings is measured on the machine dials, or on the CRT, if a computer-assisted machine is used. Balancing is not considered much of a problem if the planes in which the unbalance is located are known, as shown in the simple two-disk rotor used at the beginning of this section. However, consider a centrifugal compressor having eight impellers, all with cover discs, a balance piston, thrust collar, and coupling hub. It is extremely difficult to look at bearing readouts and decide which plane or even which impeller, notwithstanding that a given impeller cover or back shroud may be the item needing correction. This is the dilemma of the service shop.

For new rotors, where the elements have not yet been put on the rotor, other techniques can be used. First, the components can be individually balanced on a precision mandrel. Precision means that the runout is a few tenths of a mil (.001 inch). The runout high spot should be scribed on the mandrel. The new component now can be reasonably well-balanced. As the component is removed from the mandrel, the mandrel mark should be transferred to the component. When all the components are completed, the shaft is checked for runout. The high spot should be marked. As the components are stacked onto the shaft, the marks on the shaft are aligned with those transferred to the component. This works well with keyless rotors (no key between shaft and component). Experience has shown that in most cases with keyless rotors when the stacked rotor is put in the balance machine and checked, the residual unbalance is within the acceptable tolerance. If not, the rotor must be unstacked and the problem located. It must be remembered, however, if the components were properly balanced and the rotor comes out with unbalance, there must be a problem. To begin to balance the assembled rotor means unbalancing some components and covering up a problem. It should also be remembered that balancing in the wrong plane will permit operation with two unbalanced couples canceling each other out at the shaft ends, giving no apparent unbalance. With the couples still in the rotor, there are residual moments that will add unwanted stress to the shaft. If taken to the extreme, shaft breakage could result.

For rotors with keys, an alternative method also endorsed by API, is *progressive balancing*. In this method, the components are stacked onto the rotor one or two (but never more than two) at a time. The balance correction may only be made to the element added. When the rotor is completely stacked, only very minor trim balance should be required. If more balancing is indicated, unstacking is required to prevent compromising already balanced elements. Since most balancing is done using clay to determine the amount of unbalance, and then corrections performed by grinding material from the element, a poor job will leave a very sad looking element. More material in another location will need removal to correct the errors of the previous job.

With both balance methods cited, unstack if things appear to go wrong; don't grind a balanced impeller. One common problem is the manner in which a component is installed. The components are fitted to the shaft with an interference fit, even when keys are employed. To get the component on the shaft, the bore must be dilated. This is normally done by heating the component. It is tricky, at best, to get the bore to open evenly. It is even more difficult to get the cooling to take place uniformly. If it does not, the component becomes cocked and either locally distorts the shaft (bows) or has excessive run out. This can be detected before the balance step, but because humans are involved, not everything that is supposed to happen will happen. If the balancer is well-disciplined, he will detect the problem, but if not, the component gets a "butcher" job.

High Speed Balancing

The at speed balance (high speed balance) facility was originally used in Europe and is a rather sophisticated and costly installation. Time on the machine is expensive, partly due to the owner having a large investment to amortize and partly because the machine requires more than just an operator.

The installation consists of a chamber large enough to house the rotor and bearing pedestals, with ample room for the technicians to add the trial weights and do the grinding necessary. The rotor assembly is moved into the chamber on tracks. There are other versions where the chamber opens from the top to permit an overhead crane to lower the rotor assembly. There is a large vacuum pump to remove the air for the rotor runup. The vacuum is necessary to reduce the windage and friction of the elements moving in an open area at operating speed. The horsepower to spin the rotor would be significant, and the heating in a closed chamber would prohibit such an operation unless completely evacuated. The drive must be variable-speed, and there must be a lubrication system to service the bearings of the rotor undergoing balance, drive equipment, and vacuum pumps. Normally the drive is a motor with a gear. To get a wider speed range, some facilities keep several gear sets for the different speed ranges. The rotor monitoring equipment is elaborate. It includes a computer to do the data logging and the calculations necessary to determine weight locations. To make a weight change, the vacuum on the chamber must be relieved. After the change, the chamber must be re-evacuated and the run repeated.

Why does anyone need this? Not all rotating equipment practitioners agree on the use of the at speed balance. There is one fear that high speed balancing will cover up a problem. Those in favor of the high speed balance argue that since most rotors operate above a bending critical or possibly above more than one bending critical, that balancing should take into account the rotor's mode shape at its operating speed. On the positive side, repaired rotors not given a mechanical run test may be operated at rated speed. While an oblique pro, cocked components will sometimes straighten when run at speed.

Field Balancing

The third method of balancing is generally called *field balancing,* because this is where it's often done. If it appears that major balancing is required, however, this method should not be used. Normally it is a trim, or so called touch-up balance. There is always some minor adjusting of the elements after coming up to temperature and speed. The minor movement is caused by the imperfect cooling and the failure of the component to come onto the shaft in a perfectly square condition while shrink fitting. This area of consideration, referred to as the element seating, may be enough to raise the residual unbalance, but not enough to remove the rotor and unstack, leaving more vibration than desired. On axial or centrifugal gas compressors, where there is normally no provision for field balancing, this can be time-consuming because the cover must be removed to make a trial weight change. Most balancing involves reading unbalance and phase, adding a trial weight, watching the resulting change, then, if the balancer is skilled, triangulating the last weight estimate from the vectors to bring in the correction. If all worked well, it still means two cover lifts. The method is more often used on steam turbines or air compressors where provision is sometimes made for field weights that are accessible without a cover lift. This technique is also used on a vendor test floor if it is believed that the out-of-tolerance compressor can be corrected in a less costly manner with a test floor trim balance. In a vendor shop, there are normally several skilled balancers available.

Two methods for determining the location of the unbalance vector are used. One is with portable vibration pickups and a strobe light. The pickup is attached to the bearing housing with a magnet or other temporary attachment. Numbers are painted on some rotating part, generally the coupling. The numbers identify evenly spaced marks on the circumference. If keys are used, the numbers start there to establish a physical mark. When the compressor comes up to speed, the vibration signal triggers the strobe light, theoretically at the maximum point of the vibration signal. The peak-to-peak reading of the probe forms the vector length. The numbers, frozen by the strobe light, give the vector direction. The technique proceeds as stated earlier with a trial weight.

The alternate method uses the proximity probes and an oscilloscope. A Lissajous figure is established on the oscilloscope. The orbit pattern and the keyphase mark are used to generate a vector. Weights are added or removed and the changes in the orbit are noted. Triangulation is used to anticipate the next move. For more complete information or technique, the reader is referred to a book on the subject by Jackson [1].

Reciprocating Shaking Forces

Reciprocating compressors include linear translational motion as a part of their normal operation. A single cylinder compressor will have an inherent set of shaking forces as part of its operating characteristics. The forces are resolved to reaction forces on the main bearings, where they transmit through the frame and the foundation. The purpose here will be to show the reader the basics of how these forces are generated and a general philosophy on how they can be minimized by balancing or by cylinder

Table 9-1
General Normalized Functions Describing the Dynamic Characteristics of the Crank and Connecting Rod Mechanisms

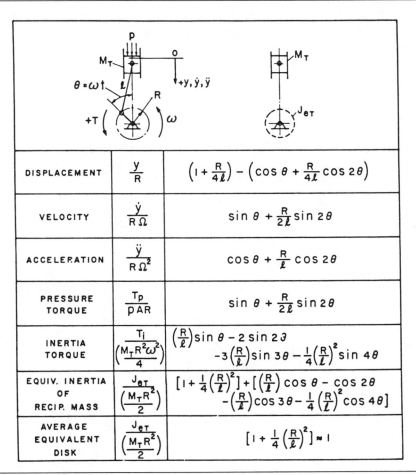

DISPLACEMENT	$\dfrac{y}{R}$	$\left(1+\dfrac{R}{4\ell}\right)-\left(\cos\theta+\dfrac{R}{4\ell}\cos 2\theta\right)$
VELOCITY	$\dfrac{\dot{y}}{R\Omega}$	$\sin\theta+\dfrac{R}{2\ell}\sin 2\theta$
ACCELERATION	$\dfrac{\ddot{y}}{R\Omega^2}$	$\cos\theta+\dfrac{R}{\ell}\cos 2\theta$
PRESSURE TORQUE	$\dfrac{T_p}{pAR}$	$\sin\theta+\dfrac{R}{2\ell}\sin 2\theta$
INERTIA TORQUE	$\dfrac{T_i}{\left(\dfrac{M_T R^2 \omega^2}{4}\right)}$	$\left(\dfrac{R}{\ell}\right)\sin\theta-2\sin 2\theta$ $-3\left(\dfrac{R}{\ell}\right)\sin 3\theta-\dfrac{1}{4}\left(\dfrac{R}{\ell}\right)^2\sin 4\theta$
EQUIV. INERTIA OF RECIP. MASS	$\dfrac{J_{eT}}{\left(\dfrac{M_T R^2}{2}\right)}$	$[1+\dfrac{1}{4}\left(\dfrac{R}{\ell}\right)^2]+[\left(\dfrac{R}{\ell}\right)\cos\theta-\cos 2\theta$ $-\left(\dfrac{R}{\ell}\right)\cos 3\theta-\dfrac{1}{4}\left(\dfrac{R}{\ell}\right)^2\cos 4\theta]$
AVERAGE EQUIVALENT DISK	$\dfrac{J_{eT}}{\left(\dfrac{M_T R^2}{2}\right)}$	$[1+\dfrac{1}{4}\left(\dfrac{R}{\ell}\right)^2]\approx 1$

Source: [4] Reprinted with permission of John Wiley and Sons, Inc.

arrangement. One compressor arrangement used is the balance opposed configuration. Cylinders are oriented in the horizontal position and located on each side of the crankshaft. See Figure 3-3. This is made more difficult with a compressor, even if the arrangement is made completely symmetrical, as the volumetric reduction due to compression requires each cylinder bore, and, therefore, piston size, to be different. This is one reason, however, lighter materials such as aluminum are preferred for the larger pistons.

Table 9-1 is a handy little chart to visualize a vertical, single-cylinder compressor and the basic functions. The functions are normalized to keep them in a dimensionless form [4]. With the following set of equations, the x and y components of the inertia forces for a single cylinder can be calculated. For the derivation, the reader is referred to references [4, 5]. Figure 9-4 depicts the generalized stage to aid in the definition of terms

$$F_{xp} = (m_{rot} + m_{rec})(r\omega^2)(\cos\theta)(\cos\phi) - (m_{rot})(r\omega^2)(\sin\theta)(\sin\phi) \qquad (9.4)$$

$$F_{xs} = (m_{rec})(r^2\omega^2/l)(\cos2\theta)(\cos\phi) \qquad (9.5)$$

$$F_{yp} = (m_{rot} + m_{rec})(r\omega^2)(\cos\theta)(\sin\phi) + (m_{rot})(r\omega^2)(\sin\theta)(\cos\phi) \qquad (9.6)$$

$$F_{ys} = (m_{rot,})(r^2\omega^2/l)(\cos 2\theta)(\sin\phi) \qquad (9.7)$$

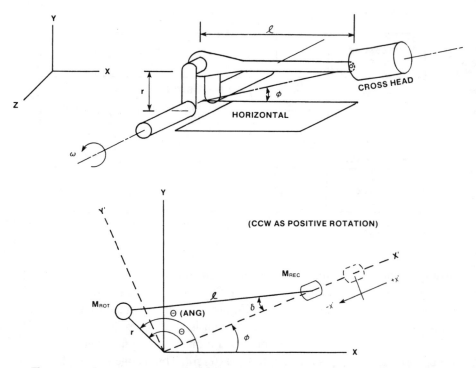

Figure 9-4. Terminology for analysis of a one-cylinder reciprocating compressor.

where

F_{xp} = primary shaking force in the horizontal direction
F_{xs} = secondary shaking force in the horizontal direction
F_{yp} = primary shaking force in the vertical direction
F_{ys} = secondary shaking force in the vertical direction
ω = angular velocity, rad/sec
θ = crankangle, also ωt
ϕ = cylinder travel angle relative to horizontal direction
t = time, sec
m_{rot} = total mass of the rotating parts, per cylinder
m_{rec} = total mass of the reciprocating parts, per cylinder
r = crankthrow, also stroke/2
l = connecting rod length

To obtain the mass of the reciprocating elements, the individual recip-rocating parts (piston, piston rod, crosshead, crosshead pin) and all other hardware such as nuts must be weighed and the weights added. One more item has to be considered. The connecting rod has both translation-al as well as rotational motion. A simple way of obtaining each part is to assume the connecting rod consists of two masses, one at each end con-nected by a massless center section. Figure 9-5 shows a direct method of obtaining these values by using two scales. The value obtained from the crank end is assigned to the rotating mass, m_{rot}.

At this point, as far as shaking forces go, the gas forces do not make a contribution. If the rod load or bearing loads are to be analyzed, the gas forces must be calculated and added vectorially to the inertia forces to

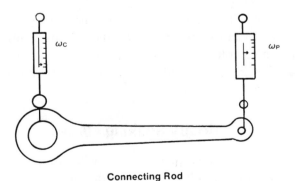

Connecting Rod

Figure 9-5. Method of obtaining weights to a two mass connecting rod model.

get a net bearing force. This is mentioned at this point because it is sometimes misunderstood.

After the forces are evaluated for each cylinder of a multistage compressor, all forces must be summed in the x and y direction. For the maximum shaking forces, the value of the crank angle, which contributes the maximum force, should be used. This involves taking the respective sine and cosine functions to their maximum. For example, a vertical cylinder will have the maximum component force at a crank angle of 0° and 180°. At this time, the horizontal components, primary and secondary, are zero.

While on occasion, the forces may cancel, the moments must also be evaluated. If the vectors are assumed to be acting at the centerline of the crankshaft at the line of travel of the piston, and the resulting vectors summed, it can be seen that on a multistage compressor, if the lateral third stage should balance vectorially, a couple still exists, which results in an unbalanced moment. If it is recalled that $\theta = \omega t$, then by referring to Equations 9.4 and 9.6, the forces (and resulting moments) are a function of θ, while being proportional to ω is acting at running speed. The secondary forces (and resulting moments) as seen in Equations 9.5 and 9.7 are a function of $2 \times \theta$, or twice ω, and thereby act at twice running speed.

The forces and moments discussed so far have included no provisions for balance. There are two methods of balancing the reciprocating compressor. One that has been mentioned is the arrangement of the cylinders. For example, a horizontally opposed design is an attempt to balance by arrangement. The other method can be used alone on a single-cylinder machine, or used together with arrangement on the multistage. Crankshaft counterweights are used, and whenever possible, they are attached to the crankshaft on either side of the connecting rod journal. Other placements are possible. Placing a mass at radius, r, for a given, ω, will generate a rotating vector whose direction is a function of the attachment angle. Intuitively, it would be 180° from the crankthrow. Before final locations can be decided, the vectors for the various crank angles must be plotted on a polar plot. The magnitude and location of the weight or weights can then be decided. Refer to Figure 9-6 for counterweight position with respect to the crankshaft.

Rotary Shaking Forces

Foundation designers, particularly ones who are familiar with reciprocating compressors, will often ask about the shaking forces exhibited by

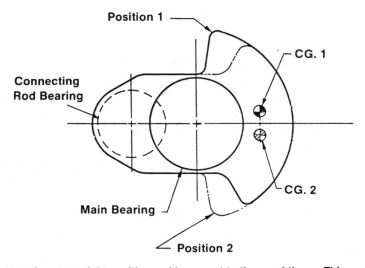

Figure 9-6. Counterweight positions with respect to the crankthrow. This applies only to the maximum counterweights.

the rotary or dynamic compressors. Normally these forces are rather insignificant relative to the weight of the compressor. To a designer, a numerical value has more meaning and allows him to decide the relative significance to the foundation design.

Rotary and dynamic compressors have in common the force due to unbalance. As discussed, the objective is to balance a rotor to a level of .1g or better, which was previously stated to be a force equal to or less than 10% of the rotor weight.

This represents a shaking force at a frequency equal to rotor speed. For foundation design, a value of 5 to 10 times the residual unbalance or ½ to 1 times rotor weight at operating speed would be a reasonable design value. The direction of the force is perpendicular to the shaft, and operates as a rotating vector which can be centered between the bearings.

The dynamic compressors, axial and centrifugal, exhibit one other non-steady force, which is often questioned by foundation designers. This is the torque reaction force at the casing feet and, thus, the foundation due to surge. As a rule of thumb, a value of ½ the maximum steady-state torque reaction at the casing feet can be used. The frequency is the surge frequency, which is less than operating speed but not readily predictable, because it is heavily dependent on the connected system.

Rotor Dynamics

Lateral critical speeds are somewhat misnamed, since they are now thought of as a damped response to some form of rotor excitation. Classically, they got named when compressor (or other rotating machine) speed, whether starting or operating, happened to coincide with a responsive lateral resonance (damped natural frequency). Since the rotor of the compressor included inherent unbalance, it would excite the rotor-bearing support system. Because this occurred long before the "speed squared" effect of unbalance should have been felt, it received the name critical speed, and was tagged as a speed to avoid. Figure 9-7 shows a typical rotor response plot.

The undamped critical speed is proportional to the static deflection of a simple shaft as seen by the following equation for a mass concentrated at a single point [6].

$$N_c = 9.55 \sqrt{g/\delta_s} \tag{9.8}$$

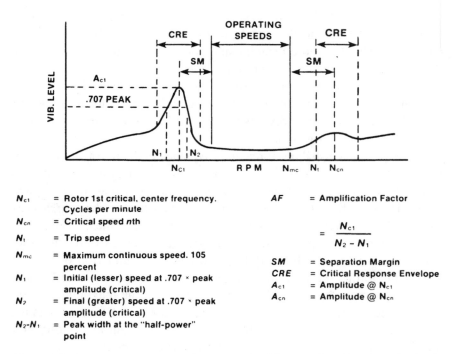

N_{c1}	= Rotor 1st critical, center frequency. Cycles per minute	AF	= Amplification Factor
N_{cn}	= Critical speed nth		$= \dfrac{N_{c1}}{N_2 - N_1}$
N_t	= Trip speed		
N_{mc}	= Maximum continuous speed. 105 percent	SM	= Separation Margin
N_1	= Initial (lesser) speed at .707 × peak amplitude (critical)	CRE	= Critical Response Envelope
		A_{c1}	= Amplitude @ N_{c1}
N_2	= Final (greater) speed at .707 × peak amplitude (critical)	A_{cn}	= Amplitude @ N_{cn}
N_2-N_1	= Peak width at the "half-power" point		

Figure 9-7. Typical rotor response plot. (*Courtesy of the American Petroleum Institute*)

where

N_c = critical speed, rpm
δ_s = static shaft deflection
g = gravitational constant

All the early calculations were based on simple beam analysis. The method was improved when the summation of moments was introduced by Myklestead [5] and Prohl [6].

The interesting fact is that the two men just mentioned were initially working independently of each other. The initial analysis used infinitely stiff bearings and, while the method improved the results, it did not match the compressor critical speed test results consistently. Later work recognized the existence of a bearing oil film with elastic and damping properties.

As a journal rotates in a lubricated bearing, the viscous nature of the lubricant causes the shaft to trap oil between the journal and the bearing surface. A wedge of oil or fluid film develops and produces a hydrodynamic pressure sufficient to carry the journal load and keep the two surfaces separated. A fully developed oil film exhibits stiffness and damping characteristics which will vary in magnitude with journal rotational speed. The bearing stiffness represents the spring rate of the oil film, while damping indicates the film's ability to dissipate vibrational energy. These properties are calculated in the form of linear, non-dimensional coefficients for use in the damped unbalanced response calculation. Stiffness alone is used in the undamped calculation (see Figure 9-8).

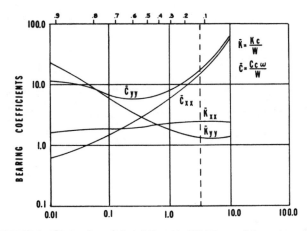

Figure 9-8. Principal non-dimensional linear stiffness and damping coefficients for short bearing. (*Courtesy of Turbocare, A Division of Demag Delaval Turbomachinery Corp., Houston facility*)

When rotor geometric data, shaft diameter and lengths, element sizes and weights are put together, the mass elastic rotor system can be modeled. To complete the model, the support spring rate (oil film) is included. When the information is available, a computer analysis would include support parameters, such as pedestal and/or foundation stiffness. The results of the solution for frequency yield the undamped critical spectrum. These results are plotted in two forms. One is a mode shape diagram (deflection vs. rotor length) (see Figure 9-9). The plots are useful to identify the various modes and to check for node points in the bearing. Damping will have little or no effect if there is no movement in the bearing. To state in positive terms, damping is a dynamic property and is a function of velocity which implies movement. The second plot is a map of the response frequencies plotted against bearing stiffness (see Figure 9-10). The response frequencies are given in speed units, rpm. The operating speed is drawn in with a straight line bounded by the API 617 margin requirement. The bearing stiffness versus speed map is cross-plotted. It was mentioned previously that bearing stiffness is a function of speed. At one time, this was the extent to which rotor analysis was pursued. The figure cited would have had all interested persons happy because the points where the bearing maps cross the three response plots do not coincide with operating speed nor do they encroach into the margin specified.

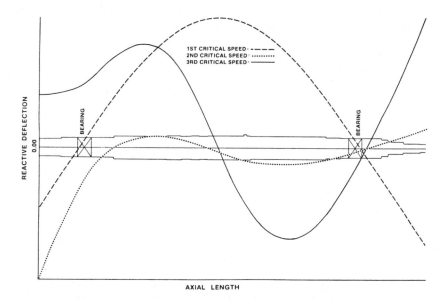

Figure 9-9. Typical mode shapes of an undamped system.

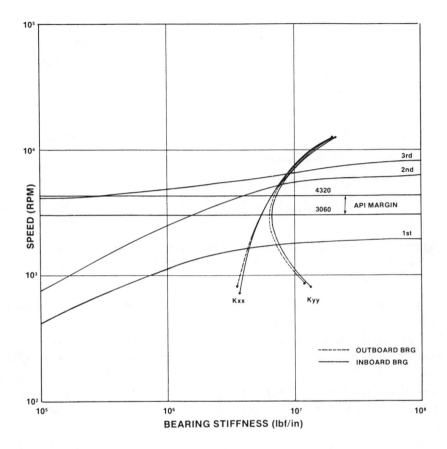

Figure 9-10. Map of response frequencies plotted against bearing stiffness.

Damped Unbalance Response

Unfortunately, particularly for the vendors, the undamped analysis opened somewhat of a Pandora's box. While the calculated frequencies were not quite accurate, they were better than any previous analysis, once the actual bearing parameters were input.

It should be mentioned in the course of the rotor analysis development, the art of tilting pad design moved from just carrying a load at speed to one of tailoring the arc of the pad by a design method called preload to achieve the stiffness and damping values needed to control the location of the response (see Figure 9-11). A stiff rotor relative to the bearings

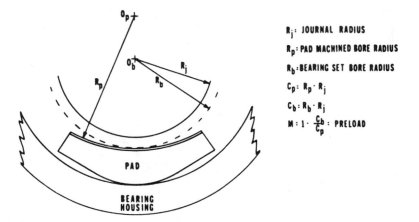

Figure 9-11. Tilting pad bearing schematic with definition of preload. (*Courtesy of Turbocare, A Division of Demag Delaval Turbomachinery Corp., Houston facility*)

(where the speeds on the three modes become close to horizontal) isn't influenced by bearing tuning.

The dismay mentioned is that undamped response finds numerous other responses. It also picks up two other modes, referred to as *rigid rotor modes*. These modes are controlled primarily by the elastic properties of the bearings with little or no rotor bending (rigid). While helicopter and aircraft designers had observed these effects, this came as a shock to quite a few of the compressor design engineers. Besides these classic modes, others would crop up from time to time with little or no apparent physical significance. Seeing these plots, eyebrows were raised and the users decided the vendors were trying to hide something. There is no way of knowing the hours expended on what amounts to phantom criticals. With the improvement of the damped response analysis and test stand data, it was determined that most of these were indeed unresponsive in real life. Both vendor and user gained with the damped analysis. The criticals, which are real, can be identified and can be the focus of attention.

Users still like to see the undamped plots because they are easier to interpret, but the vendors are afraid that they may have to design away from an unresponsive value and waste money. This should resolve in time. Figure 9-12 shows the various modes discussed in a classic form.

The damped (unbalance response) plots are not really very impressive, but they address the information every operator would like to see. Amplitude is plotted against speed at various stations such as at the probes (see

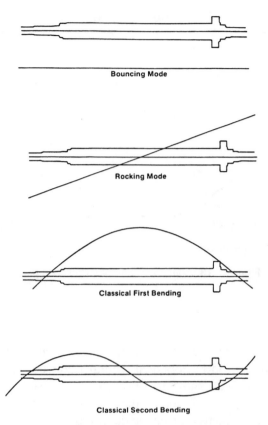

Figure 9-12. Classical mode shapes.

Figure 9-13). A response, of course, is a sudden rise in amplitude at a speed. The analysis is performed using a mathematical model that includes the effects of damping in the equations, making the model much more complex than any previous analysis. The University of Virginia carries out continuous research in which a consortium of users, vendors, and researchers provide funds, and data and interchange ideas to advance the science of rotor dynamics. Other organizations, such as the Bently Rotor Dynamics Research Corporation and Texas A&M University, are also carrying on similar work on a continuing basis.

The value of a damped response is that the areas under seals or other close clearance areas can be investigated on a dynamic basis. This can be done at any selected station on the rotor. If there is encroachment on clearance, the rotor can be tuned to avoid the problem areas. The reliability of the machine can be considered rather than arguing the need for large margins to avoid the unknown.

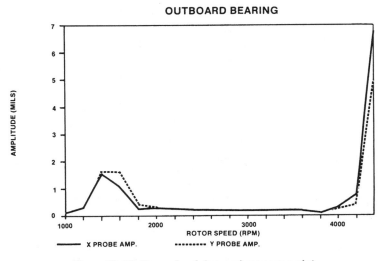

Figure 9-13. Example of damped response plot.

Torsionals

Besides the inherent lateral natural frequency characteristic, compressors are also influenced by torsional natural frequencies. All torsionally flexible drive trains are subject to non-steady or oscillatory excitation torques during normal operation of the system. These excitation torques can be an inherent function of either the driver or the driven equipment and, when superimposed on the normal operating torque, may appear to be of negligible concern. However, when combined with the high inertia loads of many turbomachinery trains and a torsional resonant frequency of the system, these diminutive ripples can result in a tidal wave of problems.

The torsional resonant response of a system is an interaction of all the components in the train. Calculation of torsional natural frequencies is based on the entire system and these frequencies are valid only for that given arrangement. If any component of the train is replaced by an item with torsional characteristics different from the original, the system torsional response must be recalculated and new torsional natural frequencies determined. Occasionally, an original equipment manufacturer is requested to calculate the torsional and lateral critical speeds of the supplied item. Unfortunately, the purchaser is unaware that this request is of limited value since the torsional response of a single item in a train is meaningless. Likewise, a torsional shop test will yield meaningless results if the train is not assembled and tested with every item destined for the field.

The interesting aspect of torsional problems in turbomachinery systems is that the first indication of a problem is usually a ruptured shaft or coupling in the field. Silent and deadly, a torsional response can lurk at synchronous or non-synchronous frequencies, and be steady or transient in nature. Once a torsional problem is found in the field and the excitations are determined to be inherent in the system, the only solution available, to put the system back on line quickly, is to decouple the excitation source or to dampen the system response.

Table 9-2 contains some sources of torsional excitations encountered in the operation of turbomachinery systems.

A quick review of system torsional response may help explain why a resilient coupling works. Figure 9-14 is a torsional single degree of freedom system with a disk having a torsional moment of inertia J connected to a massless torsional spring K.

Newton's law for a rotating body states:

$$T = J \alpha \tag{9.9}$$

Table 9-2
Typical Sources of Torsional Excitation in Turbomachinery Systems

SOURCE		FREQUENCY	MODE
Gear	Mesh	No. of Teeth × Gear RPM	Steady
	High Spot	1x Gear RPM	Steady
	Quality	2x Gear RPM	Steady
Steam Turbine	Nozzle Passing	No. of Blades × RPM	Steady
Electric Motor	Variable Frequency Drive	No. of Pulses × Motor Line Frequency	Steady
	Synchronous Start	2x Slip Frequency	Transient
Compressors	Centrifugal Surge	Broad Band	Transient
	Reciprocating	No. of Cylinders × RPM	Steady
Electrical Faults		Varies	Transient

Source: *[13]*

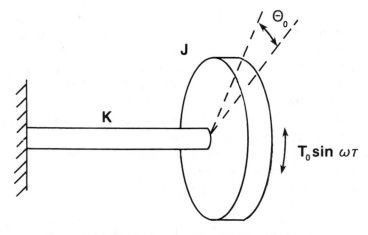

Figure 9-14. Single degree of freedom torsional model.

where

T = torque
J = mass moment of inertia
α = angular acceleration

Angular acceleration is the time rate of change of angular velocity which, in turn, is the time rate of change of angular displacement.

$$\alpha = d\omega/dt = d^2\theta/dt^2 = \ddot{\theta} \qquad (9.10)$$

where

θ = angular displacement, rad
ω = angular velocity, rad/sec

Assume that oscillatory excitation torque of $T_0 \sin \omega t$ is applied to the system in Figure 9-14. By definition, when the excitation frequency coincides with the torsional natural frequency of the model, all torques will balance and the system will be in a state of resonance.

For equilibrium, the following equation of motion must be satisfied

$$J\ddot{\theta} + C\dot{\theta} + k\theta = T_0 \sin \omega t \qquad (9.11)$$

where

$J\ddot{\theta}$ = inertial torque
$C\dot{\theta}$ = damping torque
$k\theta$ = stiffness torque

In its simplest form, damping is neglected and no external forcing function is applied, resulting in the equation

$$J\ddot{\theta} + k\theta = 0 \qquad (9.12)$$

separating the variables,

$$\ddot{\theta} = -(k/J)\theta \qquad (9.13)$$

The general solution of this second order differential equation is

$$\theta = C_1 \sin t \sqrt{k/J} + C_2 \cos t \sqrt{k/J} \qquad (9.14)$$

Assuming the disk is displaced θ_0 radians and then released, the following initial conditions apply:

$t = 0$

$\theta = \theta_0$

$\dot{\theta} = 0$

The first condition yields

$$C_2 = \theta_0$$

Differentiating and substituting, the second condition yields

$$C_1 = 0$$

resulting in the specific solution

$$\theta = \theta_0 \cos t \sqrt{k/J} \qquad (9.15)$$

The period, P, of this vibration is

$$P = 2\pi \sqrt{J/k} \qquad (9.16)$$

The reciprocal of the period is

$$1/T = \left(\frac{1}{2}\pi\right)\sqrt{k/J} = f_n \tag{9.17}$$

where

f_n = torsional natural frequency, cps

For a complex, multimass system like that shown in Figure 9-15, the equations of motion become quite complex, especially if a forcing function exists and internal damping is included. Inertial damping (damping to ground) is neglected. The equations of motion for this system would take the form:

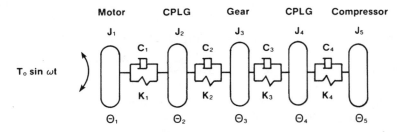

Figure 9-15. Multimass torsional model.

$$J_1\ddot{\theta}_1 + C_1(\dot{\theta}_1 - \dot{\theta}_2) + K_1(\theta_1 - \theta_2) = T_0 \sin\omega t \tag{9.18}$$

$$J_2\ddot{\theta}_2 + C_2(\dot{\theta}_2 - \dot{\theta}_3) + C_1(\dot{\theta}_1 - \dot{\theta}_2) + K_2(\theta_2 - \theta_3) - K_1(\theta_1 - \theta_2) = 0$$

$$J_3\ddot{\theta}_3 + C_3(\dot{\theta}_3 - \dot{\theta}_4) + C_2(\dot{\theta}_2 - \dot{\theta}_3) + K_3(\theta_3 - \theta_4) - K_2(\theta_2 - \theta_3) = 0$$

$$J_4\ddot{\theta}_4 + C_4(\dot{\theta}_4 - \dot{\theta}_5) + C_3(\dot{\theta}_3 - \dot{\theta}_4) + K_4(\theta_4 - \theta_5) - K_3(\theta_3 - \theta_4) = 0$$

$$J_5\ddot{\theta}_5 + C_4(\dot{\theta}_4 - \dot{\theta}_5) + K_4(\theta_4 - \theta_5) = 0$$

The solutions to a problem of this magnitude can be found in references [3, 7] and others. Figures 9-16 and 9-17 are torsional mode shape diagrams of some typical systems. While the rigorous solution to the multimass damped system is not within the scope of this book, several interesting points should be made.

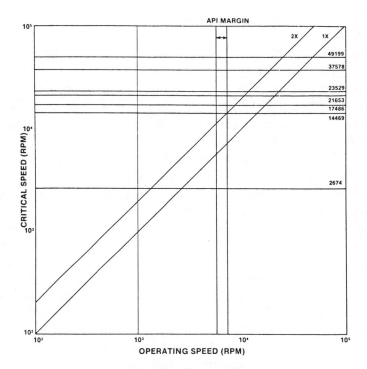

Figure 9-16. Typical torsional interference map.

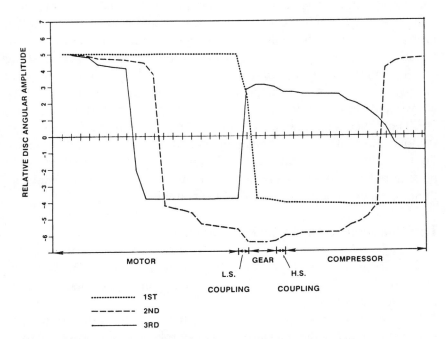

Figure 9-17. Typical plot of torsional angular deflection.

1. An nth degree of freedom system will have (n − 1) natural frequencies.
2. Equation 9.11 indicates that the torsional natural frequencies of a system are a function of the torsional inertia and stiffness of the system.
3. The natural frequencies of a damped system are essentially the same as the undamped systems for all realistic values of damping.
4. The displacements of the system at resonance will be a function of the magnitude of the driving or excitation source and damping.
5. Damping in the system represents dissipation of vibratory energy that reduces the amplitudes in the system.
6. Damping is a function of the angular velocity change across the damper.

As previously mentioned, the torsional response of a system is a function of the stiffness and inertia in the train. While some parameters of the system can be changed, the inertia J is usually fixed by the basic process. For example, the J value for a centrifugal compressor is largely a function of impeller diameters and widths. These, in turn, are set by the required head and flow. Theoretically, hub and shroud thickness could be varied to tune the system. However, any change in impeller hub thickness, shroud thickness, or shaft diameter may significantly alter the lateral response of the unit. These modifications would lead to trade-offs that probably would not be considered acceptable from a process or operational point of view. As a necessary competitive procedure, some machine manufacturing must proceed in parallel to the engineering analysis. Basic changes to impellers and shafts while torsional problems are worked out would definitely slow the process down and impact delivery. The driver cannot be designed until the system inertia is known, yet the torsional analysis cannot be completed until driver parameters are set. So it would appear that an iterative design and analysis procedure would be required and could go on for quite some time. This would complicate things considerably if it were not for the couplings.

Torsional Damping and Resilient Coupling

Many believe that couplings are only torque transmitters between different pieces of equipment with a secondary function of handling ever present misalignment.

While couplings are typically sized and chosen based on the before mentioned requirements, they can take on another role, that of a torsional fixer.

Once the driver and driven equipment have been chosen and it is determined that none of the items will be subject to any lateral vibration problems, the system torsional analysis is performed. If a calculated torsional natural frequency coincides with any possible source of excitation (Table 9-2), the system must be de-tuned in order to assure reliable operation. A good technique to add to the torsional analysis was presented by Doughty [8], and provides a means of gauging the relative sensitivity of changes in each stiffness and inertia in the system at the resonance in question.

Typically, the couplings in a turbomachinery train will be the softest torsional elements in the system. As a result, they represent the controlling factor in the system's overall stiffness since the total system spring rate cannot exceed the stiffness of the softest spring.

Should analysis indicate that a coupling is in a sensitive position, then a small amount of custom design in a relatively standard coupling can accommodate the de-tuning of the critical in question. One note of caution: while changes in stiffness or inertia may de-tune a given resonance, their effect on the other criticals must also be determined, since any change in the system will result in a new set of resonant frequencies.

If modification to coupling stiffness cannot effectively de-tune the system, or if unanticipated excitation frequencies are encountered in the field, the designer has another option to desensitize the train. This "last resort" or "shotgun" approach involves the resilient or torsionally damping coupling. These couplings add damping to the system in order to dissipate vibrational energy and effectively reduce twist amplitudes during a resonant condition. While this does not eliminate the source of the problem, it does allow the manageable operation of the unit under resonant conditions.

Once again, a resilient coupling applied during a resonant condition must be at a location sensitive to the applied damping. Since damping is a function of the relative velocities of the coupling hubs, little would be gained by placement at a node or points of small angular velocity changes.

Another alternative to consider if operation on a known resonance is anticipated is a stress analysis. It is possible that the stresses at resonance may be within an acceptable level permitting the compressor train to operate without a problem.

Specification

A guideline for specification of torsional damping and/or resilient couplings can be found in Appendix B of API Standard 671, *Special-Purpose Couplings for Refinery Service* [9].

The majority of torsional resilient or damping couplings currently in use can be classified into five major categories:

1. Quill shaft
2. Metal-metal resilient
3. Elastomer, compression
4. Elastomer, shear
5. Fluid

The torsionally soft or resilient coupling such as the quill shaft and the metal-metal resilient transmit torque can handle misalignment through springs, metal strips, coils, disks and diaphragms. They will tune the system by changing the spring constant K.

The elastomer compression coupling provides both tuning and damping to the system. In some cases, the two functions interact, that is, the stiffness K or damping C may be a function of the other. The elastomers are torsionally softer than the metal-metal resilients, but will introduce higher levels of damping into the system.

There are two major types of elastomer compression couplings. One is the *torus* type in which the elastomer is attached directly to the coupling hubs (see Figure 9-18). The other is the compression type with the elastomer held in compression by the hub geometry (see Figure 9-19).

Probably the obvious question that arises after the preceding discussion is, "Why not use a soft, highly damped coupling in every case and sidestep the torsional problem altogether?" The solution is often compromise. On the good side, these various couplings provide greater latitudes in the selection of coupling characteristics to solve the torsional problems mentioned earlier. Furthermore, if analysis fails to predict a torsional problem and one arises in the field, these couplings are a quick and inexpensive means of bringing the unit back on line.

On the bad side, many of the elastomeric types are highly nonlinear in their characteristics. The elastomeric compression-type couplings are very soft at small wind-ups under low loads, but once the elastomer has filled the available squeeze space, the coupling is effectively rigid. This makes prediction of system response difficult unless the load and coupling characteristics are well defined prior to installation.

The elastomeric couplings generally do not have a life factor equivalent to a gear or flexible element coupling. This is further complicated by the fact that if the coupling is to provide damping, the dissipated vibrational energy is converted to heat, which can further shorten the life of the ele-

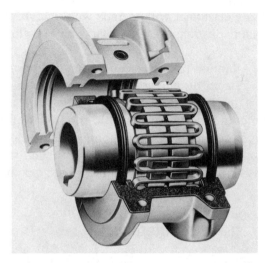

Figure 9-18. Torus type coupling. (*Courtesy of Falk Corporation*)

Figure 9-19. Elastomeric type coupling, commonly used for torsional system tuning. (*Courtesy of Kop-Flex Power Transmission Products*)

ment. Likewise, any elastomer will degrade with time, resulting in a coupling with characteristics that are nonlinear with both load and time of service.

Most torsional vibration problems can be flagged prior to installation of a compressor system by performing a thorough torsional analysis. Typically, the system can be de-tuned by varying the torsional stiffness of a spacer type coupling with very little modification to a standard design. However, if de-tuning with coupling stiffness alone turns out to be ineffective, a stress analysis can predict a problem, or if field operation results in a system failure, a resilient damping type coupling may be the answer.

References

1. Jackson, Charles, *The Practical Vibration Primer,* Houston, TX: Gulf Publishing Company, 1979.
2. Feldman, S., *Unbalance Tolerances and Criteria,* Proceedings of Balancing Seminar, Vol. IV, Report No. 58GL122, General Engineering Laboratory, General Electric, Schenectady, N.Y., 1958.
3. ISO 1940/1, *Mechanical Vibration—Balance Quality Requirements of Rigid Rotors—*Part 1: "Determination of Permissible Residual Unbalance," First Edition, International Organization for Standardization, Geneva, Switzerland, 1986.
4. Harker, Ralph J., *Generalized Methods of Vibration Analysis,* New York: John Wiley & Sons, 1983.
5. Den Hartog, J. P., *Mechanical Vibrations,* Fourth Edition, New York: McGraw-Hill Book Company, 1956.
6. Marks, Lionel Simeon, Ed., *Standard Handbook for Mechanical Engineers,* Eighth Edition, Baumeister, Theodore, *et al.,* Eds., New York: McGraw-Hill Book Company, 1978, pp. 5–73.
7. Myklestead, N. O., "A New Method of Calculating Natural Modes of Uncoupled Bending Vibration of Airplane Wings and Other Types of Beams," *Journal of the Aeronautical Sciences,* Vol. II, No. 9, April 1944, pp. 153–162.
8. Prohl, M. A., "A General Method for Calculating Critical Speeds of Flexible Rotors," *ASME, Journal of Applied Mechanics,* September 1945, pp. 142–148.
9. Kerr Wilson, W., *Practical Solution of Torsional Vibration Problems,* Vols. I & II, London: Chapman & Hall, Ltd., 1956.

10. Doughty, S., "Sensitivity of Torsional Natural Frequencies," *ASME 76-WA/DE-18,* New York: American Society of Mechanical Engineers, 1976.

11. API Standard 671, *Special-Purpose Couplings for Refinery Services,* Second Edition, Washington, D.C.: American Petroleum Institute, 1990, Reaffirmed 1993.

12. Brown, Royce N., "A Torsional Vibration Problem as Associated with Synchronous Motor Driven Machines," *ASME 59-A-141, Journal of Engineering for Power, Transactions of the ASME,* New York: American Society of Mechanical Engineers, 1960, pp. 215–225.

13. Brown, Royce N., *Torsional-Damping—Transient and Steady State,* Proceedings of the Thirteenth Turbomachinery Symposium, Texas A&M University, College Station, TX, 1984, pp. 203–207.

14. Chapman, C. W., "Zero (or Low) Torsional Stiffness Couplings," *Journal Mechanical Engineering Science,* Vol. II, No. 1, 1969, pp. 76–87.

15. Eckert, Joachim, "Transient Torques and Currents in Induction Motors Resulting from Supply Changeover and Other Transient Conditions," *System and Equipment,* pp. 103–106.

16. Hafner, K. E., "Torsional Stresses of Shafts Caused by Reciprocating Engines Running through Resonance Speeds," *ASME 74-DGP1,* New York: American Society of Mechanical Engineers, 1974.

17. Hizume, A., "Transient Torsional Vibration of Steam Turbine and Generators Shafts due to High Speed Reclosing of Electric Power Lines," *ASME 75-DET-71,* New York: American Society of Mechanical Engineers, 1975.

18. Holdrege, J. H., Subler, William, and Frasier, William E., "A.C. Induction Motor Torsional Vibration Consideration—A Case Study," *IEEE Paper No. PCI-81-2,* pp. 23–27.

19. McCormick, Doug, "Finding the Right Flexible Coupling," *Design Engineering,* October 1981, pp. 61–66.

20. Pollard, Ernest I., "Synchronous Motors . . . Avoid Torsional Vibration Problems," *Hydrocarbon Processing,* February 1980, pp. 97–102.

21. Pollard, Ernest I., *Torsional Vibration Due to Induction Motor Transient Starting Torque,* undated manuscript.

22. Pollard, Ernest I., "Transient Torsional Vibration Due to Suddenly Applied Torque," *ASME 71-VIBR-99,* New York: American Society of Mechanical Engineers, 1971.

23. Porter, B., "Critical Speeds of Torsional Oscillation of Geared-Shaft Systems Due to the Presence of Displacement Excitation," *ASME 63-WA-8,* New York: American Society of Mechanical Engineers, 1963.

24. Sohre, J. S., "Transient Torsional Criticals of Synchronous Motor Driven, High-Speed Compressor Units," *ASME 65-FE-22,* New York: American Society of Mechanical Engineers, 1965.

25. Wallis, R. R., "Flexible Shaft Couplings for Torsionally Tuned Systems," *ASME 70-PET-38,* New York: American Society of Mechanical Engineers, 1970.

10

Testing

Introduction

The word *test* is quite broad in its definition, and many of the inspection steps in the course of the compressor manufacturing cycle can appropriately be called tests. An example would be the material tests. The API mechanical equipment standards, however, attempt to narrow the test definition. This chapter will discuss testing within these narrowed definitions. The first test defined in most API mechanical equipment standards is the hydrostatic test, and it will, therefore, be the first test covered in the chapter.

Objectives

The question generally arises, "Why test?" There are reasons for testing arising from both the manufacturers' and the users' points of view. Traditionally, manufacturers have carried out factory tests in order to avoid the costly expense involved with site rectification of deficiencies. Testing is expensive, and manufacturers have no incentive to test low value items except where consequential damage can be caused by the equipment supplied. Figure 10-1 shows a typical centrifugal compressor test stand with several compressors being readied for test.

Many large petrochemical projects involve considerable investment. Plant output value can be measured in staggering amounts of money if value of lost production is calculated. An incorrect pressure switch installed on a compressor may have very little direct financial impact to a compressor manufacturer, but it can cause a loss to the user far in excess of the total value of the compressor package itself. Extra care must therefore be taken in compressor performance tests. In the planning of a project, testing has to be reviewed to:

- Verify design parameters, including performance
- Proof testing of all functions
- Correct faults found

The required lead time for ordering equipment has to take into account all the above factors. Too often the time required for possible design alteration, which may be found necessary, is overlooked. In order to min-

Figure 10-1. Compressors and steam turbines on a large modern test stand. (*Courtesy of Elliott Company*)

imize loss time, each item of equipment can be reviewed for the risk factor involved in its design. The longest time required for modification is in the ordering of raw material. Spare raw material must, therefore, be considered at the time the original order is placed.

The extent of this should be inversely proportional to the manufacturers' experience and directly proportional to the financial consequences of lost production. For example, blank gear sets will allow speed change. Material for guide vanes and impellers will accommodate modifications to aerodynamic performance. Spare material will offset the risk of material defect.

The first test to be considered is not an operational test, but the first major assembly proof test.

Hydrostatic Test

It is conventional to perform a hydrotest on all pressure-containing casings for all compressors. For the reciprocating compressor, where there is no casing, it is the cylinders that are tested individually, based on $1\frac{1}{2}$ times the maximum working gas pressure, but not less than 20 psig. The 20 psig minimum is generally recommended for the rotary, positive-displacement casing as well as dynamic compressors.

Reciprocating cylinder water jackets are tested at $1\frac{1}{2}$ times specified working pressure for the cooling fluid. API 618 recommends a pressure of no less than 115 psig.

For compressor casings designed for different pressure levels, the hydrotest is complicated by the requirement for split-level pressure testing. The casing has to be divided with special closures to permit the isolation of the pressures. It is not unusual for the closures to leak, complicating the test interpretation, but this is not a cause for rejection. Generally, these types of casing designs are discouraged, but nonetheless are furnished on occasion.

The duration of the test pressure in most cases is 30 minutes minimum; however, the test must be held for long enough beyond that time to permit a thorough examination for porosity and casing wall leakage.

Interpretation of joint leakage is somewhat more complicated because the pressure profile causing the leakage may not always be a true representation of the ultimate service conditions.

Casings are designed for controlled deflection, a more stringent criterion than stress. To avoid a misunderstanding, stress is not ignored, but because the deflection requirements inherently keep the stress relatively low, it is not the primary concern. Obviously, in those areas where deflection is not the limiting factor, the stress is then used for the design basis. This makes the testing of casing integrity somewhat more difficult. Hydrotesting is used as in pressure vessels, and the casing is inspected for leaks. Since the hydrotest is based on a value of 1.5 times the maximum allowable working pressure, the test checks the basic safety of the casing to ensure that the stress levels were not exceeded in any area. Only when design modifications are made is the casing instrumented to directly monitor deflections. The tendency is to use joint leaks as a measure of casing suitability. Of course, if the casing doesn't leak, everything is acceptable; however if the joint leaks, the problems begin. The joint may well leak from the distortion due to the hydro pressure; however, it can be argued that this is not a proper test since the joint should never experience this level of pressure during operation. If the casing is given a gas test at the maximum allowable pressure at the time of final assembly when the joint is bolted and made ready, as it is intended to operate in the field, a better judgment may be made. This little dissertation will not settle the arguments, but, if the objectives and methods are considered as the designer intended, perhaps the discussions will not be as heated.

Impeller Overspeed Test

Obviously this test applies only to the centrifugal compressor. The manufacturer normally performs this test whether specified or not. Since the user does have an interest in the test, some discussion of the parameters is in order. API 617 mandates the test, which includes the requirement to spin the impeller at a speed of 115% of the maximum continuous speed for a duration of one minute.

The tests are performed in a special overspeed chamber, which is evacuated to a minimum level to minimize windage pumping of the impeller and to control the temperature rise in the impeller.

The overspeed test is a form of proof test, in that it looks for a failure. In this case, it is a permanent deformation. Not all permanent deformations are failures. In fact, manufacturers have autofrettaged high tip speed impellers for decades. The fourth edition of API 617 was somewhat misleading as it identified the bore as the only critical area. On process compressors, particularly with a hardness limit due to suspected H_2S contamination, the cover disc (front shroud) eye on high specific speed impellers will be the more likely candidate for a permanent deformation. The semiopen impellers would have the bore critical, but are less frequently used for process applications.

This test is better left to the vendor for interpretation unless the user representative is quite experienced in the design of centrifugal impellers. The vendor can be asked to identify the critical areas prior to the test if there is time to check his analysis. Actually, the liability to the vendor is much too high for him to risk an overstress failure just to pass a test. Unfortunately, failures come in the unexpected situations, where the only hedge is "hind sight."

Operational Tests

General

The most straightforward way of proof testing the equipment would be to reproduce site conditions completely. Under such conditions, the testing would easily establish if the compressor performance requirements are met and if the driver and support systems function as intended and are adequate for the purpose.

With large power installations of 40 to 50 thousand horsepower compressing flammable gases, this is not usually possible, and each individual item has to be separately tested. A typical situation could be a 40 thousand horsepower gas turbine driving multibody process gas compressors, where the engineering and manufacture involve one contractor and six sub-contractors located in the UK, U.S.A., Japan, and Indonesia. The difficulties of separately testing each sub-unit to ensure compatibility will be compounded by distance and culture.

Establishing the interface design parameters is easy enough, but forcing designers to establish acceptable tolerance on interface boundary conditions is difficult. Operating parameters need tolerance just as much as manufactured dimensions.

After acceptable design tolerances have been agreed to by all parties, then these become the basis for system design and unit acceptance criteria.

Testing, as a topic, is usually discussed after completion of design. It is not out of order on more complex equipment to perform a detailed review of test requirements prior to placing an order.

Mechanical Running Test

The mechanical running test is generally run at a no-load condition, or if a load is used, it is done for reasons other than the establishing of a measured capacity. If API specifications are used for the compressor, a mechanical running test is mandated on all but reciprocating compressors. The running test for a reciprocating compressor must be specified on each contract. In all cases, the test duration is a minimum of four hours. Table 10-1 lists the objectives of centrifugal compressor mechanical testing.

Table 10-1
Objectives of Centrifugal Compressor Mechanical Tests

Partial Verification Of:
• The quality of overall unit assembly
• Freedom from internal rubs
• Bearing fit, alignment, and adequacy of lubrication
• Rotor-bearing system dynamic stability and calculated critical speed
• Vibration levels
• Correctness of assembly and tightness of shaft oil seals
• Drive coupling fit-up and balance
• Lubrication system cleanliness and performance
• Train component compatibility (optional)
• Noise (optional)

Source: [11]

One problem with the vendor test stands is that they fall into the "do as I say, not as I do" category. This means the test stand lubrication system bears little resemblance to the nice API 614 lube system purchased with the compressor. The user should not be surprised if he finds the vendor's test stand lube system to be somewhat on the shabby side, depending, of course, on the age of the test stand.

Objectives of Centrifugal Compressor Mechanical Tests

It is a good idea for the user to have a witness to check that someone has inspected and verified operation of all safety and warning instrumentation used in the shop. Granted, the compressor is in the vendor's shop and repairs are on the manufacturing account, but why risk damage and late shipment just because the witness was bashful? While he's at it, have him ask if the oil is being filtered to a level of at least 10 microns. It probably isn't feasible to open the filter and look, but at least ask.

For this test, as much of the contract equipment as is practicable should be installed on the compressor, such as the vibration probes, temperature sensors, even the job coupling if a match-up to the driver is possible. If the contract coupling isn't used on the test, the coupling used should duplicate the mass-moment of the contract coupling. Recall that the ideal test would be to duplicate all the field conditions. This picture should be in the user's mind, bounded only by contract scope and practical shop limitations. A little creative imagination can make a routine test a meaningful test for the user.

One objective is to verify oil flow, friction horsepower, and heat rejection to the lube oil. Very often these tests are carried out under vacuum conditions resulting in very low aerodynamic thrust loads. It should be noted that the thrust friction horsepower constitutes the greatest loss and, therefore, the results will have to be adjusted by calculation from the test thrust to full design thrust. The error found from the test compared to design is of little significance to compressor efficiency, but is serious to oil cooling capacity. It should be noted that no-load to full-load friction loss is a linear relationship for thrust bearings and for gears. Errors in predicted loss are due to variations in running clearance and oil flow that affect the churning loss. It is important that tests are carried out at site-design temperatures with the correct viscosity of oil.

If oil buffered seals are used on the compressors, the seal leakage toward the process side of the compressor must be carefully measured, as it is (and should be) a small value. While five gallons per day doesn't sound too small, in a four-hour run, this is less than two pints, making the hold-up time at the inner seal chamber and in the lines to the drain pots a significant value. This makes exact measurement quite difficult.

On compressors with gas seals, it is very desirable to use the contract buffer gas skid. Use of the skid provides the contract filtration level as well as having the regulators sized for contract conditions and checks the buffer gas skid functioning. Most shop systems are built of surplus parts and tend to use manual regulation, which may lead to serious test stand problems. If the skid is not available, it is important to have the vendor provide a comprehensive plan on how the shop buffer gas will be set up.

On those compressors where rotor dynamics can be a problem, which is on all but the standard units (even them sometimes) and reciprocating compressors, this is the point where the acceptance of the compressor for vibration and critical speed criteria should take place. Finding these in the field later is what the user and vendor want to avoid, and this is why all the elaborate and careful work is done at the running test time.

Suitable instrumentation has to be installed to monitor rotor response. This must be taken into consideration during the design stage and the required finish, concentricity, and physical properties of the instrument target areas monitored as a quality control operation during manufacture. Installation and calibration of probes during shop assembly in conjunction with manufacturing quality control avoids costly test stand delays. If magnetic tape records are obtained of the shaft vibration during shop test, these can then be used for field comparison during operation.

Shop test facilities should include instrumentation with the capability of continuously monitoring and plotting rpm, peak-to-peak displacement, and phase angle (X-Y-Y′). Presentation of vibration displacement and phase marker by use of an oscilloscope makes visualization easier.

The vibration characteristics, determined by use of the instrumentation, will serve as the basis for acceptance or rejection of the machine. API standards generally require that the equipment be operated at speed increments of approximately 10% from zero to the maximum continuous speed and run at the maximum continuous speed until bearings, lube-oil temperatures, and shaft vibrations have stabilized. Next, the speed should be increased to trip speed and the equipment run for a minimum of 15 minutes. Finally, the speed should be reduced to the maximum continuous speed and the equipment should be run for four hours. API does not require that the four hours be uninterrupted; however, it is generally interpreted that way. The interpretation is one of the many test criteria to be discussed. It would seem that a break in the test at the midpoint is not the same as having it cut short five minutes from the end because the vendor's boiler took an upset that was not related to the compressor test. The vibration during the shop test is normally specified as the API limit of 1.0 mils peak to peak, or the value from Equation 10.1, unfiltered, whichever is lower.

$$A = \sqrt{12,000/N} \qquad (10.1)$$

where

A = peak to peak amplitude, mils

N = maximum continuous speed, rpm

The limits given are for the centrifugal compressor and for the steam turbine. API 619 should be consulted for the limits of vibration of the helical lobe compressor.

The parameters to be measured during the test will be speed and shaft vibration amplitudes with corresponding phase angles. The vibration amplitudes and phase angles from each pair of X-Y vibration probes shall be vectorially summed at each response peak to determine the maximum amplitude of vibration. The major-axis amplitude of each response peak shall meet the limits specified and set by the specification. At the end of the four-hour test, the unit should be decelerated and recording measurements taken. The unit should then be accelerated to full speed and similar recordings made. The gain of the recording instrumentation used shall be predetermined and pre-set before the test so that the highest response peak is within 60–100% of the recorder's full scale on the test unit coast-down (deceleration).

Vectorial subtraction of slow-roll (300–600 rpm) total electrical and mechanical runout is permitted by the API rules. Vectorial subtraction of bearing-housing motion may be justified if it can be demonstrated to be of significance.

If the previously agreed-on test acceptance criteria are not met, then additional testing will be required.

Rotor Dynamics Verification

The API mechanical standards for the rotary and centrifugal compressor have a test specified for proof of rotor insensitivity. This would normally be the test invoked at this point to ultimately prove the rotor. The fifth edition of API 617 expanded on this test and changed the acceptance criteria from those based on amplification factor to an acceptance level based on internal seal clearances.

The results from these tests will have additional value if they are used to verify the mathematical predictions made. In this way, the accuracy of other mathematical predictions, which cannot be verified except during site operation, is confirmed.

The manufacturer will have carried out a full mathematical investigation into the system vibration response. One quality control requirement will be to establish whether the correct assumptions have been made. Rotor assembly and component weights should be obtained during manufacture and verified against mathematical data used.

The determination of the first bending critical speed is well established; however, there is also concern with regard to the rotor support system's sensitivity to exciting forces. These come from unbalance and/or gas dynamic forces arising during operation in service. Operation with dirty corrosive gas will soon cause rotor unbalance. The rotor dynamics verification test is concerned with synchronous excitation, namely unbalance. The test must also verify that the separation margins are to specification.

System response studies should include rotor residual unbalance, which can be verified during manufacture. An additional study must be made to consider the application of a deliberate unbalance weight at a location chosen by the vendor. The amount and location of the unbalance must be used in the following test. Normally the location chosen is the coupling.

After the completion of the balanced acceleration and deceleration runs, the unbalance weight must be placed at the predetermined location and the acceleration and deceleration test repeated using the same recording instrumentation. The results should be compared to the appropriate computer predictions and the API acceptance criteria applied. Acceptance criteria were added in the API 617 sixth edition.

String Testing

String testing with all the unit coupled together will verify the system response. Individual unit testing may not always give a true result because this may be affected by the test driver characteristic. However, the unit response is affected only by the moment transfer across the couplings. The moment transfer factor can be verified by deliberate unbalance of the test driver hub end and by noting the vibration response at each bearing of the compressor. The rotor response to deliberate unbalance at the compressor shaft hub must also be measured to compare the effects.

Stability

Another potential problem is due to rotor instability caused by gas dynamic forces. The frequency of this occurrence is non-synchronous. This has been described as aerodynamic forces set up within an impeller when the rotational axis is not coincident with the geometric axis. The verification of a compressor train requires a test at full pressure and speed. Aerodynamic cross-coupling, the interaction of the rotor mechanically with the gas flow in the compressor, can be predicted. A caution flag should be raised at this point because the full-pressure full-speed tests as normally conducted are not Class I ASME performance tests. This means the staging probably is mismatched and can lead to other problems [22]. It might also be appropriate to caution the reader: this test is expensive.

Helical-Lobe Compressor Test

Mechanical testing of the non-lubricated helical-lobe compressors is modified from the previously described test. For example, API 619 only requires a two-hour mechanical run. The procedures and monitoring requirements are generally the same as previously described. A run comparable to the overspeed run is the *heat run*. The compressor is run on air at the maximum allowable speed, and the discharge temperature is allowed to stabilize at a value 20°F higher than the rated discharge temperature. The compressor is then run for 30 minutes.

After the heat run, the compressor continues to run on air and the highest pressure practical is imposed, while the speed is set to the normal operating speed. The capacity and power should be noted as well as bearing temperature and the other instrumentation used during the test. If oil buffered seals are used and the test run is expected to exceed 250°F, the test procedure may have to be modified to avoid the possibility of an explosion hazard.

Reciprocating Compressor Test

Occasions arise when it is not practical to even run a no-load test on the reciprocating compressor because of certain driver arrangements. While it is recognized as being a compromise, a *bar over* test can be performed. The compressor is assembled, as if it were to be run. The machine is then turned manually through the various crank positions and stopped while clearances and runout of the parts are measured.

Spare Rotor Test

Spare rotors are frequently part of a new compressor contract. A decision must be made regarding the testing of the spare rotor. The recommended practice is to mechanically run spare rotors, using the same procedures used for the main. As a matter of logistics, the compressor is usually shipped with the last run rotor. If a performance test has been specified, it can be run before the rotors are changed and both rotors performance tested. This would not be economical, however.

As an alternative, used when the cost of the extra running does not warrant mechanical testing of both rotors, the spare rotor can be fitted to the case and only clearance measurements taken. This is philosophically similar to the reciprocating compressor bar over test.

Static Gas Test

Immediately after the running test, any compressor intended for toxic, hazardous, flammable, or hydrogen-rich service should be gas tested with an inert gas to the maximum seal design pressure. The test is held at least 30 minutes and the casing and its joints checked for leaks, using a soap bubble method or other suitable means for leak detection. When no leaks are detected, the compressor will be considered acceptable.

When the seal maximum pressure is such that the gas test must be conducted well below rated discharge pressure, then another test, or substituted test, may be invoked in which the seals are removed. It may require removing the rotor; however, with this procedure the joint makeup still remains a problem when the rotor is replaced. However, without the seals, the test described earlier is performed, using the rated discharge pressure rather than the maximum seal pressure. The acceptance test is the same also; no leaks should be detected. The latter test should help to still controversy about minor joint leaks on a hydrostatic test, if the casing is proven at rated pressure on gas.

The gas testing described will apply to all compressors with casings. For the reciprocating compressors, the gas test, as specified by API 618, is applied to the cylinder. The test is used automatically only when gas under the molecular weight of 12 is the contract gas or the gas contains in excess of .1 mol % of hydrogen sulfide. The test gas is specified to be helium. Consideration should be given to applying the test for the toxic, hazardous, and flammable gases as well.

Testing of Lubrication Systems

The hydrostatic testing is the first test used on the lube system. The system is tested while assembled or partially assembled, based on the particular system. A test pressure of 1½ times the maximum allowable working pressure, a minimum of 20 psi for the oil side, is used for the test. For the oil-wetted parts, the test fluid should be light oil, which is normally the recommended lubricant for the compressor train. The test period is the length of time needed to inspect for leaks, or a minimum of thirty minutes. Acceptance is based on the lack of leaks as visually observed or the lack of a drop in the test pressure.

The operational test of the lube system is, as the name implies, a functional test to check as many of the features as practical under running conditions. The first and last step is a demonstration of the cleanliness of the system. This is followed by a running test of a four-hour duration. The test should simulate the field operation with the compressor in every way practical. All equipment to be furnished with the lube system should be used in the test, including the standby pump start and trip switches. All other instruments should be used to demonstrate their operation. Prior to starting the four-hour run, the system should be thoroughly inspected for leaks and the leaks corrected. If no steam is available for a steam turbine (if one is used), the four-hour run can be made on the electric pump. However, every effort should be made to use an alternate source of energy such as compressed air, to operate the steam turbine.

For the standby pump start test, which is an important test to ensure the pumps transfer without large pressure swings, a check should be made to see if the relief valves lift or the pressure falls to a pressure one half the difference between the standby pump start pressure and the compressor trip pressure. The transient pressure is best measured with a multipen chart recorder. The chart speed must be high enough to fully display the pressure variation. While not as good as the chart recorder, a simple shop-made test setup can be substituted. A spare switch is temporarily connected to the same location as the other switches. This switch is calibrated to close at the threshold acceptance pressure. The contacts are wired to a test light through a seal relay. The purpose of the relay is to maintain the circuit when the pressure restores, so the light will remain on if the switch reached its setting. On the transfer, if the test light is illuminated, the console failed the test. Obviously, this test must be made with both pumps operational. The same is true for control valve response, though the balance of the control valve operation can be done with the standby pump. Because of the problems involved in simulating events, it is best to conduct the test with the main pump if at all possible. As many as practical of the operational steps should be performed. The transfer valves for the coolers and filters should be operated. The pressure should not fall to the standby pump start level in the transfer. If control oil is being supplied, a governor transient should be simulated to check the pressure level for a drop to the level of starting the standby pump. A check should be made of the transfer valves to demonstrate that the leakage is less than the filter body draining capacity. A zero leakage is preferred.

The lube system is considered acceptable when no abnormal conditions occur during the test. Instability or excessive pressure swings during one of the steps is considered an abnormal condition. Corrections to the system will be required and the system retested and demonstrated to be free of the abnormalities before it can be considered acceptable.

Shop Performance Test

To provide further assurance of compressor field performance, as specified, shop performance testing is an option for the user. A shop performance test may further be indispensable because of one or more of the following situations:

- To assure attainment of specified performance before actual installation.

- To be able to find and correct performance deficiencies quickly and effectively without production loss at the site.

- To be able to check guaranteed performance without the fouling effects caused by foreign material or impurities in actual duty gas.

- To be able to test according to accepted test codes which may be impractical or impossible at the site.

- To be able to log simultaneous readings in the shop, after stabilization is reached, of all points to evaluate correct performance.

Open-loop testing with air or with the specified gas is relatively straightforward, but there are difficulties in carrying out closed-loop tests with equivalent gases.

Selection of the test gas is an important consideration and the requirement for accurate mixture control and availability of reliable thermodynamic data really narrows the choice to industrial gases freely available in the market, for example, R22, carbon dioxide, and nitrogen. When compressors are designed for low molecular weight gases, helium nitrogen mixtures may have to be used. When mixed gases are used, the gas should be bought premixed and certified in gas bottles ready for charging the loop, or mixed in a separate monitored mixing chamber whenever possible. This precludes problems with stratification which can occur if the gases are mixed directly in the loop.

In order to maintain gas purity, it is desirable to select a positive suction pressure in order to minimize air leakage into the loop. Apart from the need to minimize inlet density and, hence, test power, the selection of test inlet pressure is a matter of convenience.

Test Codes

The basis for code testing is the ASME Power Test Code, PTC 10 [1] or ASME PTC 9 [2] as applicable. Several specific points made in the code were intended as guiding principles, yet are often misunderstood. The following facts must be considered.

- The codes establish the rules for a test, including the definitions of code or non-code.
- The codes are not a textbook on testing.
- The codes have to assume that the properties of the gases involved are known. It recognizes that this is not always the case, but must place the burden of knowledge of gas on the persons performing the tests.
- The codes establish a basis on which to agree or disagree. Ultimately the final test procedure and method must be agreed on by the purchaser and vendor.

The PTC 9 establishes the deviation limits from the contract conditions that the test may use and still meet the code (see Table 10-2).

The PTC 10 defines three classes of testing. The code attempts to categorize testing and thereby establish an inherent degree of accuracy. These categories are based on methods of test and methods of analysis.

Table 10-2
Maximum Allowable Variation in Operating Conditions

Variable	Deviation of test from value specified (plus or minus)	Fluctuations from average during any test run (plus or minus)
(a) Inlet pressure	2% of abs pressure	1%
(b) Pressure ratio[1]	1%	
(c) Discharge pressure[1]		1%
(d) Inlet temperature		1°F
(e) Inlet temperature deviation for any stage	15°F	
(f) Speed	3%	1%
(g) Cooling water inlet temperature		2°F
(h) Cooling water flow rate		3%
(i) Metering temperature		3°F
(j) Primary element differential pressure		2%
(k) Voltage	5%	2%
(l) Frequency	3%	1%
(m) Power factor	1%	1%
(n) Belt slip	3%	None

[1]*Discharge pressure shall be adjusted to maintain the pressure ratio within the limits stated.*
Source: [2], Reprinted by permission of the American Society of Mechanical Engineers.

Class I includes all tests made on the specified gas (whether treated as perfect or real) at the speed, inlet pressure, inlet temperature, and cooling (if applicable) conditions for which the compressor is designed and is intended to operate, that is, an air machine or a gas-loop test on the specified gas within the limit set by Table 10-3.

Classes II and III include all tests in which the specified gas and/or the specified operating conditions cannot be met. Class II and Class III basically differ only in method of analysis of data and computation of results. The Class II test may use perfect gas laws in the calculation, while Class III must use the more complex "real gas" equations. An example of a Class II test might be a suction throttled air compressor. An example of a Class III test might be a CO_2 loop test of a hydrocarbon compressor. Table 10-4 shows code allowable departure from specified design parameters for Class II and Class III tests.

Table 10-3
Allowable Departure From Specified Operating Conditions for Class I Test

Variable	Symbol	Unit	Departure (%) (1)	
(a) Inlet pressure	P_i	psia	5	(2)
(b) Inlet temperature	T_i	°R	8	(2)
(c) Specific gravity of gas	G	ratio	2	(2)
(d) Speed	N	rpm	2	—
(e) Capacity	q_i	cfm	4	(3)
(f) Cooling temperature difference	—	°F	5	(4)
(g) Cooling water flow rate	—	gpm	3	—

(1) Departures are based on the specified value where pressures and temperatures are absolute.
(2) The combined effect of items (a), (b) and (c) shall not produce more than 8 per cent departure in inlet gas density.
(3) See Par. 3.13 of PTC 10 for limitations on range of capacity.
(4) Difference is defined as inlet gas temperature minus inlet cooling water temperature.
Source: [1] Reprinted by permission of the American Society of Mechanical Engineers.

Table 10-4
Allowable Departure From Specified Design Parameters
for Class II and Class III Tests

Variable	Symbol	Range of Test Values Limits—% of Design Value Min	Max
Volume ratio	q_i/q_d	95	105
Capacity-speed ratio	q_i/N	96	104
Machine Mach number	M_m		
0 to 0.8		50	105
Above 0.8		95	105
Machine Reynolds number where the design value is	Re		
Below 200,000 centrifugal		90	105
Above 200,000 centrifugal		10*	200
Below 100,000 axial compressor		90	105
Above 100,000 axial compressor		10**	200

Mechanical losses shall not exceed 10% of the total shaft power input at test conditions.

* Minimum allowable test Machine Reynolds number is 180,000
** Minimum allowable test Machine Reynolds number is 90,000
Source: [1] Reprinted by permission of the American Society of Mechanical Engineers.

It becomes apparent, the farther the test gas parameter deviates from those of the contract gas, the more difficult the correlation. This leads to test error.

In a multistage compressor, these errors cumulate and play a significant role in mismatching the inlet conditions at every successive stage. If a multistage compressor has side steams, testing becomes quite complex because each section may require a different test speed. A compromise must be made if sections are to be tested at the same time. This compromise is sometimes not feasible, and the sections must be tested separately.

An equation to determine the equivalent speed that would simulate the test conditions for testing the compressor on a gas different from the design process (contract) gas is given as:

$$N_t = N_s \sqrt{\frac{T_{1t} Z_t (n/n - 1)_t (r_p^{n-1/n} - 1)_t}{T_{1s} Z_s (n/n - 1)_s (r_p^{n-1/n} - 1)_s}} \tag{10.2}$$

where

   N = rotative speed, rpm
   T_1 = inlet absolute temperature, °R
   Z = average compressibility
   n = polytropic exponent
   r_p = pressure ratio
   s = contract gas constants (process gas)
   t = test gas constants

Loop Testing

A closed-loop test probably will be necessary to performance tests within the code. There is more accuracy in the loop test than in the air test because volume ratio matching can be more closely achieved. The loop test has several limitations that may not be obvious. Loop tests are generally expensive and time-consuming. They may become complicated with complex compressor configurations and become impractical to set up. Finally, the number of gases available for shop-loop testing is quite limited. It should be noted that most gases must be recovered after the test because of environmental regulations. Air, which is quite available for normal testing, is too dangerous for use in loop testing. In a loop test, air and oil may come into contact causing the danger of explosion. Figure 10-2 shows a typical shop test schematic of a loop arrangement. A two-body loop test arrangement is shown in Figure 10-3.

Because of safety concerns, all combustible and/or toxic gases must be used in outdoor test loops or in a special indoor test building with the required safety monitoring equipment. The gas cost factor makes the problem even more difficult. The problem of known gas properties adds another complication. Despite all the negative aspects just mentioned, most performance tests are closed-loop tested.

Test Loop Design

After calculating the test point conditions at the equivalent design rating, similar calculations should be carried out at surge and overload. These points can then be checked for significant deviation from equivalence. The operating envelope, so defined, can now be drawn on a Mollier diagram. This is especially important when the test gas used is a refrigerant in order to establish if condensing can occur in any part of the loop. Although the gas discharged is hot, any of this gas in static lines will, of course, be at ambient temperature and the possibility of condensation must be checked. When oil film seals are used, condensation in the

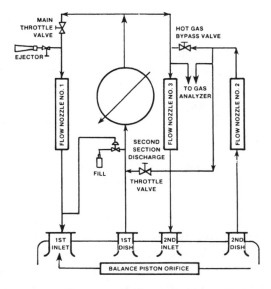

Figure 10-2. Schematic of a typical shop test-loop arrangement.

Figure 10-3. Compressors set up for both a closed-loop and string test on the test stand. (*Courtesy of A-C Compressor Corporation*)

overhead tank will cause serious problems. On the other hand, the use of mechanical contact seals on compressors will avoid this problem except for the effect on static instrument lines because there is no overhead tank. When manometers or transducers are used on the discharge side and pressure lines are arranged to be short and self-draining, it is still possible for an acceptable test to be carried out.

In order to avoid the need to measure velocity head, the loop piping must be sized to have a velocity pressure less than 5% of the static pressure. Flow conditions at the required overload capacity should be checked for critical pressure drop to ensure that valves are adequately sized. For ease of control, the loop gas cooler is usually placed downstream of the discharge throttle valve. Care should be taken to check that choke flow will not occur in the cooler tubes. Another cause of concern is cooler heat capacity and/or cooling water approach temperature. A check of these items, especially with regard to expected ambient conditions, should be carried out. It is surprising how often tests are abandoned because of these problems, causing delays in shipment.

The code requires the submission of an instrument and piping flow sheet. All instruments should be numbered for identification and the actual instruments labeled. The code is very helpful in recommending the range and sensitivity required, and close adherence is recommended.

Currently, most manufacturers use automatic data gathering. Here, pressure signals will be obtained by the use of transducers. Under these circumstances, the transducers should be calibrated per code and certified.

If data gathering is also computer-linked, then the problem is to check out the correctness of the program. Hand calculation of one rated point for flow head, efficiency, and horsepower will serve as verification.

Gas Purity

Significant errors will arise if gas purity is not accounted for. It should be noted that the code lays down no conditions for this, and a figure of 99% or better should be targeted. In order to obtain a good purity at the start, all pipe joints should be taped and the system evacuated to a low vacuum several times with intermediated purging with the test gas to remove the residual contaminants.

Change in molecular weight affects the equivalent speed-measured head and density. This means that monitoring of gas purity must be carried out throughout the test. In any event, loop capacity is usually small and gas should be added and vented during the course of the test as the test discharge pressure changes. To ensure adequate mixing and to avoid stratification in the case of refrigerant gases, the gas should be injected up stream of the gas cooler.

Sidestream Compressors

Sidestream compressors are used in refrigeration processes where, for economy, the refrigerant is flashed off at different pressure levels. Ideally, separate compressors could be used to successively compress the gas back up to the condensing pressure level. The pressure ratio for each stage is low enough to enable this to be done with only one or two impellers in each section. Because of this, compressors can be made with all sections in one casing so that mixing of the streams takes place internally.

The inlet conditions to the suction of an intermediate section are those obtained by the mixing of the side stream entering the nozzle and the gas discharged from the preceding section. In designing the compressor, the manufacturer will assume suction pressure and temperature at the section based on complete mixing of the streams with an allowance for the losses in total pressure between the side load flanges and the section inlet. The test conditions must be based on the inlet condition at the side load flanges, but the pressure loss between flange and section inlet has to be verified in order to establish machine performance. If these losses are significantly different, then this can change the validity of the test results and some adjustment to the test speed may be required.

The measurement of the actual interstage test conditions is also difficult due to turbulence and stratification of the gas at the measuring points. However, much depends on the mixing arrangement adopted, and most manufacturers are coming to the view that the two streams should be brought to the same velocity prior to mixing, if only to ease the problem of computation.

Test Arrangement for Sidestream Compressors

The easiest, but most expensive, arrangement is to test each section independently by installing one section at a time. This involves three separate tests followed by dismantling and reassembly after each test. Test inlet conditions can be accurately maintained, and the problems of installing instruments for measuring pressure profiles across inlet passages are overcome by the space available. Verifying the losses between side nozzle and inlet to the section is less certain as the flow arrangement is not exactly identical to design.

Another method of test is to test all sections simultaneously. In this method, the weight flow through each section is maintained in the same proportion. A separate gas cooler is required for each stream and the loop must have sufficient capacity to enable stable conditions to be obtained. In this system, the machine is tested as fully built but requires extensive shop test space to accommodate the multiple gas coolers and large test-loop piping required, if there are several sidestreams.

Loop testing the sidestream compressor is probably the most difficult, if more than two sidestreams are involved. Reviewing the sidestream problem should give an insight to the various configurations and the test arrangement.

Instrumentation

Test instrumentation has been touched on, but a few additional comments are appropriate at this point. The code provides guidance test arrangements and instrumentation. It includes details on sensor point location as well as pressure tap construction. Flow measurement is defined in detail.

One of the critical measurements is torque or shaft power. A variety of methods is recognized: direct methods such as torque meters or reaction mounted drivers (dynamometers) and indirect methods such as electrical power input to drive motors, heat balance, or heat input to a loop cooler. See Part 7, Measurement of Shaft Power, PTC 19.7 1961 [3] for additional information.

If a preference were to be given, it would be listed in the order stated above, recognizing that the dynamometer is limited in size, but works well with the PTC 9 class of equipment.

Test Correlation

In the discussion of gases and their correlation calculation, the point was made that the code can not be the final authority on gases. Do not expect the vendor to be all-knowing in the areas of all gas properties. The contract gas properties are the responsibility of the user. Much data are published on gases which, together with high-speed computers, make the job of defining gas properties somewhat easier today than it was when the code was written. This is not to say that all properties of all gases are as well defined as they might be. Gas mixtures, in particular, are always a problem.

The use of a gas mixture presents a two-part problem. If the state of the mixture is such that it may be considered a mixture of perfect gases, classical thermodynamic methods can be applied to determine the state of each gas constituent. If, however, the state of the mixture is such that the mixture and constituents deviate from the perfect gas laws, other methods must be used that recognize this deviation. In any case, it is important that accurate thermodynamic data for the gases are used.

Presently, sophisticated computer programs are available based on properties published by various scientific organizations. These programs are normally based upon the equations of state derived by Benedict Webb-Rubin [4] and Starling [5] for use with hydrocarbons. Modifications are often applied to cover other industrial gases. There are alternative equations of state that are more appropriate. Differences in values still may occur due to variations in thermodynamic properties published by different authorities and to differing mathematical techniques and assumptions used in programming. Therefore, an agreement must be reached prior to testing as to which authority and mathematical technique will be used. Three major parameters affecting performance are

- Reynolds number
- Mach number
- Volume ratio

Reynolds Number

PTC 10 has one correlation that has been found to be incorrect. Equation 5.27-1 will permit a Reynolds number correction for high Reynolds number gas (above 10^6), which is much too optimistic. ISO standards allowed no corrections, which is more nearly correct.

Work done by Wiesner [6] is a much more accurate approach. The subject has also been reported on more recently by Simon and Bulskamper [7]. They generally agree with Wiesner that the variance of performance with Reynolds number was more true at low value that at high values. The additional influence above a Reynolds number of 10^6 is not much. It would appear that if a very close guarantee depended on the Reynolds number to get the compressor within the acceptance range (if the Reynolds number was high to begin with), the vendor would be rather desperate.

The other factors mentioned come into play with the inherent mismatch that occurs when an overall factor must be applied to correlate the flange-to-flange method. The correlation proposed in the code is based on work by Schultz [8] and was quite good for its day. When combined with modern calculation methods and equations of state, the philosophy is still valid.

A final tool available is the vendor's ability to generate a set of off-design curves. By performing a stage-by-stage analysis of the proposed test gas, correlations are much easier to make. This is probably best illustrated by an example reported by Wong [9].

Abnormalities in Testing

Even if the test is conducted in accordance with code requirements, some errors can still occur. A test was carried out with a four-impeller gas booster which gave an unexpected result. As an experiment, the test was repeated with different test gases.

All of the tests were with code conditions except for CO_2 where the Mach number variance was more than plus 5%. The results are shown in Figure 10-4. It was concluded that the R12 test provided the more nearly

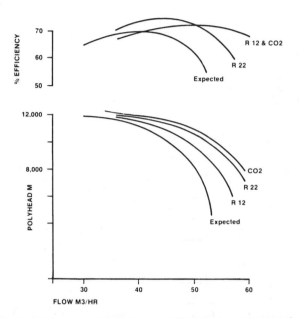

Figure 10-4. Test results from different gases. [9] (*Reprinted by permission of The Institution of Mechanical Engineers*)

correct correlation. Table 10-5 shows a comparison of testing with different gases. It should be noted that R-12 is no longer available as a test gas due to environmental regulations. It is still included in the comparison to show the effects of correlation with different gas characteristics.

The preceding experience did lead to a lack of confidence, and it was concluded that an impeller-by-impeller performance check should be carried out theoretically at test speed with the test gas. This idea was carried out on a cold methane compressor with nine impellers. The results of this were fruitful as can be seen from Figure 10-5, and the expected performance at various test speeds calculated in accordance with the code is shown in Figure 10-6. As a result of this work, two test speeds were

Table 10-5
A Comparison of Testing with Different Gases

Item	Design	1	2	3
Gas	Feed Gas	R12	R22	CO_2
Mol wt	32.36	120	86.48	44
Rpm	11,000	5350	7010	10987
Inlet density	11.8	17.9	4.58	3.59
Outlet density	59.4	47.85	9.42	10.67
Machine Mach no.	0.816	0.792	0.86	0.92
Reynolds no.	3.207	1.5114	1.95	2.71

Note: Reynolds number is $\times 10^{-6}$
Source: [9] Reprinted by permission of the Institution of Mechanical Engineers.

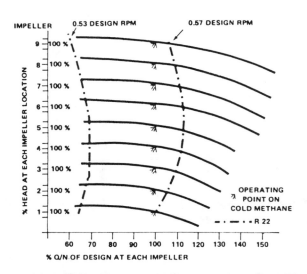

Figure 10-5. Effect of deviation difference R22 test gas and cold methane [9]. *(Reprinted by permission of The Institution of Mechanical Engineers)*

adopted and an agreed correction factor applied on test head and efficiency for flow test point.

Field Testing

After all the problems of shop testing have been discussed, one would wonder why anyone would want to consider field testing. Here are a few reasons.

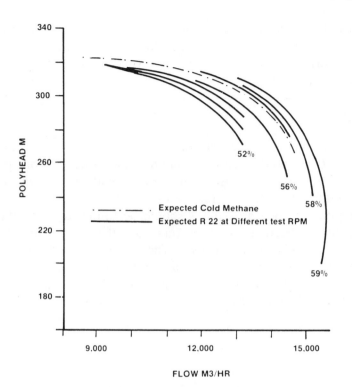

Figure 10-6. Effect of test results due to interstage deviation [9]. (*Reprinted by permission of The Institution of Mechanical Engineers*)

- The process gas may be used (Class I test).
- Sidestreams can all be working.
- Pressures and temperatures are the process values.
- Performance of existing compressors can be evaluated.

With the constant competitive pressures on the petrochemical industry, it is important to operate all plant equipment at the best performance level possible. This is certainly true for compressors.

The many problems with correlation and good shop tests discussed in this chapter would seem to lead to the conclusion that one should field test. It is still better, however, if at all possible, to test in the shop. The new compressor field tests should be limited to only those units where performance is in doubt and shop test correlation is just too difficult. A four sidestream multi-component hydrocarbon gas would probably qualify as difficult to shop test.

Because there is always the possibility of shop testing, there probably is more incentive to take some of the older equipment and evaluate its current performance or possibly the performance after a "turnaround" with a field test.

Planning

One thing that will become quickly evident, as aptly stated by Matthews [10] is "production plants are not test stands." As one pursues the task of getting an existing compressor ready for a field test, it becomes quite clear as to how true the statement is.

The applicable code procedures should be followed, particularly in regard to instrumentation. On an existing plant, this will require planning in advance of a scheduled shutdown. The pressure and temperature tap locations and installation details can all be developed with the compressor running. All details must be followed. When a drawing states, "deburr, do not chamfer" that is exactly what it means (see Figure 10-7).

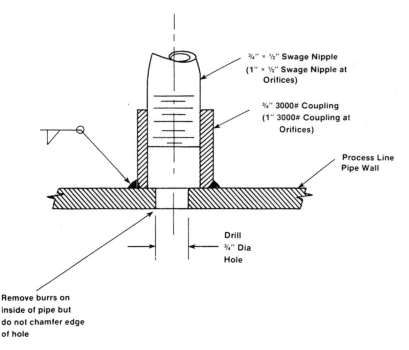

¾" × ½" Swage Nipple
(1" × ½" Swage Nipple at Orifices)

¾" 3000# Coupling
(1" 3000# Coupling at Orifices)

Process Line
Pipe Wall

Drill
¾" Dia
Hole

Remove burrs on
inside of pipe but
do not chamfer edge
of hole

Figure 10-7. Typical pressure tap detail.

The necessary steps are a lot easier to accomplish if the compressor installation is still on the drawing board. This author's best field tests were run on units where the instrumentation was installed during plant construction. Sometimes, as plants are going through revisions for other reasons, it is possible to add the necessary compressor test equipment.

Part of the planning should include the evaluation of test uncertainty. This evaluation can be limited to a common sense approach based on available instrumentation and the locations relative to the ideal. A more sophisticated study can be made in which instrumentation accuracy and the impact of any inaccuracy on the measured parameters is evaluated. This is a complex task with the need being based on the motivation for the test. If the test is being performed to settle a dispute, a formal understanding of the uncertainty should be developed. Methods for evaluation of test uncertainty are found in ANSI/ASME PTC19.1 [11].

Flow Meters

The most difficult part of a field test is the flow meter, if it wasn't planned in the construction phase. There is no way to simulate a meter run if you don't have the proper pipe length. Figure 10-8 is an example of the requirements. An ASME long radius flow nozzle is preferred by the author, though a short throat venturi will do. The probability is that an orifice is all that will be available. It should be examined before and after the test to verify not only the bore diameter, but the finish. The bore should

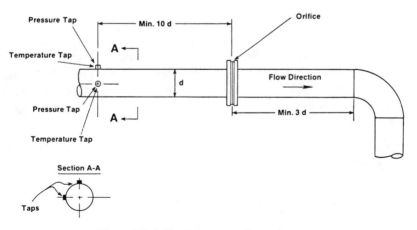

Figure 10-8. Requirements of a meter run.

still be smooth, free of fouling material, and the edges sharp and square. It will be difficult to achieve, but remember the better the instrumentation performs, the less the data scatter that will require rationalization later.

Gas Composition

If the planning has been positive to this point and all objectives have been realized, at least within reason, the rest is almost all down hill, at least until the next stumbling block—gas composition. It would be a real shame to get a good meter run with a first-class primary element and get nothing but a puzzled look when asking about gas composition measurement. This is not to indict plant operating people and accuse them of not knowing their business. Over a 24-hour period, they can probably tell you to a gram just how much of what went through the compressor. No one ever asks for this information every 15 minutes. This problem will have to be handled somewhere, perhaps by drawing samples every time a test point is reached and taking them to the analytical laboratory for analysis. If the samples are carefully marked, and do not react with the sample bottle, or decompose with time, the problem may be solved. With a little luck, an online analyzer may be available as part of the plant monitoring system.

If this appears to be somewhat difficult, then communication is taking place. These statements are not to discourage, but rather to make the test engineer aware of things to expect and to encourage a little research.

Location

The location as well as the quantity of the instruments is important. Four are preferred, two can be used, but voting is not possible. If space for three can be found, then there is probably room for four. Figure 10-9 is a sketch taken from a field test where only two sets of instruments could be installed.

Thermowells must be used for safety. The wells slow things down. The wells should be filled with a conducting oil like kerosene, or if not fillable, with conducting grease. No one wants to measure thermowell air ambient temperature. See Figure 10-10 for a thermowell detail.

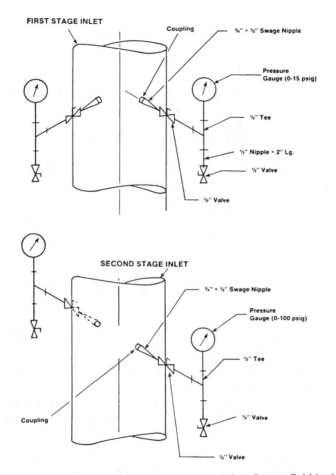

Figure 10-9. Pressure indicator piping arrangements taken from a field test where only two sets of instruments could physically be installed.

Power Measurement

If the instruments are in, the meter run was available, and gas composition can be accurately determined as needed, one last minor hurdle should be addressed. This is power measurement. The indirect method, such as heat balance can be used. In fact, it should be used as a redundant method.

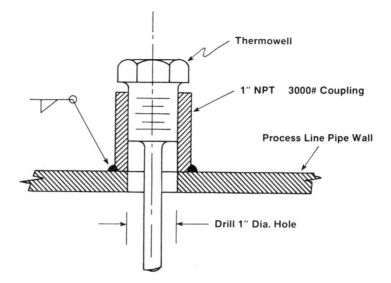

Figure 10-10. Typical temperature tap detail.

Field testing should make use of any redundant method available. However, while a somewhat personal preference, shaft torque is preferred. Couplings instrumented for torque measurement are becoming more common. The reliability of the torque measurement coupling has improved considerably. Possible detractors are cost and spacer length. If the spacer is not long enough to provide for the necessary deflection, the torque output readings are not reliable.

If not available and the driver is a motor, the local electrical engineers may be coerced into procuring accurate current (CT) and potential (PT) transformers to permit the use of watt meters. The everyday plant CTs and PTs aren't suitable for a quantitative field test. When having to solve a compressor's performance, one will use what is available, but only because it is expedient.

Speed

One of the most important parameters to measure is speed, yet it is often overlooked. The fan law in Chapter 5 states that for a centrifugal compressor, the head is proportional to speed squared. For an axial, the sensitivity to speed is as great or greater. Capacity is directly proportional to speed on all compressors, both the positive displacement and the dynamic type. Because of the significance of the speed parameter, it is important that accurate speed measurements be made during a test.

Conducting the Test

The problem with field testing is getting the test gear in place. If the process has been suitably defined, with the computer terminal turned on to the local favorite equation of state, the testing may begin. If the bodies are available to take the data, an all-manual set up works well. If getting to the point to read the data is a problem, then a transducer should be used. It should be good quality and calibrated before and after the test run. Automation by way of a data logger does help keep the scatter down since all points can be read in synchronism or at very close timing.

The test codes offer a good guide to use for the analysis. One technique to smooth the data is to screen off what appear to be bad points. The code suggests one bad reading out of a set of four instruments at any point is cause to discard the point. Since the data are hard to acquire, if three of the four readings look all right, keep the point and throw away the bad reading.

When the data have been screened and performance calculated, don't be surprised if all the points plot like a target in a shooting range, perhaps with a tighter circle rather than describing the complete compressor curve.

Using a cooled bypass around the compressor, or by any means possible, try to establish spread between the points. It is amazing how data points can look different and still come back to the same point on the curve when the calculations are complete. If the compressor is tied to a reactor, the points may have to be taken over quite a period of time which can be done if the instruments are maintained. Be careful of fouling when using this method.

It's better to run a test after a shutdown and prior to putting the compressor on stream. This will allow the test to take place in a day or two. Unlike the shop test, a field test is rarely completed in four to six hours.

References

1. "Compressors and Exhausters," *ASME PTC 10-1965,* New York: American Society of Mechanical Engineers, 1965.
2. "Displacement Compressors, Vacuum Pumps and Blowers," *ASME PTC 9-1970, ANSI PTC 9-1974,* New York: American Society of Mechanical Engineers, 1970. New York: American National Standards Institute, 1974.

3. "Measurement of Shaft Horsepower," *ASME PTC 19.7-1961,* New York: American Society of Mechanical Engineers, 1961.

4. Benedict, M., Webb, G. B., and Rubin, L. C., *Chemical Engineering Progress,* Vol. 47, No. 8, 1951.

5. Starling, K. E., *Fluid Thermodynamic Properties for Light Petroleum Systems,* Houston, TX: Gulf Publishing Company, 1973.

6. Wiesner, F. J., "A New Appraisal of Reynolds Number Effects on Centrifugal Compressor Performance," *ASME 78-GT-149,* New York: American Society of Mechanical Engineers, 1978.

7. Simon, H. and Bulskamper, A., "On the Evaluation of Reynolds Number and Relative Surface Roughness Effects on Centrifugal Compressor Performance Based on Systematic Experimental Investigation," *ASME 83-GT- 118,* New York: American Society of Mechanical Engineers, 1983.

8. Schultz, J. M., "The Polytropic Analysis of Centrifugal Compressors," *Journal of Engineering for Power,* January 1962, pp. 69–82.

9. Wong, W., "Acceptance Testing of Centrifugal Compressors and the Application of *ASME Power Test Code PTC 10-1965,"* I. Mech. Conference Publication 1978-1, Mechanical Engineering Publications, Ltd., for the Institution of Mechanical Engineers, 1978, pp. 123–129.

10. Matthews, Terryl, *Field Performance Testing to Improve Compressor Reliability,* Proceedings of the 10th Turbomachinery Symposium, College Station, TX: Texas A&M University, 1981, pp. 165–167.

11. Neal, D. F., *Centrifugal Compressor Mechanical Testing,* Presented at The Third Compressor Train Reliability Symposium sponsored by Engineering Advisory Committee, Manufacturing Chemists Association, April, 1973.

12. Brown, Royce N., "Test Compressor Performance—In The Shop," *Hydrocarbon Processing,* April 1974, pp. 133–135.

13. Brown, Royce N., *Centrifugal Compressor Testing,* Presented at The Third Compressor Train Reliability Symposium sponsored by Engineering Advisory Committee, Manufacturing Chemists Association, April, 1973.

14. Sayyed, S., "Aerodynamic Shop Testing Multistage Centrifugal Compressor and Predicting Gas Performance," *ASME 78-PET-28,* New York: American Society of Mechanical Engineers, 1978.

15. API Standard 613, *Special-Purpose Gear Units for Refinery Services,* Second Edition, Washington, D.C.: American Petroleum Institute, 1977.

16. API Standard 617, *Centrifugal Compressors for General Refinery Services,* Fourth Edition, Washington, D.C.: American Petroleum Institute, 1979.

17. API Standard 618, *Reciprocating Compressors for General Refinery Services,* Fourth Edition, Washington, D.C.: American Petroleum Institute, 1995.

18. API Standard 619, *Rotary-Type Positive Displacement Compressors for General Refinery Services,* Second Edition, Washington, D.C.: American Petroleum Institute, 1985.

19. "Flow Measurement," Chapter 4, "Instruments and Apparatus," *ASME PTC 19.5; 4-1959,* New York: American Society of Mechanical Engineers, 1959.

20. Lock, Jack A., *Techniques for More Accurate Centrifugal Compressor Performance Evaluation,* ASME 29th Annual Petroleum Mechanical Engineering Conference, New York: American Society of Mechanical Engineers, 1974.

21. Bultzo, C., *Turbomachinery Acceptance Testing for Conformance to Design Criteria and Overall Performance Evaluation,* ASME Petroleum Division Conference, New Orleans, LA, September 14, 1972

22. "Measurement Uncertainty," Part 1 Instruments and Apparatus, *ASME PTC 19.1,* New York: American Society of Mechanical Engineers, 1985.

11

Negotiation and Purchasing

Introduction

After the estimates are made, process design has taken shape, and the funds to build the plant are authorized, it's time to buy the equipment. To the uninitiated, the task can appear quite formidable. Even to the veteran, the chore is great with a multitude of details to remember. Note, the title of the chapter is not how to, because that is left to each engineer and his client or company to decide. What will be presented here can be more nearly described as a checklist and outline to help the purchaser recall all the details and avoid as many surprises as possible.

Procurement Steps

This chapter will present a list of steps that represent the general industry practice in the procurement of a major piece of equipment such as a large compressor. This list would be modified by a specific user as required for a given job, but most of the steps are recommended for purchasing a large compressor, regardless of the type. As the size of the machine decreases, some of the steps can probably be dropped. The order of the list is in the sequence in which the events take place. After presenting the list, a general description of what each step entails will be given.

The steps in the procedure include:

1. Preliminary sizing
2. Inquiry specification
3. Bid/quotation
4. Bid evaluation
5. Pre-award meeting
6. Purchase specification
7. Award contract
8. Coordination meeting
9. Engineering reviews
10. Inspections
11. Witness tests
12. Shipment
13. Site arrival
14. Installation and startup
15. Successful operation

Supplier Partnerships

Supplier partnerships or alliances have been used to control the cost of procurement. They are based on methods developed in Japan, initially in the automobile industry. These methods have been extended into the process industry and for the purchase of such items as compressor trains. Generally as originally conceived, they were intended mainly for commodity items. By using some innovative approaches, the concept of partnerships or alliances has been extended into the purchase of custom equipment.

In this text, partnerships include alliances. Some companies prefer the alternate name. Partnerships are structured in many different ways. They may be project specific, commodity specific, or even company specific. Each company has its own qualification criteria. It is beyond the scope of this book to go into detail on these because that is material for a text of its own. The intent here is to introduce the reader to the concept if it is not known and meld in the negotiation modifiers for both neophyte and veteran.

There is basically one goal for the partnership, with possible side benefits. The goal is to save money. When all the dust settles, that is it. However, there are associated benefits, the best of which is to minimize or even do away with adversarial relationships. It may be wishful thinking to believe that these will be completely eliminated. A related, but very important, benefit is to build trust between the parties. This is necessary to achieve the goal.

One outcome of the partnering arrangement is the minimizing of the supplier base. Some advocates call for single sourcing, others take a more conservative approach and keep two or three potential suppliers, and use criteria to select from the individual companies. The purpose is to lower the bid and bid evaluation cost. Going out for bid to many suppliers is costly to the company and wasteful for the industry and eventually shows up as a cost to the user community. The experienced practitioners of partnering attempt to eliminate bidding and directly negotiate the purchase of the compressor train for the custom-type equipment. For more standard commodity items, a blanket purchasing agreement is worked out for a fixed period of time, such as a year. The elimination of bidding removes a significant step in the procurement process, and the purchaser expects to benefit both from the elimination of the bid evaluation as well as in the negotiated price due to the supplier's savings in bid preparation. Additional savings may be realized in the form of specification standardization. The reduction of variations in specification from project to project can produce significant savings.

The concept of partnering is quite interesting and, when carefully done, does produce benefits for the purchaser and the supplier. It is heavily dependent on people and their ability to engage in a trusting relationship. There are many interesting variations that, while tempting to write about, must be left for another book. While the following steps are still valid, it is hoped that the potential for eliminating or streamlining some steps is possible with a partnering arrangement.

Preliminary Sizing

The preceding chapters offered some general guidance on the sizing of the various types of compressors. Before the start of the specification, the equipment should be sized, at least in a preliminary manner. Actually, part of this may be integrated into the process calculation. However, after the process calculations are complete, a review of the equipment best suited to perform the task at hand should be made.

One reason for the sizing step is that it saves time and embarrassment if bids are collected on the correct type and style of compressor. It just doesn't look good to use a centrifugal compressor specification if a reciprocating compressor is needed. In fact, even the specifying of a single stage when a multistage is needed because of temperature limits, doesn't start the procedure on a proper note.

It is not unusual to find more than one compressor suited to the job. Additional factors, such as inherent reliability and efficiency of the various types, should be considered. First cost will be an unavoidable factor that must be addressed. If, after consideration of all pertinent factors, there is still an overlap, then it may be advisable to inquire about more than one type of compressor. Normally this doesn't happen, because once the types are reviewed, some factor sways the decision. If, however, more than one compressor will be inquired, a specification should be prepared for each. It is quite difficult to write a comprehensive multipurpose specification. While there are some areas that will be the same, a repeat of each document will make evaluation much easier and cleaner later on.

Specifications

A specification should outline to the bidding vendors just what it is you want to accomplish with the equipment and how reliable it must be. Compressor reliability can be achieved if it begins prior to vendor design and fabrication. Therefore, a truly reliable piece of equipment begins in the specification.

Basic Data

Enough data must be supplied to rate the compressor. For example, the mass flows, inlet and outlet pressure, inlet temperatures, type of gas or gas physical constants need to be itemized. The basic data must be very clearly stated and complete to the extent necessary to achieve a common understanding between the user and vendor.

Operations

Very few pieces of equipment operate at a single set of conditions. Thus, operating conditions must be divided into a set of normal conditions and those other than normal. The entire anticipated range of operating conditions should be defined either by range limits or alternate operating conditions. To establish guarantee points and nameplate rating points, unique conditions within these parameters must be selected. After defining the normal operating conditions, consideration should be given to "off-design" conditions. For example, these would include abnormal molecular weight deviations as might be encountered in the startup of a machine on a gas other than the design gas. Another example might be encountered in an equipment holding period where part of a process is not in operation.

In describing abnormal, unusual, or off-design operating conditions, even insignificant items should be mentioned, if for no other reason than to establish conversation at a later date. While these conditions may frighten the vendors at the time, it is better to point them out in the specifications rather than to use the "Oh, by the way" method to the service representative at startup. Examples of little potential vendor surprises are hydrogen sulfide or chlorides in the gas, even though they are in the ppm levels. Many times it is erroneously assumed that a vendor's knowledge of the process is more than your own, and therefore, the vendor should have been aware of these unusual conditions. The same reasoning can be applied to known potential process upsets. If the vendor has had an unfortunate experience with a process similar to yours and recognizes it in your process, he may well anticipate the problem. However, you should be much more aware of your potential upsets than the vendor. An example of this might be the temperature runaway potential in the gas being fed to a hot gas expander. At least the specification could be used to question the vendor on the degree of over temperature tolerance that his machine can handle. Another example is the potential of the sudden dead ending of a centrifugal or axial compressor during switching operations of batch-type reactors.

Some consideration of surge characteristics should be made, and unless your own plant design includes antisurge controls, it should definitely be mentioned in the specification. Possibly, the vendor can quote antisurge equipment with his offering. At a later date, the specification will serve as a good reminder to both the vendor and purchaser if some mention of compressor control is made. The details may be worked out at any convenient point, such as at the bid review or possibly later in a coordination meeting.

One way to decrease your reliability and cause problems is to not mention fouling potential in centrifugal compressor specifications if process experience exists. The experience may exist but not be recognized because it did not come directly from operating centrifugal compressors but from a smaller plant operation with reciprocating equipment. By mentioning a fouling potential in the specification, it may be possible for the vendor to include some form of washing in his offering. Alternatively, he may be able to allow additional head margin in his design. By working together, the period of time before cleaning a unit can be maximized.

Writing the Specification

Up to this point there has been no mention of how you begin to write a specification, or even the basic organization thereof. Since most user companies have established formats, the foregoing comments are intended to be general enough to fit an existing format. For equipment covered by an API standard, the past practice has been to write an "overlay" specification to supplement the API document. These added specifications address the exceptions the specification writer wishes to take to the API document and also allows him to select options and provide the site specific as well as environmental information called for in the API document. There has been an industry trend to minimize the "overlay" specifications and to use the API document without exception. Because of user input, later revisions of the API documents attempt to address this desire of the industry. User information is placed on data sheets, making it more convenient to select options and provide user data. For non API compressors, the procedures outlined here are still applicable. While not as great a task, the options selections for API-based specification can also benefit.

Another useful approach is to attempt to get a broad review of the specification draft by operations, maintenance, and engineering. The specification must have an author who can write the basic document and who later edits the draft into the scope and language of the final specification. Single authorship, with committee review, is preferred over committee authorship. Only in rare instances is vendor participation in the specification authorship really justified, except in partnership relationships.

Specification Outline

The items mentioned above are basic design considerations that must be considered for all specifications. Before proceeding, a list of items to be covered in the specification will help establish a format for a specification draft. These items include:

- General
- Basic design
- Accessories
- Inspection and testing
- Vendor data
- Guarantee and warranty

General

The general section covers a broad group of items. These items may apply to any of the specific areas to be included in the following sections. This section normally starts with a description of the equipment being inquired, or later purchased, such as a reciprocating, centrifugal, or screw compressor. This is followed by a brief description of the scope of supply. If item numbers are used, the item numbers that will designate the equipment should be included. Injection should be specified at this point if needed. Code requirements, such as area classification and statements concerning reliability should be included. Even rules on how the specification should be answered, such as the addressing of each paragraph in the proposal, can be mentioned early, rather than waiting until the end where specific proposal items are stated.

Other items related to the site conditions that should be communicated are temperature range, if outdoor and type of protection, if outdoor (Is there a roof? Are there sides? Is the entire installation completely open?). The requirement for heat tracing and other details of this nature that the vendor is to include in his scope of supply should be specified. Whether the machine is at grade or is elevated must also be specified.

For many of the details, the use of a data sheet is helpful. The API standards for compressors have data sheets that help act as a checklist. It is suggested that the major items, those just mentioned, be included in the verbage in the body of the specification as well as on the data sheet. The data sheet collects details of the items, especially those that are more clearly conveyed in tabular form.

Basic Design

At first glance, it may appear unnecessary to specify design parameters. This might appear to be the vendor's responsibility. While the vendor does take care of the basic design, there are optional areas where he must receive instructions (by way of the specification) to be able to complete the design.

The data sheet is used to convey the process design parameters, capacity, temperature, pressure, and gas data. The verbage can explain or better define any of the specifics. For example, if gas is to be removed from the compressor and cleansed, cooled, and otherwise modified by the owner's equipment, this is the place to provide the interfacing parameters and brief explanation for the vendor as to where or how to react. This information can be used to establish the diaphragm pressure differential for a centrifugal compressor. This in turn affects the mechanical design. Also, statements concerning momentary flow interruptions that may occur with switching exchangers should be conveyed.

If the decision for split-case or barrel construction is to be left to the vendor, enough details of the gas must be outlined. There is no reason the user can't make these determinations on his own and specify them in the scope.

If casing limitations are fixed by user-supplied relief valves, this information should be conveyed to keep the vendor from rating the compressors on other data. Evaluations can be more of a problem if the same design basis isn't universal with all vendors. Startup and shutdown consideration influence various components, shaft end seals, seal system pressures, and even thrust bearings in some instances. The use of an alternate startup gas, or the desire to operate a gas compressor on air to aid in plant piping dryout should be covered.

If there are speed limitations, such as piston speed in reciprocating compressors, these must be stated. While it would seem a range would be good to give latitude, it really works in another manner. Whatever is given as the high number will be the target, with the vendor testing the approximate value given and exceeding it by some amount to see if a concession can be gained. As a practical step, values should be given as maximum allowable and then held at that value to keep the bidding compatible.

Other options, such as replaceable cylinder liners, must be covered. API mandates replaceable liners with some variations allowed when they are not available. It would be beneficial to state the acceptable alternatives. This approach is probably useful in all areas.

Cylinder connection locations should be defined, as well as the use of air cooling. API 618 requires jacket cooling unless the user specifically states otherwise. Rider rings or wear bands, which are useful in helping extend ring life, are an option in API 618 and normally will not be furnished unless specified. If there is any doubt about a feature, rather than blindly specifying, a call for vendor comments can be made at that point in the specification to bring out later discussion. The price can be included as a breakout or adder.

For centrifugal compressors, nozzle orientation must be specified. If the compressor is on an elevated foundation, the down-nozzle orientation makes maintenance much easier, because major piping does not have to be removed to lift the casing upper half. Figure 11-1 shows an unusual case where all the nozzles are in the upper casing half.

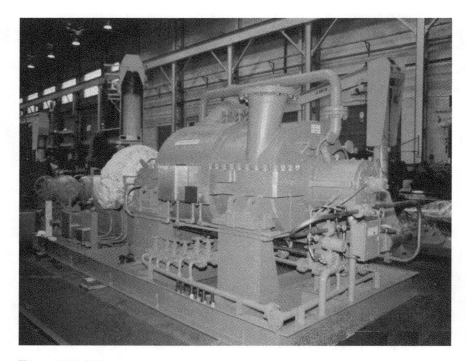

Figure 11-1. This steam turbine-driven centrifugal compressor, shown just prior to shipping, has all nozzles in the upper half of the casing. Note the non-vertical orientation of the inlet nozzle. (*Courtesy of A-C Compressor Corporation*)

In reciprocating compressors, the use of explosion-relief devices must be stated if desired. The type and style of distance piece must be specified together with the connections for the collection and disposal of leakage gas. API 618 gives a rather complete coverage of all options.

A basic item on the reciprocating compressor is the lubricated cylinder. The vendor will need guidance from the user advising whether lubrication can be used.

For the screw compressor, in many cases the application will dictate a flooded or dry type compressor. The user can either state this specifically if possible or give enough information for the vendor to make a choice. Care must be exercised in that all vendors cannot furnish both. The user may be required to evaluate the two types for the same service.

There are many other options that can be specified. The appropriate API standard can be consulted. Paragraphs that contain user decisions or choices are marked to help the user in writing the specification. Vendor literature generally gives guidance in those areas where the vendor has options available, which is a help for non-API equipment.

Materials

Past experience by operations provides a guide to materials specifications. It is unwise to assume that the vendor is completely knowledgeable about materials for the user's process. Stating operations' experience with materials or defining materials in the specifications, will help the vendor in his design and will prevent his having to gain experience the hard way at your expense. However, you do not have to make the specification so rigorous that you prohibit vendor experience with materials. By proper wording of the specification, you can invite comments that reveal vendor experiences in materials. Specifying a non-standard material will call the vendor's attention to a possible non-routine situation. A large number of forced outages are traced to vendor design, but they really reflect the lack of application knowledge, which could have been prevented if the purchaser communicated the specifications properly.

Some of the most obvious examples of problems with gas and materials are frequently found in refining or petrochemical applications. One is the presence of hydrogen sulfide. Austenitic stainless steel, normally a premium material, cannot be used if chlorides are present due to intergranular corrosion and subsequent cracking problems. The material choice is influenced by hardness limitations as well as operating stresses that may limit certain performance parameters.

Material and welding quality inspections, based on industry standards, are desirable. Again, API gives some guidance. Beyond this, the inspection method and acceptance criteria must be clearly stated. In preparing his bid, the vendor must understand exactly what is desired.

Bolting materials must be considered. While not strictly a material problem, bolting also includes the requirement for studs, the tapping requirements, bolt head style and many aspects, as well as plating type if allowed. In the various compressor standards, API provides guidance, but particular plant experience may add to or modify those requirements.

Bearings

On some compressors (of any type) there may be a choice of bearings. For high reliability compressors in critical service or where outage is costly, the choice would be for pressure-lubricated journal and thrust bearings. Babbitted liners or pad type are preferred to rolling element bearings when available. For the higher speed compressor, tilting-pad bearings are preferred to plain journal bearings. For thrust bearings, the self-leveling, tilting-pad type is preferred. If imbedded bearing temperature sensors are desired, they are generally specified in the bearing section of the design portion of the specification rather than later when peripheral instrumentation is covered.

Shaft End Seals

By appropriate notation, the vendor may be requested to offer alternatives for the specific application, if a particular type of seal has not been selected at the time of the specification writing. It is very difficult to generalize the seal selection, because of the many requirements and numerous methods available. For specification purposes, the single most important factor to pass on to the vendor is the requirement for a positive seal, which would be a mechanical contact of the oil buffered or dry gas type or oil film seal. All other types have significant process gas leakage associated, which can be controlled if required, but at best cannot be considered in the positive class. For oil buffered seals, a quantity of acceptable inner leakage should be established for use later at test time. In the inquiry phase, it can be given as a target value for vendor comment or acceptance or as a requested value from the vendor, which would then be made a contract value for the order specification.

Seals, particularly the restricted leakage type, require auxiliary piping, control valves, and other items. If more than a suggested schematic is to be included in the vendor's scope of supply, a statement to that effect should be made. For the oil buffered systems, if a separate oil system is required, this must be stated, and it can be detailed in the accessory section with the lube system. For dry gas seals, the type of support system must be specified. The vendor's proposal should be requested to outline materials and to permit user review of the standard materials for compatibility. As always, if a problem is known to exist or is suspected, acceptable materials may be specified. The vendor can use the list to review his ability to conform or to propose alternatives. This method tends to save some time.

Accessories

Often, more time is spent specifying and discussing accessories than is spent on the basic compressor. This is because the open options are much more numerous. Also, the average user is more familiar with piping and instrumentation, both of which fall into the accessory area.

Lube and Seal System

The specification must recognize this very important auxiliary part of the compressor. API 614 is a very complete specification and should be used when applicable. When direct use is not applicable, it makes an excellent guide. The seal system in many cases is combined with the lube system. When separate, many of the seal system components are fundamentally the same, with some specifics, such as operating pressure and control, being unique. For those applications where the sealant is other than lubricating oil, the particular fluid will need discussion. The system basic components, on the other hand, will probably be quite similar. General items that must be considered include:

- *Piping.* Specify materials—stainless steel has numerous advantages. Specify welded construction and the minimum use of fitting. Tape-type thread lubricants should be avoided.
- *Pumps.* Specify the type and number of pumps required. Minimum is two full-size pumps.
- *Driver.* If steam is available, a common choice is a steam turbine driver for the main pump and an electric motor for standby. Some plants prefer two electric motor-driven pumps. Give the minimum steam condition. Give the expected voltage drop on the electric system if more than 10% is expected, especially on large systems.
- *Accumulator.* Required for multivalve steam turbine servo control.
- *Filter.* Specify twin filters with a non-interrupting transfer valve. Specify the degree of filtration (10 microns or lower, possibly even one micron in some applications).

• *Cooler.* Specify twin coolers with non-interrupting transfer valve. Specify materials of construction, utilities, and fouling factor.

• *Reservoir.* Retention time should be 5–8 minutes. Stainless steel material is recommended.

Drivers

Chapter 7 attempted to give sizing criteria for the various driver types as well as some descriptive detail. If the vendor is to furnish the driver, the type of driver should be specified in the compressor specification. Driver specifics should be placed in a separate specification, making sure that areas of similarity are described consistent with the compressor specification. Even if the vendor is capable of supplying the driver from within his own organization, the specifics should stand alone because a different group of people will be working with it. It is safe to assume that the two groups do not communicate any better than if they were two independent organizations. Should they by chance communicate, everyone would be ahead, but nothing is really lost in preparing two documents. API covers steam and gas turbine drivers quite well. Motor standards are available for both induction and synchronous motors.

Many large users have a tendency to not include the driver as a part of the compressor vendor's scope of supply. If this is the option selected, the compressor vendor should be so informed. Not only will the compressor vendor need to know the type of driver, but also information relative to the compressor vendor's responsibility regarding the driver must be established. There are numerous interfaces that must be delegated either to one or the other, the vendor or the user. Responsibility for drawings of the complete train, such as the arrangement drawing, lubrication interface, and other similar drawings must be stated. Also, system response analysis, such as torsionals, should be assigned. Details that are easy to forget are items at the interface, such as coupling, coupling guards, and items of this type. If the equipment is to arrive at the site ready-to-assemble and on-time, someone must be assigned each item of responsibility. Sometimes the compressor vendor is assigned coordinative responsibility, for a fee, of course.

Gear Units

Many of the compressors require some form of speed matching to the driver. For most drivers to reciprocating compressors, this will be a reducer. For gas turbines to almost all compressors, a gear unit is required, which could be a reducer or an increaser. Motors to centrifugals are almost universally associated with increasers. While basically simple devices, gears have been the cause of unscheduled shutdowns and problems. As with all components, the gear must be given specification treatment commensurate with the rest of the compressor train to achieve the same level of reliability desired for the compressor and driver. As pointed out in the accessory chapter, the gear is covered by API 613 or API 677. Normally, the compressor vendor will furnish the gear, so the gear specification is a part of the compressor specification package. There is no reason it can't be a stand-alone document. There was a time when gears were strictly purchased on price. If the compressor vendor is not checked, this may still happen. This is not good. If nothing else, the specification for the gear, regardless of how documented, should give the user some selection power regarding the gear. In the overall competitive bidding, the gear should not be made a factor as such. This has proven to be poor economics in the past.

Couplings

It is impossible to specify a completely foolproof coupling, particularly by manufacturer and model number. Still, in the interest of reliability, the specification should cover the subject of couplings. This requires the use of plant feedback and a lot of research on the part of the specification writer to determine the latest developments in coupling design. Unfortunately, when a vendor furnishes a coupling, it often becomes just another outside purchased item that he may purchase with a minimum specification and strictly on low price.

API again offers some help. API 671 covers the special purpose coupling as either a specification or guideline. It has been common practice for the compressor vendor to furnish the couplings. Most compressor vendors have no problem accepting this responsibility, while driver vendors sometimes prefer not to furnish them. In some compressor types, the coupling style will be somewhat influenced by the user. When torsionals become a major consideration, the vendor may have to make the choice based on system need. In all cases, the user should establish his right of approval if the vendor performs the selection.

Mounting Plates

Either soleplates or baseplates should be furnished for most of the compressor types. Reciprocating compressors are sometimes directly mounted to the foundation and the frame grouted. Most other compressors are mounted on an intermediate surface in the form of soleplates or a structural steel base. The user must make a number of decisions, beginning with the basic decision of sole or baseplate. This decision must be revealed to the vendor in the specification. If a baseplate is chosen, the extent of the baseplate must be established, such as a full length base to be used for all equipment on the train. If soleplates, who is to furnish the soleplates? It may be each individual vendor, the compressor vendor, or the user. Original setting conveniences may be further specified by calling for adjustment screws or subsoleplates and shims to help level either the soleplate or baseplate. It is recommended that the soleplates be machined on both sides to facilitate original set up. The soleplate can be pre-coated with an epoxy primer on those surfaces to be covered with grout if an epoxy grout is to be used. These decisions are left to user or user and contractor, as opinions differ widely. For centrifugal, axial, and large screw machines of the dry type, soleplates are recommended over baseplates. This statement is the author's preference, recognizing there is a wide variation of opinions.

Controls and Instrumentation

Each user has a favorite list of vendors that conform to plant standards for the various control and instrument items. These include gauges, transmitters, control valves, and the host of various hardware used. In the past, the compressor vendor was involved with supplying vibration and bearing temperature sensors and monitors. Larger users now prefer the purchase of monitors to be a user function, allowing the compressor vendor to supply the sensors in most instances. It is probably best to allow the compressor vendor to furnish the control loop used locally or the lube and seal system, primarily to give overall responsibility on vendor furnished items. On those items interfacing to user control loops, the user should probably maintain responsibility. These thoughts are offered in the form of suggestions rather than recommendations. It certainly is in order for a user to include a list of preferred vendors with the specification. However, it is good to keep an open mind, particularly where the compressor vendor, who is being held responsible, has a strong feeling for a given instrument vendor.

For centrifugal and axial compressors, some form of override control is recommended for constant speed motor drivers to sense motor overload and override the process control until the cause of overload has passed. The override can take the form of closing, moveable inlet guide vanes, when these are purchased, or a suction throttle valve. The compressor vendor frequently furnishes them.

Control panels usually become quite a complex item because of the great amount of coordination necessary in having them conform to plant panel standards, including non-compressor vendor items that usually end up there, and generally can be quite a problem. It is recommended that a great deal of planning take place prior to specifying a panel in the compressor specification. If it can be fully specified early, it may not be a total nightmare. Generally, a user is better off to take on these items himself. This does not include the highly standard panels that are offered with plant air packaged compressors.

Inspection and Testing

If the API standards are used, the inspection requirements are reasonably well outlined. At this point in the specification, the type of inspection to be used as well as any special inspection should be stated. The details can be resolved later, but the vendor should be given some idea of what inspection methods are used, such as visiting, resident, or no inspection. The tests that are to be witnessed should be covered. API defines two categories of test witness. One is the formal witness test, generally where the vendor runs a pre-test prior to running the customer witness test. Since this is double testing, it is more expensive to the vendor who may, if he feels the market will allow, pass the extra cost along in his bid price. The user is usually kept guessing at that point. The other test is the observed test. This works reasonably well with resident inspectors who are technically able to act as test witnesses. The problem is the schedule of the first test is not too predictable, as it depends on vendor shop loading, last minute problems, and dozens of little irritating items totally unpredictable yet ever present. The attraction to the observed test is that it is the only test and, therefore, should not carry an extra charge to the customer. With the witness test, there is some advance notice required. Because the test has been run, the formal test should be of a minimum duration. The evaluation for which type of test to use is strictly a user decision. If engineers must attend the witness test (where the engineer's time is critical), then the best choice is to take the witness test. While extra performance testing can be specified, outside the code performance test and the selection of the ASME testing, there is not too much incentive for the vendors to go further. The ASME type code tests are expensive, but when reliable operation and performance are important, they do help determine field performance and give a checkpoint from which to proceed if problems develop in the field. Also, they do catch problems while the compressor is still in the shop. There is a significant increase in cost to repairing problems in the field compared to doing it in the shop.

Full power, string tests and the other optional tests, as used in centrifugals, must be evaluated for each application. There is a strong feeling that string tests do find problems, particularly since contract couplings are used and criticals or vibrations due to overhung weight can be uncovered.

Vendor Data

All along, suggestions have been made for the proposal. This is the place in the specification to summarize the data, curves, and drawings expected with the proposal. At inquiry time, it is also time to anticipate what documents will be needed from the selected vendor. This helps the bidder to properly plan the amount of documentation and to evaluate it against his standard. Don't expect a large extra amount of paperwork to be free or that the vendor will comply where it's not stated. Spell out all documents logically so that proper evaluation of the purchased equipment is possible and all the data needed to complete the site engineering are provided. The final document list should be covered again to be sure it is clear and reasonable. Probably the most important document to single out is the vendor's portion of the data sheet, which should be filled in and returned with the proposal and then later completed for an as-built, or at least, as engineered record of the compressor. This one document is very handy when questions about the compressor are raised years after shipment.

Guarantee and Warranty

Current policy in API is to avoid commercial language in the API standard. API standards do not include a guarantee and warranty section. In the revisions, as they occur, other paragraphs deemed commercial are disallowed. This puts the burden of guarantee and warranty verbage on the user. A practical solution, if the user is not familiar with industry practices, is to solicit from the vendor a proposed guarantee that should be a zero tolerance on head and capacity. This would only be good practice. Then the negotiation can take place on the input power guarantee. As a design acceptance value, U.S. industry has used plus or minus 4%. However, the vendor should so state in his proposal, and this can then be negotiated or accepted, and made part of the contract. The warranty period can be treated the same way. Large users have negotiated this item through their own terms and conditions for some time and have rarely relied on API.

Bid or Quotation

After the inquiry specification has been issued to a group of vendors by the Purchasing Department or company equipment group, the vendors start the bid preparation. There is very little for the user to do but wait. Questions that arise must be funneled through communication channels as set by each organization. It is suggested that a neutral atmosphere reign during this period. Questions should be answered as asked. Don't tell the fellow how to build a watch if he only asks the time. If the question raised reveals a specification defect or a general problem, an addendum, or letter, should be sent to all vendors clearing that specific point. Good business would seem to dictate that a due date be set and adhered to. If an extension is granted to one vendor, all vendors should be granted the extension. This, of course, is again dictated by individual company policy.

Bid Evaluation

When all bids are received, one engineer should be assigned the job to review the bids for completeness and to prepare a bid tabulation. On a large job, this job may be divided into several tasks and may take a considerable length of time. On smaller jobs, it may be a relatively short project.

Once the bids are tabulated for specification compliance in the form of a chart for easy review by all others involved in the project, an overall evaluation should be made, factoring in energy cost, first cost, and time value of money using an established economic equation. Most companies have a standardized formula. If the data are available, total cost of ownership can be estimated, which for larger equipment is considered a good measure for evaluation.

The vendor order can be established on a bid-factor basis. To this list can be added pluses or minuses based on experience with the vendor. In some companies, a spare parts package cost is obtained with the bids.

Any resemblance to the first package cost and any later spare costs, unfortunately, is a coincidence. It is suggested, if funds permit, that at least the first spares be purchased with the compressor because they can be expected to come at a good price.

Use the pluses and minuses only when vendors of equal experience are available. It is a trap to assume that the vendor for whom there are no minuses might be great if the lack of minuses is due to lack of experience. Time will surely correct this.

Pre-Award Meeting

Once the evaluation is done, a preferred vendor should have surfaced. Beyond the evaluation, it is up to the user, his company policy, and conscience to make the selection. Once made, a meeting with the vendor is recommended. It may be good practice to keep the second choice vendor in the background at this time. Also, under no circumstances should anyone leak the selection to the vendor. This is the first of a trading session, and until the vendor is sure he has an order, he will stay in a trading posture. If two vendors were tied or too close to call, both vendors may be called in individually.

The vendor should be informed that a decision is imminent and that he should bring technical personnel capable of fielding in-depth questions.

The specification and the proposal should be reviewed in depth. Options and their cost should be covered. Completeness is important because it is now time for the vendor to commit to any remaining items. If there was a single vendor in the lead, then, once the open questions are resolved, it would be safe to place the order. If, at the meeting, all questions were settled in a satisfactory manner, it can be quite good psychologically to place the order on the spot.

On the other hand, if there appears to be a problem, particularly where there may be more issues open than resolved, a second-place vendor may be interrogated. After the two meetings, those involved can make the selection or recommendation, as policy dictates, as to the final candidate.

One item to mention that may seem trivial is that in many cases a somewhat adversary relationship often develops between vendor and user. This is really counterproductive. It may be inherent in the role of the user to be somewhat paranoid. It is worth spending time doing attitude adjustment to continue the project as partners working to get a joint successful project. Actually, the project is really more fun. There is a lot of work, but there is no point in making it other than pleasant by having poor human relations. If people problems seem to appear early on, do something to get them resolved.

Purchase Specification

While it is somewhat tedious to document the myriad details that have at this point been agreed to, it is time to rewrite the inquiry specifications. Care should be taken regarding the vendor's cardiac health to use the same verbage so as not to appear to be springing surprises. The best way to upset the vendor is to write into the specification something completely new or different from all the previous agreements.

Award Contract

While the vendor may have been told that he is the successful bidder, he becomes the vendor when a contract is written and accepted by him. This is important because the clock is started at this time and all future dates will be referenced back to this date. This also is the date from which delivery is counted. The so-many-months-after-approval-of-drawings method is pretty obsolete. Drawings are reviewed, not approved. Vendor manufacture begins officially with the contract acceptance.

Coordination Meeting

Approximately six weeks after placing the order, a coordination meeting should be held with the vendor. While such a meeting is probably not beneficial for a standard compressor or standard package unit, it can make a significant difference in time saving and understanding between the user and vendor on custom engineered equipment. If the compressor system is of the custom design or if significant changes are required of a standard design, then this meeting will be beneficial in those instances. Compressors that come into this category are process compressors of the centrifugal, axial, and larger screw types. Large reciprocating compressors, particularly of the API 618 type, qualify. Small rotaries or small standardized packaged air or refrigeration compressors do not.

An important factor, endorsed by many users, is the meeting location. This meeting should be held at the vendor's plant. It is important for the vendor's personnel to have some exposure to the user. This meeting is intended for the actual vendor working people to find out, first hand, exactly what the user wants. Detail specifications get direct answers. Most of the people at the plant involved in following the various parts of the compressor through the manufacturing process would normally not travel to the user location. At this point, the user has fewer people involved. Also, the user, who is not familiar with the vendor's facilities, gets a chance to become more acquainted. Items to be reviewed must be set up in an agenda and should be available to the user a week or more ahead of the meeting. A partial list of the items is as follows:

- Purchase order, scope of supply, and subvendor items
- Data sheets
- Schedules for transmittal of drawings, production, and testing
- Inspection, expediting and testing
- The physical orientation of the equipment
- The lube-oil system
- Seal system
- A review of applicable specifications and previously agreed-upon exceptions to the specifications
- Drivers
- Couplings

Additional items may be covered as warranted. Additional meetings may be planned, with the location based on meeting type, for purposes of reviews. For example, with compressors where subcontracted lube systems are used, a meeting at the subvendor's location to review a CAD model and the overall layout would be appropriate. With reciprocating compressors, a meeting at the location where the analog study is being performed should be scheduled. A review of the rotor dynamics study on a centrifugal or axial compressor, at the point where the analysis was made should be considered. Foundation and piping model layout review would take place at the engineering location. These are just a few examples, but all meetings should be attended by those people actually involved, with a project representative from both parties to monitor the discussion and keep everything within the contract scope.

Engineering Reviews

After the coordination meeting, the user will begin receiving drawings, reports, and all the documents called for in the contract. These are reviewed and returned to the vendor as "reviewed without comment," or if a discrepancy is noted, this is marked and returned "with comment." Some of the larger users use an in-house design audit procedure, particularly on the larger, critical equipment trains. In the design audit, as the name implies, the designs are reviewed rather extensively. The idea here is that any error caught early saves both user and vendor time and money. It probably has a further benefit that may be the more valuable: it gives the user a better understanding of the compressor he will be receiving, making subsequent questions or problems easier to answer or solve.

The additional meetings, if planned, take place during this period. Unanticipated problems of any nature that arise during this period should be given immediate attention. Allowing a problem area to go without an adequate solution will only appear again later when time is running short and more components have been manufactured, making rework more expensive.

Inspections

Each company has its own inspection system, so it is difficult to generalize. The inspections planned should have been discussed as part of the coordination meeting. For large projects, additional inspection meetings should be planned. During the engineering phase, inspections will begin as parts are being manufactured, depending on the extent of the user's inspection.

During this period, the work schedule should be monitored to highlight any possible delays. Late shipments may be avoided if any materials, subvendor shipments, or manufacturing problems can be discovered early and if both parties work together to solve them.

Tests

A certain number of the tests will have been designated as witness tests. One is usually the hydrostatic test, which does not require an engineer but can be witnessed by an inspector. As a cost-saving measure, if engineers from another project happen to be in the plant at the time of the hydrostatic test, they may be able to arrange to witness the test, because the duration of the test is normally not long. It may take a total of two hours.

If a gas leak test is called for, the test can take place shortly after the hydrotest, especially on the reciprocator cylinder or centrifugal or rotary compressors, where the gas is tested without end seals.

The running tests require time and are generally witnessed by one or more user-engineers. A good team is one design engineer and one maintenance engineer. Of course, more can attend depending on company policy. Agreement on specific test procedures should be reviewed before the test. At times, a performance and mechanical run test may be combined if all conditions for each coincide. More normally, they don't coincide and must be run sequentially. For the additional testing, like string tests and other multiple unit tests, there may be enough delay, because changeovers on the test stand are made, so that it is more economical for the witness to make multiple trips rather than take up an extended residency (see Figures 11-2 and 11-3). The witness must be familiar with the contract, the specifications, and all agreements to properly administer the agreed-upon acceptance criteria. It is counterproductive to have a witness unknowingly approve the test outside the acceptance limits or conversely, to insist on a value more stringent than that for which the contract calls.

A test frequently overlooked is the lube system test. Because of the complexity of separate lube systems and the fact that they may be manu-

Figure 11-2. Changeover on a test stand can be quite time-consuming considering the complexity of the piping and other setup requirements. (*Courtesy of Elliott Company*)

Figure 11-3. Testing of the more standard integrally geared air compressors can be somewhat complex. (*Courtesy of Cooper Turbocompressor*)

factured at another facility, a separate test is called for. If the timing and logistics permit, the contract lube system can be run with the compressor and driver on the compressor test stand. For any number of reasons, this is normally just not feasible, and a separate test is needed. The witnessed test should carry the same qualifications as required for the compressor.

Shipment

Once testing and final inspection are complete, the unit is prepared for shipment. The specification should have detailed the type of storage anticipated and expected time. It should have specified indoor, warehouse, outside storage, or whatever is planned. An anticipated time point should also have been stated. The contract should have stated the mode of shipment and destination instructions. Upon notice of shipment, the traffic department (if there is one) may wish to monitor the shipment.

Site Arrival

Upon site arrival, the compressor and associated components should be carefully unloaded. A review of the shipping manifest and the components should be made to see that all parts are accounted for and were not damaged in shipment. If the unit is to be stored until a future date, care should be taken to provide as much protection as possible from the elements. Details for proper care should be coordinated with the vendor.

Installation and Startup

Commissioning the Compressor

Ideally the compressor can be placed on the foundation, aligned, grouted, and piping connected. Shortly thereafter, the unit can be started. Most of the time there are lags between the various steps. User carelessness here can delay ultimate startup and may mar what otherwise might have been a very successful installation.

At the installation, alignment between the various elements should be made prior to grouting, primarily to ensure that all parts will align when the predrilled holes are used. Once successful, the unit can be grouted and given a second alignment. The couplings should not be made up until the piping is brought into place. Indicators should be used at the shaft to verify that the alignment didn't change by more than .002 inch in any plane as the piping flanges are made up.

Once done, the couplings can be put in place and the compressor readied for startup. When the piping is in place, lubrication, seal supplies and all utilities are complete, the compressor can be run. Alignment should be monitored, as the temperatures rise to the operating level. There are several methods used to determine alignment changes. The changes should be noted and alignment corrected as required.

If time and circumstances permit, data should be recorded when the compressor is run on process gas to establish a new base reference point. During operation, all monitoring equipment should be observed to establish a signature of vibration and temperature, and to be sure these data are all within permissible limits.

Each installation will have a series of startup checks that should be made prior to going into operation. These will vary for each compressor. A little extra time to record the overall compressor operation will be valuable if problems arise.

Commissioning the Lube Oil System

When the lube system arrives in the field, it will normally see the compressor and the interconnecting compressor piping for the first time. When the system first arrives, a check should be made to ensure that all the parts have been received. The system should be kept closed as much as possible to maintain cleanliness, while the field piping is being made up. This is where the stainless begins to bring in a return on investment because it should only require keeping dirt out. The carbon steel system will require keeping oil in the lines and keeping the moisture out. Condensation is a problem outdoors or even in an unheated building. With any system, the piping should be made up and filled with oil as soon as possible. Again, this is more critical for the carbon steel system.

An inspection for shipping damage should be made of the lube console and reservoir. Shipping blocks should be removed and, where permanent supports are used, they should be installed. The drivers should be uncoupled and connected to their respective power sources. As they are operated, uncoupled direction of rotation should be noted. When all checks out, the pumps should be aligned and connected to their drivers.

There is one word that should be repeated over and over: cleanliness. As each piping spool is brought in for connections, it should be checked for cleanliness. Any dirt removed prior to connection to the system will not have to be removed in the flushing procedure.

In preparation for flushing, the vendor's recommended procedures should be followed to the extent practical. In most installations, jumpers are installed around the bearings, going from pressure side to drain to keep from depositing debris in the bearing housings. Some vendors want the bearings removed. Any orifices in the oil piping should be removed, marked, and stored in a manner that will permit replacement without mixing up the locations.

The flushing oil should be the same as the oil used for operation or one with similar characteristics. Flushing can begin when the system is filled, and the trapped air is bled from the coolers, filters, pump casings, and other areas that can accumulate air. While circulating, oil is alternately heated for a time and cooled for a time. The exact procedure varies from vendor to vendor, and also with the user companies. The procedure cited here is abbreviated to present an overview. Additional strainers should have been installed, which can be easily removed to check the progress of the flushing. The stainless has no mill scale to remove and should not use nearly the time of carbon steel for the piping system to show clean.

The procedure with carbon steel piping reminds one of some ancient rite, with many people walking around the pipe beating it with anything imaginable. The temperature is cycled as often as necessary to obtain a clean system. Care must be taken during the heating cycle to keep from exceeding any temperature limits. The viscosity of the oil will be low during the hot cycle, and should not go below the minimum allowable viscosity for the pumps. It takes time for the system to clean up, and depending on the piping material, this could be days for stainless and weeks for carbon steel. There are stories about large carbon steel system flushing duration estimates that should have been quoted in months instead of weeks. Times as high as 12 weeks have not been very unusual.

Once the system is clean and the system charged with new oil, it will be ready to run. A run with the compressor not operational should be made to check the switch settings and operation. Even though the system was tested at the factory, a rerun similar to the shop test should be made, because the system is now joined up to field pipe and some of the factors may change, requiring a few trim adjustments. If nothing was damaged in shipment and a good shop test was run, the touch-up work should be minimal. Review as many items as possible beforehand to make compressor startup a routine event.

Successful Operation

When the user and the vendor have worked together as a team, and if all people included give sufficient attention to details, the on-stream operation should be very smooth. While there are many details and, at times, endless questions to resolve, the conscientious pursuit of each item, no matter how trivial it may appear at the time, should result in an achievement of the objective. The unit should start and come on-stream with a minimum of effort.

References

1. Brown, Royce N., "Can Specifications Improve Compressor Reliability?" *Hydrocarbon Processing,* July 1972, pp. 89–91.

2. Brown, Royce N., "Selection and Specification of Process Compressors," in ASME 36th Petroleum Division Conference Publication, *Enhanced Recovery and Rotating Equipment—A Workbook for Petroleum Engineers,* New York: American Society of Mechanical Engineers, 1980, pp. 57–64.

3. Rassman, F. H., "Design Specification and Accessories," *Compressor Handbook for the Hydrocarbon Processing Industries,* Houston, TX: Gulf Publishing Company, 1979, pp. 40–41.

4. API Standard 617, *Centrifugal Compressors for General Refinery Services,* Sixth Edition, Washington, D.C.: American Petroleum Institute, 1995.

5. API Standard 618, *Reciprocating Compressors for General Refinery Services,* Fourth Edition, Washington, D.C.: American Petroleum Institute, 1995.

6. API Standard 619, *Rotary-Type Displacement Compressors for General Refinery Services,* Second Edition, Washington, D.C.: American Petroleum Institute, 1985, Reaffirmed 1991.

7. API Standard 613, *Special-Purpose Gear Units for Refinery Services,* Fourth Edition, Washington, D.C.: American Petroleum Institute, 1995.

8. API Standard 671, *Special-Purpose Couplings for Refinery Services,* Second Edition, Washington, D.C.: American Petroleum Institute, 1990, Reaffirmed 1993.

9. API Standard 672, *Packaged, Integrally Geared, Centrifugal Plant and Instrument Air Compressors for General Refinery Services,* Second Edition, Washington, D.C.: American Petroleum Institute, 1988, Reaffirmed 1991.

12

Reliability Issues

General

Overview

The author has often wondered how often machines are redesigned and rebuilt or even replaced for reasons that were later found to be incorrect. Based on observations over a considerable number of years, the number might be staggering. Ironically, the incorrect diagnosis sometimes results in a solution that will at least for the time being appear to fix the problem. This type of solution may possibly lead to a future problem because an incorrect symptom-cause relationship is established that will not hold true on another application at another time. All of this is being said to stress the need for proper and careful problem solving that accurately determines the real cause for the problem. It is difficult to address the common sense side of reliability and not make some type of plea for cool heads and calculating minds to overcome the hysteria of the moment on the occurrence of a failure. There are many good problem-solving techniques available, but they must be used in a calm, clear mindset. There certainly is no room for prejudice and "finger pointing." The correct solution to a compressor train problem certainly would do much to aid the cause of reliability.

Today there are many tools available to aid in problem solving or failure analysis. These include the Weibull Analysis, Failure Mode and Effect Analysis, and Fault Tree Analysis, to name a few. One of the most widely accepted is the Weibull analysis. This method can provide an accurate engineering analysis based on extraordinary small samples [1].

This chapter will discuss various reliability issues. The discussion will be kept at a philosophical level rather than getting into a statistical analysis. The statistical analysis is best left to others equipped with the training, tools, and the data. Hopefully, this material will give the reader some "common sense" insight into the various considerations involved in the selection and application of reliable compressors, their drivers, and auxiliary systems.

Reliability for compressors is defined in many different ways, but the most widely accepted definition states that it is the ability or capability of the equipment to perform the specified function in the designated environment for a minimum length of time [2]. The length of time, "run time," cannot be determined except by operating the compressor for the desired length of time or until it fails. It is not too practical to use this criteria as the lone measure of success. While a historical database can provide valuable information to the later design efforts, this knowledge must be supplemented with other measures to gauge the probability of success.

It is very important right from the start to bear in mind the current axiom offered in the KISS principal, "keep it simple stupid." While this may seem a little offensive, nonetheless, from time to time, one sees systems that give the impression the originators went out of their way to make things unnecessarily complicated. An example of this is controls put on systems that with a little added thought could have been made self-regulating.

For compressors in general and for some types in particular, the cleanliness of the gas stream is the key factor in a reliable operation. Moisture or liquids in various forms may be the cause of an early failure or in some cases a catastrophic failure. Corrosive gases require material considerations and yet even this may not entirely solve the loss of material issue that can certainly cause early shutdowns or failures and high maintenance cost. Fouling due to contaminants or reactions taking place internal to the compressor can cause capacity loss and the need for frequent shutdowns.

Another significant factor in a reliable operation is the basic application of the equipment. One could hope that the previous chapters have helped you avoid some of the application pitfalls. It is certainly true that no matter how well a compressor or its ancillary equipment is designed, if it is not properly applied, the results can be disastrous.

Robust Design

In the early 1970s, Sohre [3, 4] proposed a numerical approach to reliability. He developed a series of charts (see Figures 12-1, 12-2, 12-3 and

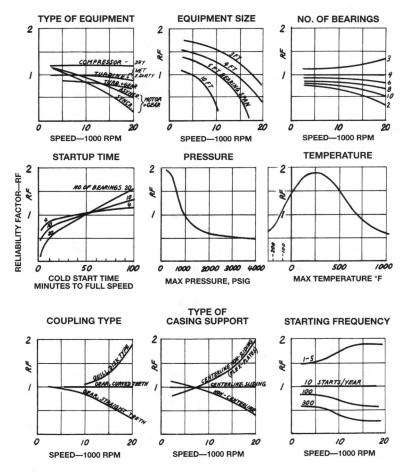

Figure 12-1. Reliability—design of plant and equipment. (*Courtesy of Sohre Turbomachinery*)

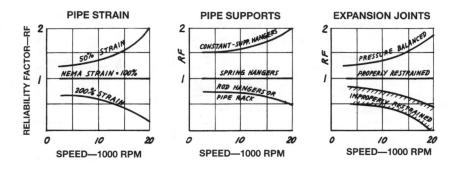

Figure 12-2. Reliability—piping. (*Courtesy of Sohre Turbomachinery*)

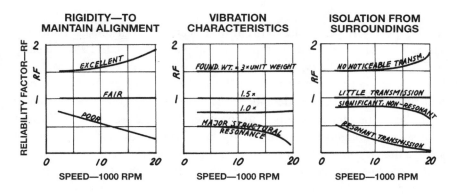

Figure 12-3. Reliability—foundation. (*Courtesy of Sohre Turbomachinery*)

12-4) on which he plotted a key parameter such as speed against a reliability factor. The reliability factor was normalized to 1. He generated a curve for each key parameter, selected from equipment design aspects, plant design, operations, and maintenance. The base value of 1 represented a normal or average industry run time. A value greater or less than 1 represented either a multiple or fraction of the run time. The curves could be used in conjunction with each other to obtain an overall effect for an installation. To normalize the resulting product of the component factors the nth root was extracted, where n was the sum of all components used to develop the product. This then produced a multiplier for a comparative average industry run time for the entire system. This approach was unique in that it was the first time to the author's knowledge this type of analysis

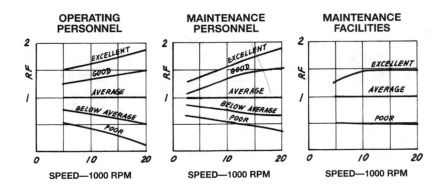

Figure 12-4. Reliability—operation and maintenance. (*Courtesy of Sohre Turbomachinery*)

was proposed. It was conceived out of a need to aid in problem solving. The advent of large single train plants took place in the late 1960s to early 1970s. Needless to say, there were many machinery-related problems. Many of the monitoring and diagnostic tools taken for granted at the current time were just being developed. The data being generated tended to overwhelm the analyst's ability to interpret it. There was no baseline for comparison. Diagnostics were strictly based on intuition, which was not always that accurate. With time, the current base of data began to evolve.

If the Sohre approach is taken in a philosophical light, it clearly indicates the need for a robust design. This is true for the basic equipment as well as the installation of foundations and piping. All of the factors were considered in the approach Sohre presented, even if in some cases, they may have been considered subjective. The key to the robust approach can be simplified as low deflection. On a rotor, the shorter the bearing span and/or the heavier the shaft cross section, the lower the deflection. This tends to increase the reliability aspect. On a foundation, heavier sections tend to have lower deflection, which contributes to the ability to maintain train alignment. Better alignment is one of the factors that contributes to higher reliability.

The Installation

Foundations

The prime function of the foundations, relative to equipment reliability, is to hold the compressor train in alignment during all operating phases for the life of the equipment. To perform this function, the foundation must be rigid. It is difficult to establish and maintain alignment between the machine components if the foundation is prone to excessive deflection. A foundation may well be able to support the machine train physically, but if not massive enough to prevent excessive differential deflection between the equipment bodies, problems with long-term operation are going to occur. Ideally the couplings, while flexible, will function much more trouble free if not operated at the extreme misalignment limit for long periods of time. It goes without saying that the basic foundation design should be robust enough to avoid or at least minimize settling so as not to affect the compressor train alignment over the life of the equipment.

Another aspect of deflection is the foundation's natural frequency. The foundation must be tuned in such a way that any resonances are not in coincidence with any of the compressor train natural frequencies. Relative to equipment considerations, it would be desirable to have all foundation natural frequencies well above any equipment critical speeds.

Suction Drums

All types of compressors will benefit if suction drums are part of the inlet system design. However, as a practical consideration, it would be difficult to justify a drum on small spared compressors, particularly the "off-the-shelf" type. If the compressor is in critical service, even when spared and/or replacement cost is deemed significant, suction drums should be considered.

The purpose of the drum is to exclude liquid from entering the compressor while operating. Ideally, the drum should be designed to operate "dry." If the drum is normally devoid of liquid, the instrumentation of the drum is straightforward. The instrumentation does not have to define a level, which can be difficult to monitor should the level be unsteady and possibly foaming. Also there should be no temptation to accept a "temporary" buildup. Finally, there is more of a volumetric safety factor to stop a liquid "carryover" for a given physical size drum if there is no resident liquid. A caution should be noted: the drum should not be a substitute for sound engineering of the compressor inlet system. Care should be taken to design the system to be free of "carryover." The drums are there for a backup system should there be that process upset that should not have even happened.

The piping configuration from the outlet of the drum is important. If possible, there should be no liquid traps in the compressor inlet line.

These are locations where condensation can accumulate. If allowed to accumulate, at some time it will enter the compressor as a slug of liquid and cause damage. If the piping configuration must contain a loop, the lower portion of the loop should be instrumented to detect the presence of liquid accumulation. Both for the drums and for the piping loops, provision must be made to drain any liquid that may be present.

Check Valves

Check valves are required in the piping system at any point where backflow of gas after a shutdown has the ability to restart the compressor, running it backwards or, for that matter, even in the normal direction. Reverse rotation is totally bad, as many components of the various compressor types are not designed for reverse rotation, and there is some possibility, generally remote, that the compressor could reach a destructive over speed. Forward rotation is bad primarily because the intent was to stop the compressor, and it is now operating out of control. This is a problem, particularly if the shutdown was caused by a compressor failure indication, and the need to stop was to prevent further damage. In this mode, it is unlikely that the compressor can attain an overspeed condition. An application with a high potential for backflow is the parallel operation of two or more compressors.

Another application with a risk of compressor uncontrolled restart is refrigeration service, where condensed liquid is normally present. On shutdown, the vapor from the boiling liquid has a high probability of producing a significant vapor flow, either for reverse or possibly forward rotation depending on the geometry of the piping and compressor nozzles.

Having presented all of the reasons for appropriately located check valves, what about the reliability of the check valve? There are some schools of thought that would argue the use of a check valve as detracting from reliability. Without an argument, check valves do fail if not properly engineered for the intended service. If the check valve is subjected to pulsating flow as is expected with many of the positive displacement compressors, it is certainly subject to fatigue failure. If a dynamic compressor is surged for extended periods, the stress on the check valve may become excessive and the valve may fail. The check valve should be of a rugged design and be properly selected for the application. It may need power assist or damping.

In some cases, power-operated block valves may be substituted for the classic check valve. However, these valves depend on an outside motive fluid and a signal from an external source to operate. It goes without saying that these ancillary items in themselves must be highly reliable. In a strict sense, overall reliability may be reduced because of the higher complexity; however, the entire system must be considered in making this type of judgment. In cases where forward flow can cause a problem, the power-operated block valve may be the only choice, because a common check won't work. The flow in this situation is in the normal direction.

Piping

Piping is required to bring the gas to the compressor and take it away. If it were not for this somewhat basic need, compressors would do well without it. Piping brings with it pipe strain. If one were to ask the maintenance department why pipes are used, the answer would undoubtedly be that piping was installed to the compressor to destroy the fine alignment just completed by the millwright.

Excessive pipe strains on equipment can cause both internal as well as external alignment problems. Some of the compressor designs are more vulnerable to internal misalignment problems from external forces than others. If the internal components become misaligned, rubbing can occur. Accelerated wear and early failure may occur. The effects of external misalignment may not be as obvious as those from internal problems but will in time reduce the length of operations. The vibration levels increase as the couplings become misaligned and the high vibration trip system may cause an unscheduled outage. Extended operation at high levels of misalignment may cause coupling failures and possibly bearing damage or, at the worst, a catastrophic failure.

The original equipment supplier should be contacted for the recommended allowable forces and moments. Equipment purchased to API standards specify a design level for allowable external forces and moments to be used in the equipment design. These values are frequently used by piping designers as field allowable values. Only when the equipment supplier echoes these values should they be used in piping design. Even under these circumstances, the entire value should not be used for piping design. Most of the piping stress programs do not account for field tolerances in fabrication, such as flange squareness and length variations. While not totally in concert with normal field fabrication practice, some of the tolerance problems can be minimized by piping away from the compressor and using some field-fit piping. It is very important to check the piping for strains by aligning the equipment and then connecting the piping while observing the shaft-to-shaft alignment change. An alignment change of no more than 0.002 inch is considered acceptable for most helical-lobe and dynamic compressors.

Compressors

Type Comparison

Comparing compressors by type is a somewhat shaky endeavor, because it involves some opinion, some prejudice, and industry-accepted perceptions. Hopefully, the prejudice has been kept under control and the perceptions of industry are based on knowledge, experience, and, at least, some data.

There are two issues to consider in the comparison of the different types of compressors relative to reliability. One is the number of parts required to perform the function. It is generally accepted that reliability is an inverse function of the parts count. This does not necessarily include the number of bolts, for example. The other is the use of wearing parts. On this basis, one would give the centrifugal compressor a higher rating than the reciprocating compressor. To come to a closer comparison, the centrifugal would rate higher in reliability when compared to the helical-lobe compressor, strictly on parts count. To temper this, add that comparisons must include other factors such as the application. The helical-lobe compressor is generally more tolerant of fouling gases than the centrifugal. To take this one step farther, the centrifugal is more tolerant of fouling gases than the axial compressor. The axial again also has a higher parts count, but does have other redeeming factors in its favor such as high capacity for the size and inherent high efficiency. As is typical with all engineering decisions, an overall evaluation is required that applies the appropriate compromises or tradeoffs.

Reciprocating Compressors

The reciprocating compressor has several strikes against it when it comes to reliability, if the main consideration of reliability is long run times as would normally be expected from the dynamic type machines. It has a lot of parts and the parts are subject to wear. Before one jumps to the conclusion and totally excludes this compressor from consideration because of these factors, the positive aspects of this machine should be reviewed. For low-volume-high-pressure ratio service and particularly for low molecular weight gases it is difficult to match. When compared to the centrifugal for this service, the centrifugal will have to operate at a high speed and require many impellers. This end of the spectrum is not the best from a reliability point of view for the centrifugal.

If turndown is a consideration, again the reciprocating compressor, with its many unloading options, does meet the challenge. This flexibility is difficult to achieve reliably in the dynamic machine and may require the compressor to operate at an unfavorable point on its characteristic curve.

Several steps can be taken to maximize the run time for the reciprocating compressor. Since wear is a function of rubbing speed, the piston speed can be kept to a minimum. Chapter 3 made recommendations for piston speed. Reliability problems due to valves are reputed to account for 40% of the maintenance cost of the compressor. Valves are the single largest cause for unplanned shutdowns. Basically, valve life can be increased by keeping the speed of the compressor as low as practical. At 360 rpm, the valves are operated six times a second. At 1,200 rpm, the valves operate 20 times a second or 1,728,000 times in a day. It is not difficult to understand why the valves are considered critical. To keep the reliability in mind, valve type, material selection and application considerations such as volume ratio, gas corrosiveness, and gas cleanliness need attention by the experts. One final note is that while lubrication is an asset to the rubbing parts, it is not necessarily good for valve reliability.

While not necessarily improving the reciprocating compressor's individual reliability, using spared units does improve the overall plant reliability. With proper monitoring, the compressors may be removed from service in a timely manner for maintenance. If the program is properly administered, unplanned shutdowns can be avoided and a higher plant reliability achieved.

Positive Displacement Rotary Compressors

The helical-lobe compressor is the more robust of the rotary compressor types. Its positive reliability factor is the absence of wearing parts if the timing gears are neglected. Timing gears are not used on the flooded type. However, on the negative side, is the parts count. There are two rotors, four bearings, four seals, and, on the nonlubricated machine, there are timing gears. Another negative aspect of the helical-lobe compressor, particularly the nonlubricated version, is the presence of high frequency pulsations. If untreated, these pulsations can contribute significantly to failures of system components. While this problem may be solved, the solution adds to the system complexity, detracting from the overall reliability. To add a side note, the location of the silencer, used to treat the acoustic pulsations, should be directly attached to the discharge flange if at all possible. There are reported cases of acoustic problems when a spool piece was used between the silencer and the compressor flange. As mentioned before, the nature of the application may alter the overall reliability picture, as the nonlubricated helical-lobe compressor will tolerate fouling gas. It tends to be unique in that arena.

Taking the balance of the lobe-type machines as a group, the biggest single factor leading to poor reliability is excessive rotor deflection. If the rotors are allowed to touch each other, generally a failure occurs. Excessive rotor deflection is caused by a higher differential pressure across the compressor than the design limit. Since most of the rotary compressors, other than the helical lobe are not intended for continuous service, their use in continuous service applications may cause the application to experience reliability problems. Sparing and monitoring for maintenance intervals commensurate with true expected run time will certainly keep the system's reliability up, if the unplanned shutdowns are kept low. Maintenance cost, included in an evaluation of life cycle cost, may well direct usage to a more suitable compressor selection.

Centrifugal Compressors

Centrifugal compressors can be very reliable, but having said that, they can also have a miserable reliability record. It is tempting to relate some of the horror stories encountered on poorly designed, poorly applied centrifugal compressors. The reader, wanting to learn from this, is directed to the literature, as there have been many volumes written on this subject. At this writing, it can be said that improvements in reliability for this compressor over the past 30 or so years have been significant.

Probably the single largest contributor to problems in the centrifugal compressor is related to rotor dynamics. Long slender rotors are the Achilles heel of this machine. It is in this design that the term robust has a great deal of meaning. Rotors are subject to critical speeds that must be encountered during startup. The longer the rotor and the higher the speed, the more the number of critical speeds that must be dealt with. To this can be added the sensitivity to unbalance. In field operation and as time passes, the compressors do degrade, which usually manifests itself in ever-increasing levels of unbalance. All this boils down to the simple fact, the more sensitive the rotor, the shorter the runtime.

Another area of concern is rotor stability. Two factors enter into the stability considerations: flexibility, just discussed, and damping. Most of the rotor damping is generated by the bearings. The design must consist of bearings with adequate damping capability and rotor motion in the bearings to generate the damping. One of the significant destabilizing forces is aerodynamic cross coupling from the impellers. This is of greatest concern with high density applications. Squeeze film dampers may be added to the bearing to generate more damping but detract because this adds a degree of complexity. Designs are very empirical and are difficult to evaluate: when they are good they are good and when they are bad they are terrible.

Individual designs must cope with the tradeoffs or compromises necessary for good reliability and yet maintain good efficiency. Unfortunately, these forces are somewhat in the opposite direction from each other. One of the first issues is speed; high speed is generally good for efficiency but can cause problems with reliability. High speed raises stresses, which must be limited for reliability. High surface speed may cause bearing and seal problems if allowed to encroach on the limits of experience. One solution is to add more impellers, but that may lengthen the bearing span to a degree where the rotor dynamics may be compromised. Another consideration is the use of multiple cases connected in series and tandem. While this is sometimes the safe option, it does increase the parts count—more bearings, more seals, and more couplings, as well as increasing the cost.

While the previous statements tend to paint a rather bleak picture, the purpose of the material is not to discourage but to help the reader understand the aspects of the reliability decisions. Also as mentioned before, the centrifugal has progressed considerably in the last 30 years. More progress is yet in the works with new analytical and diagnostic tools. This, coupled with improvements in machine health monitoring, make the centrifugal compressor outlook much brighter. It would be safe to say that it is the compressor of choice whenever the application allows.

Axial Compressors

Axial compressors offer a high volume capability in a relatively compact case. As can be said for any of the dynamic compressors, when properly applied, it can be a very reliable compressor. The compressor does not have any wearing parts, if one keeps tongue in cheek. The caveat is the blading. If parts count is used on the blading, the count is high and is a detractor from reliability. This is probably not a fair comparison, yet it is a factor to be reckoned with. The blading is prone to fouling and becomes a wearing part if particulate is allowed to enter the machine unchecked. For air service, a high level of filtration is very much in order and certainly does a great deal toward improving reliability. The blading is subject to resonance problems and exciting sources must be controlled, or conversely, exciting sources recognized and the blading tuned appropriately.

The large flow and small case do make the compressor somewhat more sensitive to piping forces and moments.

Drivers

Turbines

Gas turbine drivers generally consist of an axial compressor air section, and an axial flow expander for the rotating parts. The items discussed for axial compressors apply to the compressor section of the gas turbine and need not be repeated. The gas turbine is a complex machine, and because of this, would tend toward the unreliable. Because it is a production machine, as contrasted to the custom engineered concept used in many of the other machines, the designs are quite mature before being sold for general use. One of the more obvious areas of concern is the hot section parts. Fuel is burned in the turbine to generate the hot gas used to motivate the turbine. Here again is an area of compromise. For high efficiency, high firing temperatures are desirable. For parts life, the hotter the gas, the shorter the life. But as mentioned, because of the production nature of these units, the research to find the best compromise is not done in the installation but by off line research. Still, the limiting items for long run times are the hot section parts.

Steam turbines have more of the reliability factors in common with the centrifugal and axial compressor. Blading design considerations are quite similar to those discussed at the axial compressor. The rotor dynamics aspects are similar to the centrifugal compressor. Condensing steam turbines tend to have long bearing spans. These bearing spans are further extended if the design includes an extraction point and a condensing exhaust end, or worse, if double extraction is included with the condensing exhaust end. Steam turbines tend to exhibit more flexibility in the bearing support system that must be included in the rotor dynamic analysis, adding to the complexity. Improvements in rotor dynamics analysis have contributed considerably to the solution of these problems.

Other items unique to steam turbines that must be considered are water carryover and fouling. Carryover can leave deposits on the blading, causing loss in performance and increased thrust loads. If carried to an extreme, thrust bearing failure could occur. If a large amount of water is carried over as a slug of liquid, catastrophic failure can occur.

To minimize the effects of deposit buildup on valve stems of trip or trip and throttle valves, a pull-to-close design is preferred. While somewhat more expensive, the added feature of hydraulic-assisted closure is desirable, particularly for the larger sizes.

Motors

Motors of the four-pole design have proven to be quite reliable. The two-pole design, as used in most of the centrifugal compressor drives, does not enjoy the same reputation. Sleeve bearings are the bearings of choice for compressor drives. Ring oil lubrication is most common on these motors, but a degree of reliability can be added by using pressurized oil from the compressor lubrication system.

One area of concern that applies to induction motors is the potential of a high torque if the motor is re-energized at the wrong time after a trip. The torque capability of the motor is 500 to 700% of normal when re-connected at the wrong part of the coast down cycle. The level of torque potential is of a magnitude that a failure of the couplings, shafts, and/or gears will occur. It is not practical on motor-compressor systems to design for such high stresses. The safest procedure is to let the drive train coast to a stop before re-energizing. Alternately, though somewhat costly, is to use electric relaying that can detect the point where re-energizing is safe and automatically reclose the breaker at the proper time.

Gears

It has been said that since the publication of API 613, Second Edition, the reliability of gears used in compressor drives has moved up several orders. This edition of API 613 mandated a conservative rating formula that was widely adopted in North America and to a more limited extent globally. It is credited with a significant reduction of new installation gear failures. Prior to that time, the feeling among operating companies was that reliability and gear drives were not compatible combinations.

Expanders

Expanders have not been the essence of reliability. It is not that the expander design in itself has any significant problems. The problems for the most part seem to be related to the application. Most of the failures have been the result of the expander ingesting foreign substances, such as the catalyst in a catalytic cracking unit heat recovery application. Unlike the expansion section of the gas turbine, the inlet temperature is not as high, therefore, temperature is not a significant factor in reliability reduction.

Another vulnerable area of concern is the result of the design requirement for high temperature service. To accommodate the thermal growth potential, the various sections of the expander are quite compliant. As a result, piping and forces need to be kept to a minimum.

Applications

Process

The process application can have a significant impact on the reliability of a compressor. The reliability aspects take on two paths: the nature of the gas being compressed and the correct compressor selection for the service.

When fitted with appropriate resistant materials, the compressors can reliably compress corrosive gases. The key is the appropriate resistance. This may be achieved to varying degrees of success. Most material corrosion resistance is temperature dependent. It is important to understand the compressor temperature profile. This includes making allowance for off-design performance. In the extreme, some gases can reach a very high level of corrosion by crossing a critical temperature point, which is more commonly called burning. Thankfully, most compressor services are not that dramatic. The point is that during the application or selection phase, the subject of corrosion must be fully explored. Factors such as the presence of moisture, or degree thereof, is significant for many of the corrosive gases.

Gases that tend to polymerize with temperature must be recognized and, as mentioned for corrosion, temperature limits must be imposed at the application phase to prevent problems later in operation. The only compressor, of the types covered, that is not as sensitive to the presence of polymers is the helical-lobe compressor. It should be recognized that even this machine does have limits.

The proper compressor selection for the application is probably the most important single factor to achieving long-term reliable operation. For all the factors that make the centrifugal one of the most reliable compressors, a misapplication will negate these most dramatically.

One of the most common pitfalls for misapplication is the size of the compressor itself. Because of the centrifugal compressor's good record as a reliable machine, there is a tendency to attempt to use this compressor for all applications. The centrifugal compressor does have a lower size limit. Application ranges were covered in Chapter 1. The lower limits are not sharply defined lines, and are heavily influenced by the specific application. For example, a multistage centrifugal with an inlet capacity of 800 acfm, on the lower end of the application range, may perform reliably on a single component clean gas and fail to operate well on a dirty or corrosive gas.

Gas sonic velocities must be kept in mind when selecting the compressor type. The helical-lobe compressor is tolerant of many items that can plague the other types, but some caution must be exercised with low sonic velocity gases. The nature of the operation of this compressor tends to produce a wide spectrum of pulsations. The higher molecular weight gases, due to their lower sonic velocity, are prone to excite acoustic and mechanical resonances in the piping associated with the machine. An interesting symptom of the presence of resonant responses is that the bolts tend to loosen and fall out. This certainly is not a good situation. The application must make certain that the appropriate pulsation treatment is used. Serious vibration problems have been experienced and reliability reduced in applications where the acoustic problem was not appropriately addressed.

Reciprocating compressor pulsations were covered in Chapter 3, but need to be mentioned with the discussions on reliability. Problems with reciprocating compressor pulsations and the potential for acoustic and mechanical resonances are very similar to those experienced with helical-lobe compressors. The significant difference is the frequencies are much lower and the number of discrete frequencies per compressor are much less. However, piping vibrations can occur and there is always a danger that pipe breakage can occur. Aside from the piping acoustic and mechanical resonances, the presence of standing waves in the piping at the cylinder can reduce the output of the compressor. When the valve action is influenced by the standing wave, serious problems can occur.

For higher molecular weight gases, the centrifugal tip speeds must be reduced appropriately. The operating range of the compressor will be reduced significantly if operated at high Mach numbers. This is another example of the influence of speed in reliability.

High pressure operation, while generally quite successful, is another criteria that must be carefully considered. The increased pressure contributes to a higher density. As mentioned before, the higher density will increase the destabilizing forces on the rotor and can lead to rotor stability problems. This is a problem unique to the centrifugal; however, in the helical-lobe compressor, high differential pressure contributes to rotor deflection, which, if carried to extreme, can certainly cause early failures. In axial compressors, high pressure, high density increases the bending load of the blading, which raises the stresses and potentially will shorten blade life and contribute to early failures.

Experience

Since many of the compressors covered in the earlier chapters fall into the category of custom-engineered units, the influence of experience on reliability is worth a few words. Experience may well be considered as being "viewed through the eyes of the beholder." How can one establish a base of experience if the unit under discussion is a custom unit?

With custom-engineered compressors, since exact duplicates are rare and prototyping of the exact unit is not feasible, the first consideration is the history of the component parts. Another aspect to consider is whether the application is an interpolation of existing designs or an extension or extrapolation. The latter, while a fact of life, needs to be understood at an early stage of the overall plant design so that the appropriate risk factors may be considered prior to the final commitment. Failure to properly evaluate the risk of an extension has resulted in serious compressor and driver problems, with long, expensive outages.

The critical components should be listed and checked to see that prior usage is within the range of experience. It should be determined if the impellers have been stage tested and whether this performance has been reflected in actual field experience. Prior use history of the blading geometry being proposed should be reviewed. This should include the chord length and blade length and, hopefully, experience with both longer and shorter applications. The rotor length (bearing span), which applies to all rotary and dynamic compressors, should have a successful history on similar frame size machines. All parameters relative to the list of critical items should be carefully scrutinized.

The supplier's general experience should be used to round out the review. Has the vendor used this corrosive gas at these concentrations? Does the supplier have experience with the specified gas at the trace levels of reactive gas, preferably at both higher and lower concentrations? An example of the trace gas found in hydrocarbons is H_2S. It is very important that the supplier has experience at the anticipated pressure or density levels. The successful operation with gases in the low Mach number range should be demonstrated by the referencing of previous installations. The compression temperatures should be well within the experience range, because this is particularly critical for reliability with reciprocating compressor valves. Hopefully, these few examples are adequate to guide the reader in the proper direction, as a more comprehensive list would become quite lengthy and may not cover all the area needing review.

Operations

General Comments

It should be stated that operation should also be included in the list of considerations. It cannot be stressed enough that proven operating procedures be used. Maintenance should be performed in a timely manner as required. The manufacturer's instruction book is a good starting point for information on the last two items. The words "starting point" are important in that to this advice should be added the other considerations needed to form a well-established plant operating discipline that takes into account process, past experience, and industry knowledge. A machinery monitoring system should be in place to aid in early detection and diagnosis of possible problems, should they occur.

Gas Considerations

The use of drums for compressors was discussed earlier in the chapter. At that time, the recommendation was made that the drum be dry. This is still the prime consideration. However, if liquid carryover must be a way of life, then the question of particle size must be addressed. For the centrifugal compressor, a rule of thumb with the usual disclaimers and exceptions is a particle size of 10 microns is safe, 100 microns marginal, and 1000 microns too large. The most significant disclaimer for this rule is that the gas with particles is acting in the erosion mode and not corrosion mode. The rule can be applied to the reciprocating compressor as well, with the note that the reciprocating compressor is less tolerant of liquid than the centrifugal. The helical-lobe compressor is somewhat of an exception in that it can better tolerate liquid particles. Liquids that are near the flash point and vaporize on impact are more prone to cause problems than particles that move through the compressor as mist or flash due to the heat of compression in the gas path. None of the compressors can tolerate liquid in large slugs.

Fouling and dirty gases have been discussed several times. These must again be mentioned in the realm of operations. While the potential of dirty gases should be considered and provided for in the design of the installation, it is not that unusual to discover this was not done. One reason may be that "off-design" operation that causes carryover of foreign material either as liquids or solids is not anticipated. While solutions to this type of problem must involve both engineering and operations, it is up to operations to recognize the problem. Some fouling problems are difficult to recognize. An example that illustrates this case consisted of a gas with extremely fine particles that were able to pass through the inlet filter and were not in themselves considered a problem. Superimposed on the gas was moisture, not high enough to cause concern in itself. As the gas was compressed, and in the application, intercooled, the moisture became saturated. The moisture mixed with the apparent harmless particles to form a paste. The paste was deposited on the blade surfaces of the downstream compressor stage and dried by the heat of compression. A classic case of fouling took place due to factors that were known but considered benign. If one can draw a moral from this example, it would be that apparent insignificant factors when taken singularly may not be so insignificant when occurring in unique combination.

Operating Envelope

All compressors have an operating envelope. The size is dependent on the compressor type and the application. If the appropriate design considerations were implemented, then operation within the envelope should be problem-free. The first problem stems from a communication gap. Engineering and the equipment supplier do not always adequately communicate the bounds of the operating envelope to operations. Operations cannot be expected to operate inside an envelope that is not well-defined. Of course, operations may consciously choose to operate outside this envelope due to production pressures. It is important to consider the risks involved with this mode of operation. It may only reduce efficiency and no significant problem with equipment reliability may occur. For this, the only consideration is operating economics that may or may not be significant. However, accompanying "off-design" operation outside the envelope are factors that may well impact reliability and the life of the equipment. It is relatively universal that all of the compressors will tend to increase in operating temperature when operating "off-design." This was illustrated in the previous chapters where performance was discussed. In most cases, the operating temperature has an upper limit after which deterioration may take place. Significant changes in pressure may well cause problems by exceeding deflection ratings. In dynamic compressors, molecular weight swings can cause unexpected surging. In the rotary compressors, changing the pressure ratio will cause over or under compression, which at the least, causes efficiency problems and, at worst, accentuates an acoustic response. These are a few examples to help illustrate the point; however, many other possible scenarios do exist. Reliable operation may well hinge on the ability to recognize and properly design for these eventualities.

System Components

Lubrication

Lubrication is a rather basic requirement for any operating machinery, which certainly includes compressors. Because it is so basic, it seems rather intuitive that reliability would be highly dependent on it [4]. In some cases, such as the basic rotary compressors covered in Chapter 4, rolling element bearings are standard and many of these require grease lubrication. It is important to remember that while the bearings should be greased at regular intervals, too much is just as serious as not enough.

Most of the compressors are of the pressure lubricated type and need some sort of lubrication system. For critical service, the API 614 lubrication system should be used. These systems were covered in Chapter 8. Since lubrication systems are key to the reliable compressor operation, a few things important to a reliable lubrication system seem in order. First and foremost is to keep the system as simple and basic as practical. Parts count is also a factor at this point. Direct-operated regulators (control valves) should be used whenever they are technically feasible. Computer control adds needless complication, particularly if the system is a basic lubrication system. It does not appear to add value to take the measured lube system pressures as transmitted signals to the control room. First it adds transmitters, usually in multiple form. Then the signals are put into the plant computer for processing, which requires someone to write the appropriate control code. The output signal is sent back to the lube oil console, where it has to be converted to a pneumatic signal to operate a control valve. If the parts count parameter has any merit, it should certainly be valid in this example. This is contrasted to using a direct-operated control valve that operates directly from a lube oil pressure sensing line to position the valve to control the lube oil pressure. Combined lube and seal systems sometimes become complex. With the dry gas seal, covered in Chapter 5, the need for these complex systems should be minimized. The rotary screw pump should be used on all but the large sized systems. A centrifugal pump is sometimes used when at the upper end of the positive displacement rotary screw pump range. Because of the shape of the centrifugal pump curve, selection is somewhat more critical and should be backed by proven field experience. The upper and lower viscosity range is important for the selection of the rotary screw pump as well as the centrifugal. An operations function should be mentioned at this point. Lubricating oil condition monitoring is a vital step to ensure good quality oil is sent to the compressor. Finally, shop testing is an important step in the procurement of a reliable lubrication system. While it may not be as exciting as a full-load compressor performance test, it may well influence the reliability of that very compressor train.

Couplings

Couplings are discussed with reliability partly because in the past, as the single-train compressor installations were going through early growing pains, couplings were certainly a part of the pain. Not too many words need be used here, but possibly a bit of a reminder of the past will avoid problems in the future.

Not all that long ago, the main drive coupling of choice for compressor trains was the gear coupling. At first, the gear coupling was not a large contributor of problems. One guess is that on the early smaller trains, the couplings were oversized for the service. With oversizing, the lubrication quality, which was not good, did not cause any significant problem. It could also be said that any problems were not recognized because they were masked by all the other problems being experienced at the time. As the power density was increased, oversizing was slowly diminished, if for no other reason than economics, as the larger couplings were of a size that represented a significant cost to the equipment supplier. Also, the past oversizing had been more of a catalog rating thing and definitely was not recognized as such. The catalogs list the various sizes of couplings in a given manufacturer's line. The number of different sizes is limited by economics and the range of torque transmission coverage. Each catalog rating will cover a range of applications. The smaller sizes, when selected from the catalog, normally will fall somewhere between two successive sizes. It is normal to pick the larger size. This inherently oversizes the coupling. However, on the smaller sizes, this was not considered a problem because the cost benefit outweighs a custom design. As the power ratings went up, economics dictated a coupling selection sized closer to the application, resulting in the lubrication quality becoming more of a major factor. The fact that continuously lubricated couplings tended to centrifuge out the residual impurities in the oil (sludging) became quite significant. The presence of sludge prevented the lubricant from reaching the sliding area of the coupling and properly performing its function. Also the impact of misalignment became a factor. Highly loaded couplings with marginal lubrication tended not to operate well at high misalignment levels. The sensitivity to misalignment brought about better alignment techniques and helped minimize misalignment problems. Also, to aid in maintenance and alignment, longer spacers were used. Grease lubrication was improved, with the hopes of solving the sludging problem, but it never has achieved the run time being demanded of the ever larger trains. The advent of the flexible metallic element coupling solved the lubrication problem. It did not need lubrication. The flexible element coupling does require a discipline in machine-to-machine alignment, but given reasonable misalignments functions quite reliably.

Quality

Methodology

Quality and reliability are not free, but poor quality and reliability cost more than good quality [2]. Costs include excessive down time and associated loss of production, cost of repairs both in time and materials, and endless meetings. The list could go on but, hopefully, the point is made.

There are two general concepts of quality:

• Quality of design
• Quality of conformance

The type of compressor selected sets the quality of design. The API-based compressor is generally regarded as a higher quality than a catalog blower. Most of the chapter to this point has attempted to emphasize the many factors involved in design. Quality of conformance is a measure of how well the compressor met the specification. One of these measures is the achievement of the desired "run time." Another measure is the inspection for conformance of the parts to the drawings.

The role of quality in reliability would seem obvious, and yet at times has been rather elusive. While it seems intuitively correct, it is difficult to measure. Since much of the equipment discussed in this book is built as a custom engineered product, the classic statistical methods do not readily apply. Even for the smaller, more standardized rotary units discussed in Chapter 4, the production runs are not high, keeping the sample size too small for a classical statistical analysis. Run adjustments are difficult if the run is complete before the data can be analyzed. However, modified methods have been developed that do provide useful statistical information. These data can be used to determine a machine tool's capability, which must be known for proper machine selection to match the required precision of a part. The information can also be used to test for continuous improvement in the work process.

The advent of ISO 9000 certification has helped bring the quality issue to a focus. While ISO 9000 emphasis is on documentation of methodology and performance to the documented methods, the discipline that results is a direct step toward improvements in reliability. Out of this seems to come a goal of continuous improvement. This is a sharp contrast to the high growth era that took place in the early 1980s when poor quality on the part of the equipment suppliers was a significant factor in the lack of reliability at the field operation level.

Manufacturing Tolerances

One measure of quality is the conformation of the compressor parts to drawing dimensions. While it is customary to assign tolerances to a given dimension and base acceptance on the part being within the tolerance, there is a better way. The alternative requires a shift from the old paradigm and has been difficult to achieve in many of the compressor manufacturing locations. It has been proven in the Japanese automobile industry. The concept is to make the part to the target dimension with measurements based on the deviation from target.

The benefits are many, though the basic cost justification is more elusive. The first benefit is that assembly is easier as the variations between parts is lower and less selective fitting is required. It inherently makes for better part interchangability. For the maintenance people, parts should fit without rework or hand fitting.

The inherent contribution to reliability should be compressor assemblies that truly reflect dimensions intended by the designer. This in turn should make the compressor perform to the level intended. It should offset the desire on the part of maintenance people to improvise when parts don't fit. This improvising can certainly become a problem as the original design intent is not normally known or understood at the field maintenance level.

Summary

The preceding discussion covered quite a number of topics relative to compressor reliability. A brief summary of the items covered may help the reader recap some of the thoughts presented. Issues that commonly play a major role in the search for reliability of compressors include:

- The concept of a robust design, in the compressor component and in the installation
- The hazards of liquid carryover and the need for suction drums
- The proper application of check valves
- Problems caused by pipe strain
- The significance of proper application, selection and sizing
- The impact of high molecular weight gases
- Corrosive and dirty gas effects
- The role of experience in avoiding problems
- Operating procedures
- Ancillary systems, such as lube oil systems and couplings
- Impact of quality

It is hoped that this material will give the reader an appreciation for the many different factors that interact to cause a compressor installation to be successful or a failure. It is sincerely hoped that when all the various aspects are critically addressed that any proposed installation will achieve the high reliability that a user may expect.

References

1. Abernethy, R. B, *Weibull Analysis for Improved Reliability and Maintainability for the Chemical Industries,* 3rd International Conference and Exhibition on Improving Reliability in Petroleum Refineries and Chemical Plants, Organized by Gulf Publishing Co. and *Hydrocarbon Processing,* Houston, TX, November 1994.

2. Coombs, C. F., Ireson, W. G. and Moss, R. Y., *Handbook of Reliability Engineering and Management,* New York: McGraw-Hill, 1996.

3. Sohre, J. S., *Tubomachinery Analysis and Protection,* Proceedings of the First Turbomachinery Symposium, Texas A&M University, College Station, TX, 1970, pp. 1–9.

4. Sohre, J. S., *Reliability—Evaluation for Troubleshooting of High-Speed Turbomachinery,* ASME Petroleum Mechanical Engineering Conference, 1970.

5. Smith, J. B., *Predicting Future Failure Risk with Weibull Analysis,* First International Conference on Improving Reliability in Petroleum Refineries and Chemical Plants and Natural Gas Plants, Organized by Gulf Publishing Co. and *Hydrocarbon Processing,* Houston, TX, November, 1992.

6. Corley, J. E., *Troubleshooting Turbomachinery Problems Using a Statistical Analysis of Failure Data,* Proceedings of the 19th Turbomachinery Symposium, Texas A&M University, College Station, TX, 1990, pp. 149–158.

7. Brown, R. N., *ASM Handbook: Friction and Wear Technology,* "Friction and Wear of Compressors," Vol 18, ASM International, pp. 602–608.

Appendix A

Conversion Factors

Abbreviations. A = angstrom, atm = standard atmosphere, 760 mm of Hg at 0°C, cal = calorie (gram), cm = centimeter, deg = degree, gal = gallon, U.S. liquid, gm (and g) = gram, gmole = gram-mole, J = joule, kcal = kilocalorie, kg = kilogram, kJ = kilojoule, km = kilometer, kW = kilowatt, l = liter, lb = avoirdupois pound, m = meter, mi = mile (U.S.) mm = millimeter, N = newton, oz = avoirdupois ounce, pmole = pound mole, pt = pint, rad = radian, rev = revolution, s = second, ton = short U.S. ton, V = volt, W = watt. Others are as usual.

LENGTH

$$12 \, \frac{\text{in.}}{\text{ft}} \qquad 0.3937 \, \frac{\text{in.}}{\text{cm}} \qquad 30.48 \, \frac{\text{cm}}{\text{ft}} \qquad 2.54 \, \frac{\text{cm}}{\text{in.}} \qquad 3.28 \, \frac{\text{ft}}{\text{m}}$$

AREA

$$144 \, \frac{\text{in.}^2}{\text{ft}^2} \qquad 929 \, \frac{\text{cm}^2}{\text{ft}^2} \qquad 6.452 \, \frac{\text{cm}^2}{\text{in.}^2}$$

VOLUME

$$1728 \frac{\text{in.}^3}{\text{ft}^3} \qquad 7.481 \frac{\text{gal}}{\text{ft}^3} \qquad 3.7854 \frac{\text{l}}{\text{gal}} \qquad 28.317 \frac{\text{l}}{\text{ft}^3}$$

$$231 \frac{\text{in.}^3}{\text{gal}} \qquad 8 \frac{\text{pt}}{\text{gal}} \qquad 10^3 \frac{\text{l}}{\text{m}^3} \qquad 61.025 \frac{\text{in.}^3}{\text{l}}$$

DENSITY

$$1728 \frac{\text{lb/ft}^3}{\text{lb/in.}^3} \qquad 16.018 \frac{\text{kg/m}^3}{\text{lb/ft}^3} \qquad 1000 \frac{\text{kg/m}^3}{\text{gm/cm}^3}$$

ANGULAR

$$2\pi = 6.2832 \frac{\text{rad}}{\text{rev}} \qquad 57.3 \frac{\text{deg}}{\text{rad}} \qquad 9.549 \frac{\text{rpm}}{\text{rad/sec}}$$

TIME

$$60 \frac{\text{s}}{\text{min}} \qquad 3600 \frac{\text{s}}{\text{hr}} \qquad 60 \frac{\text{min}}{\text{hr}}$$

SPEED

$$0.3048 \frac{\text{m/s}}{\text{fps}} \qquad 152.4 \frac{\text{cm/min}}{\text{ips}}$$

FORCE, MASS

$$16 \frac{\text{oz}}{\text{lb}_m} \qquad 2.205 \frac{\text{lb}_m}{\text{kg}} \qquad 1000 \frac{\text{lb}_f}{\text{kip}} \qquad 2000 \frac{\text{lb}_m}{\text{ton}} \qquad 453.6 \frac{\text{gm}}{\text{lb}_m}$$

$$10^5 \frac{\text{dynes}}{\text{N}} \qquad 28.35 \frac{\text{gm}}{\text{oz}} \qquad 907.18 \frac{\text{kg}}{\text{ton}} \qquad 1000 \frac{\text{kg}}{\text{metric ton}}$$

PRESSURE

$$14.696 \frac{\text{psi}}{\text{atm}} \qquad 101,325 \frac{\text{N/m}^2}{\text{atm}} \qquad 51.715 \frac{\text{mm Hg(0°C)}}{\text{psi}} \qquad 47.88 \frac{\text{N/m}^2}{\text{psf}} \qquad 29.921 \frac{\text{in. Hg(0°C)}}{\text{atm}}$$

$$13.57 \frac{\text{in. H}_2\text{O(60°F)}}{\text{in. Hg(60°F)}} \qquad 703.07 \frac{\text{kg/m}^2}{\text{psi}} \qquad 6894.8 \frac{\text{N/m}^2}{\text{psi}} \qquad 14.504 \frac{\text{psi}}{\text{bar}} \qquad 0.0361 \frac{\text{psi}}{\text{in. H}_2\text{O(60°F)}}$$

$$1.01325 \frac{\text{bar}}{\text{atm}} \qquad 0.4898 \frac{\text{psi}}{\text{in. Hg(60°F)}} \qquad 760 \frac{\text{mm Hg(0°C)}}{\text{atm}} \qquad 406.79 \frac{\text{in. H}_2\text{O(39.2°F)}}{\text{atm}}$$

$$10^5 \frac{\text{N/m}^2}{\text{bar}} \qquad 0.0731 \frac{\text{kg/cm}^2}{\text{psi}} \qquad 1.0332 \frac{\text{kg/cm}^2}{\text{atm}}$$

ENERGY AND POWER

$$778.16 \, \frac{\text{ft-lb}}{\text{Btu}} \qquad 2544.4 \, \frac{\text{Btu}}{\text{hp-hr}} \qquad 5050 \, \frac{\text{hp-hr}}{\text{ft-lb}} \qquad 550 \, \frac{\text{ft-lb}}{\text{hp-s}} \qquad 42.4 \, \frac{\text{Btu}}{\text{hp-min}} \qquad 33{,}000 \, \frac{\text{ft-lb}}{\text{hp-min}}$$

$$3412.2 \, \frac{\text{Btu}}{\text{kW-hr}} \qquad 737.562 \, \frac{\text{ft-lb}}{\text{kW-s}} \qquad 56.87 \, \frac{\text{Btu}}{\text{kW-min}} \qquad 3600 \, \frac{\text{kJ}}{\text{kW-hr}} \qquad 0.746 \, \frac{\text{kW}}{\text{hp}}$$

UNIVERSAL GAS CONSTANT

$$1545.32 \, \frac{\text{ft-lb}}{\text{pmole-}^\circ\text{R}} \qquad 8.3143 \, \frac{\text{kJ}}{\text{kgmole-K}} \qquad 1.9859 \, \frac{\text{Btu}}{\text{pmole-}^\circ\text{R}}$$

$$1.9859 \, \frac{\text{cal}}{\text{gmole-K}} \qquad 10.731 \, \frac{\text{psi-ft}^3}{\text{pmole-}^\circ\text{R}}$$

Appendix B

Pressure-Enthalpy and Compressibility

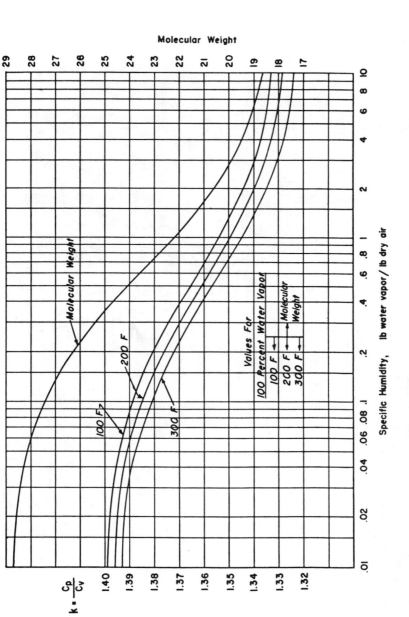

Figure B-1. The adiabatic exponent and molecular weight of air-vapor mixtures. (Reprinted by permission from *Compressed Air and Gas Handbook*, third edition, Compressed Air and Gas Institute.)

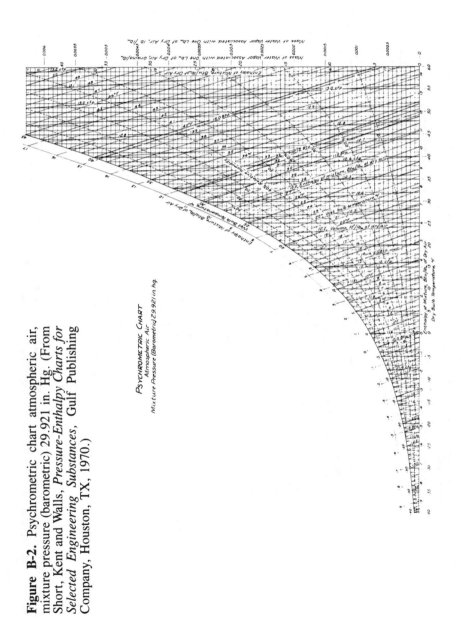

Figure B-2. Psychrometric chart atmospheric air, mixture pressure (barometric) 29.921 in. Hg. (From Short, Kent and Walls, *Pressure-Enthalpy Charts for Selected Engineering Substances,* Gulf Publishing Company, Houston, TX, 1970.)

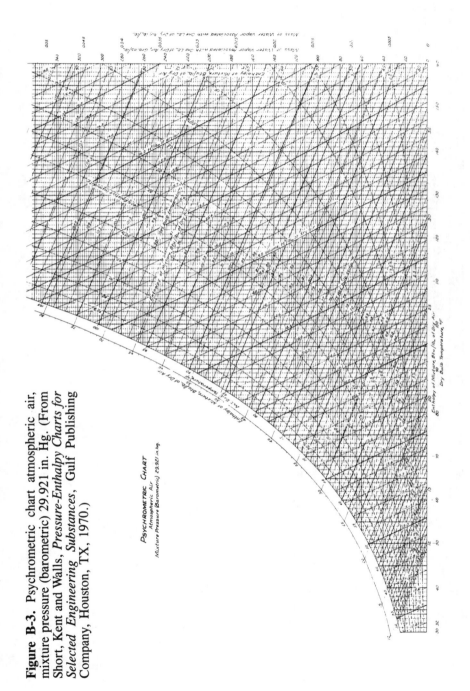

Figure B-3. Psychrometric chart atmospheric air, mixture pressure (barometric) 29.921 in. Hg. (From Short, Kent and Walls, *Pressure-Enthalpy Charts for Selected Engineering Substances*, Gulf Publishing Company, Houston, TX, 1970.)

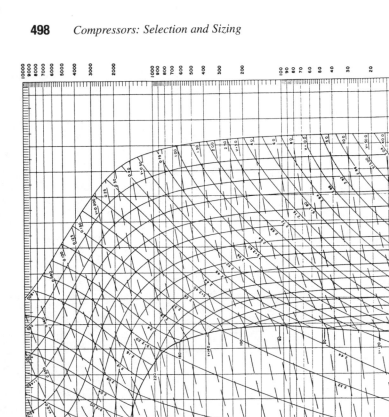

Figure B-4. Ammonia pressure-enthalpy diagram. (From Short, Kent and Walls, *Pressure-Enthalpy Charts for Selected Engineering Substances*, Gulf Publishing Company, Houston, TX, 1970.)

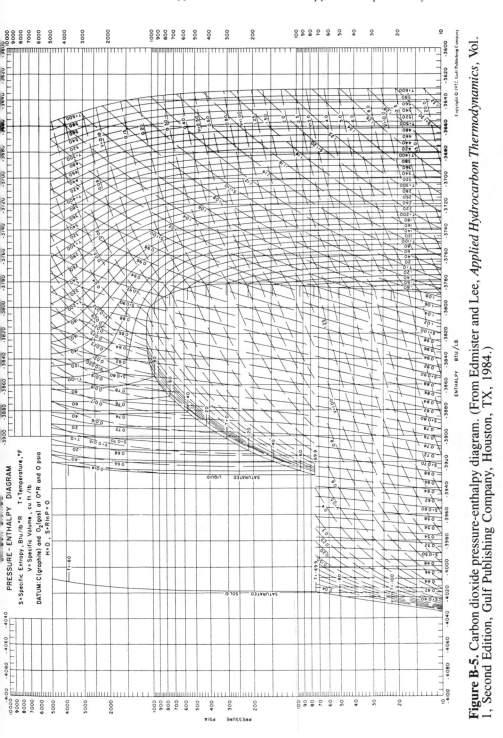

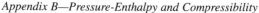

Figure B-5. Carbon dioxide pressure-enthalpy diagram. (From Edmister and Lee, *Applied Hydrocarbon Thermodynamics*, Vol. 1, Second Edition, Gulf Publishing Company, Houston, TX, 1984.)

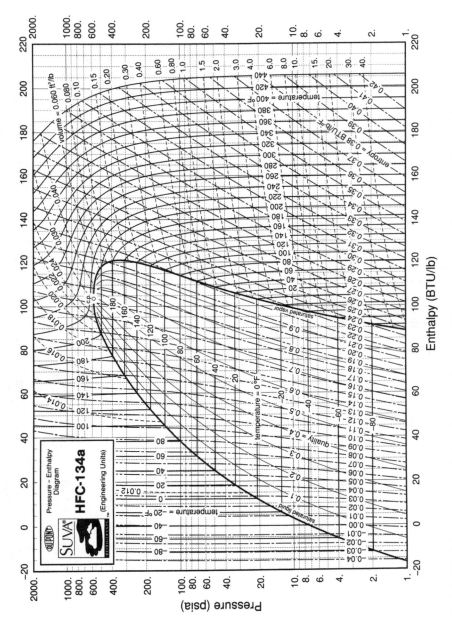

Figure B-6. HFC–134a pressure-enthalpy diagram. *(Reprinted by permission and courtesy of Dupont.)*

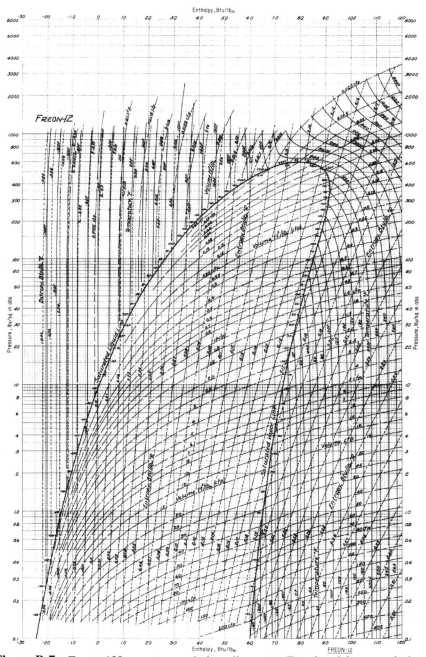

Figure B-7. Freon-12* pressure-enthalpy diagram. (Reprinted by permission and courtesy of E. I. DuPont De Nemours and Co.) *Freon and Freon—followed by numerals are DuPont trademarks.

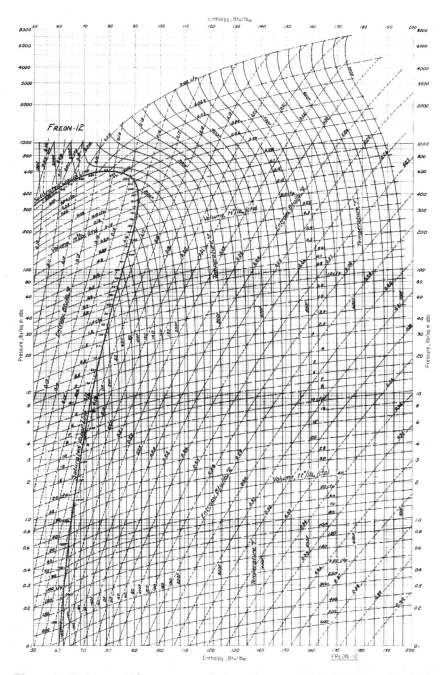

Figure B-8. Freon-12* pressure-enthalpy diagram. (Reprinted by permission and courtesy of E. I. DuPont De Nemours and Co.) *Freon and Freon—followed by numerals are DuPont trademarks.

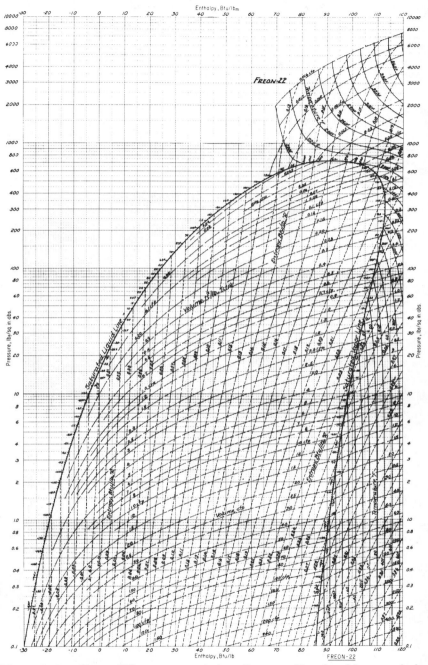

Figure B-9. Freon-22* pressure-enthalpy diagram. (Reprinted by permission and courtesy of E. I. DuPont De Nemours and Co.) *Freon and Freon—followed by numerals are DuPont trademarks.

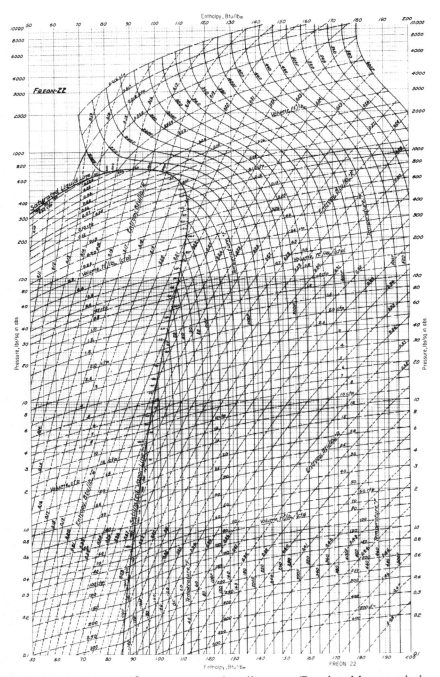

Figure B-10. Freon-22* pressure-enthalpy diagram. (Reprinted by permission and courtesy of E. I. DuPont De Nemours and Co.) *Freon and Freon—followed by numerals are DuPont trademarks.

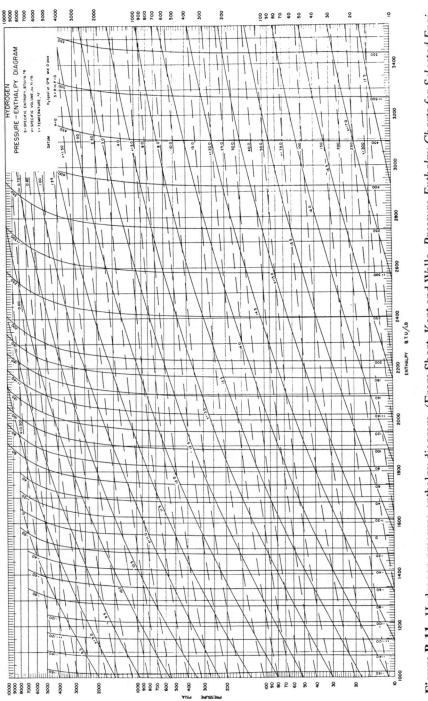

Figure B-11. Hydrogen pressure-enthalpy diagram. (From Short, Kent and Walls, *Pressure-Enthalpy Charts for Selected Engineering Substances*, Gulf Publishing Company, Houston, TX, 1970.)

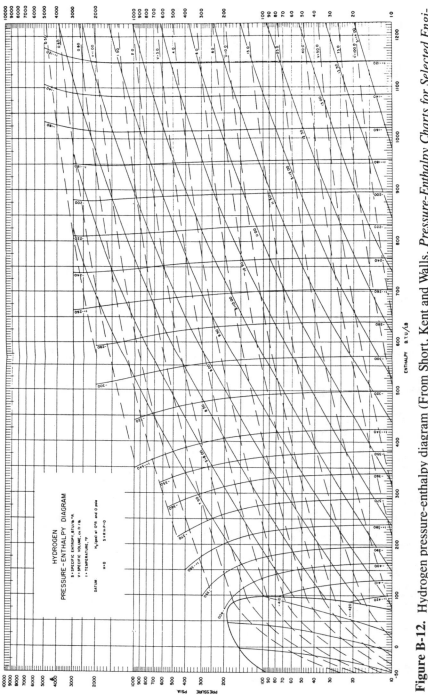

Figure B-12. Hydrogen pressure-enthalpy diagram (From Short, Kent and Walls, *Pressure-Enthalpy Charts for Selected Engineering Substances*, Gulf Publishing Company, Houston, TX, 1970.)

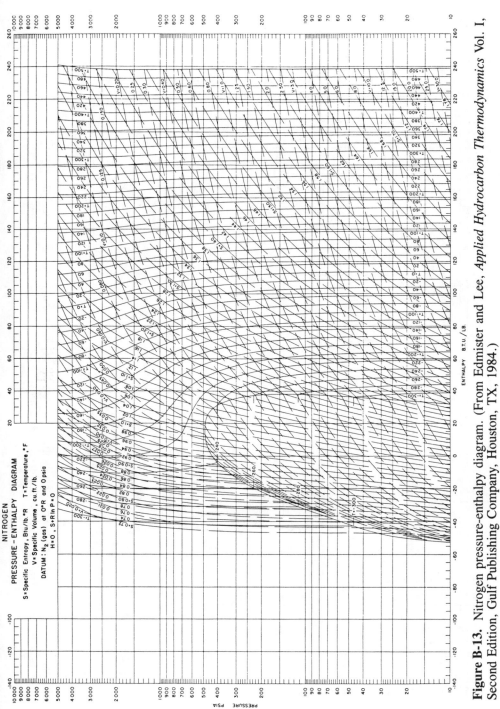

Figure B-13. Nitrogen pressure-enthalpy diagram. (From Edmister and Lee, *Applied Hydrocarbon Thermodynamics* Vol. I, Second Edition, Gulf Publishing Company, Houston, TX, 1984.)

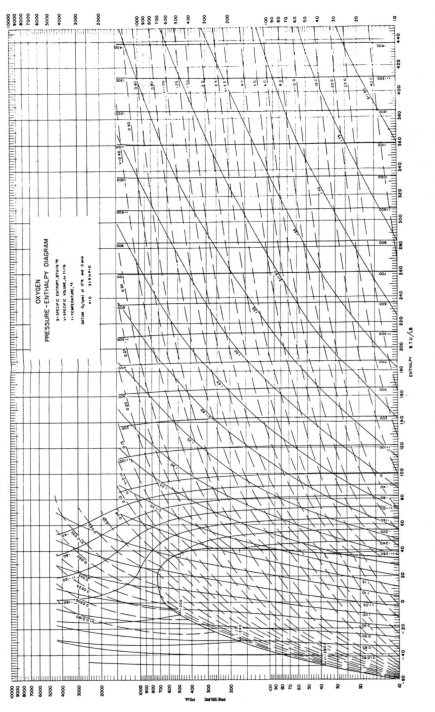

Figure B-14. Oxygen pressure-enthalpy diagram. (From Short, Kent and Walls, *Pressure-Enthalpy Charts for Selected Engineering Substances*, Gulf Publishing Company; Houston, TX, 1970.)

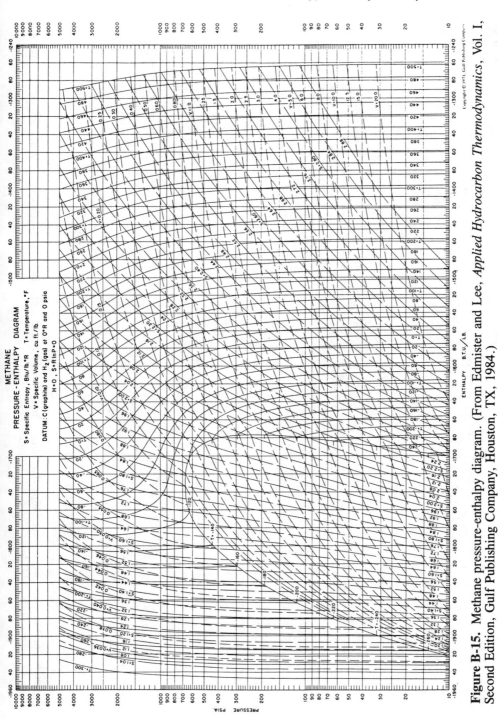

Figure B-15. Methane pressure-enthalpy diagram. (From Edmister and Lee, *Applied Hydrocarbon Thermodynamics*, Vol. I, Second Edition, Gulf Publishing Company, Houston, TX, 1984.)

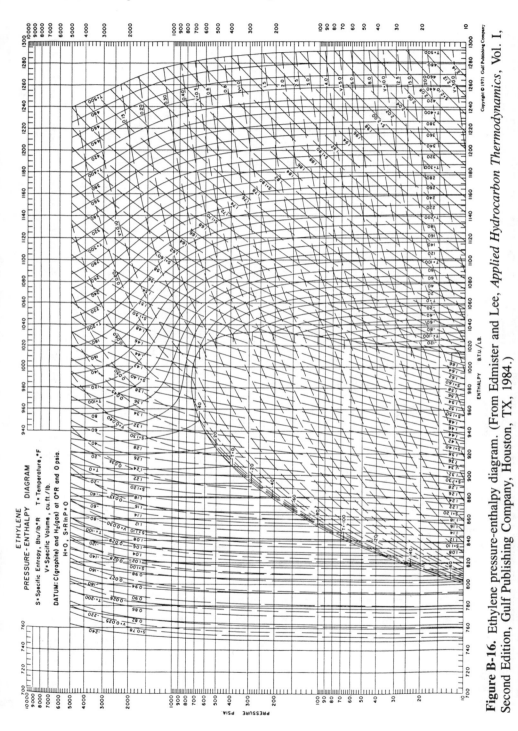

Figure B-16. Ethylene pressure-enthalpy diagram. (From Edmister and Lee, *Applied Hydrocarbon Thermodynamics*, Vol. I, Second Edition, Gulf Publishing Company, Houston, TX, 1984.)

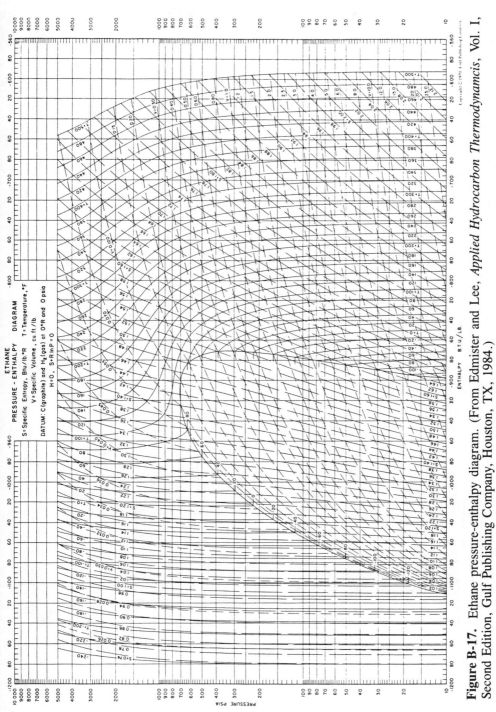

Figure B-17. Ethane pressure-enthalpy diagram. (From Edmister and Lee, *Applied Hydrocarbon Thermodynamcis*, Vol. I, Second Edition, Gulf Publishing Company, Houston, TX, 1984.)

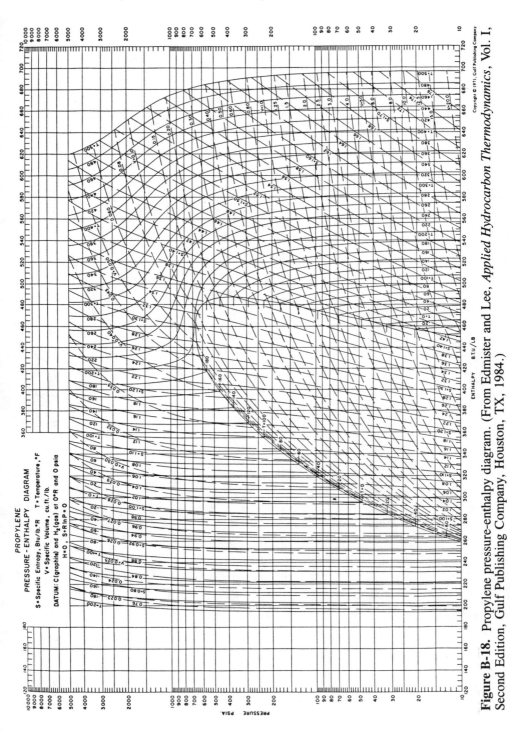

Figure B-18. Propylene pressure-enthalpy diagram. (From Edmister and Lee, *Applied Hydrocarbon Thermodynamics*, Vol. I, Second Edition, Gulf Publishing Company, Houston, TX, 1984.)

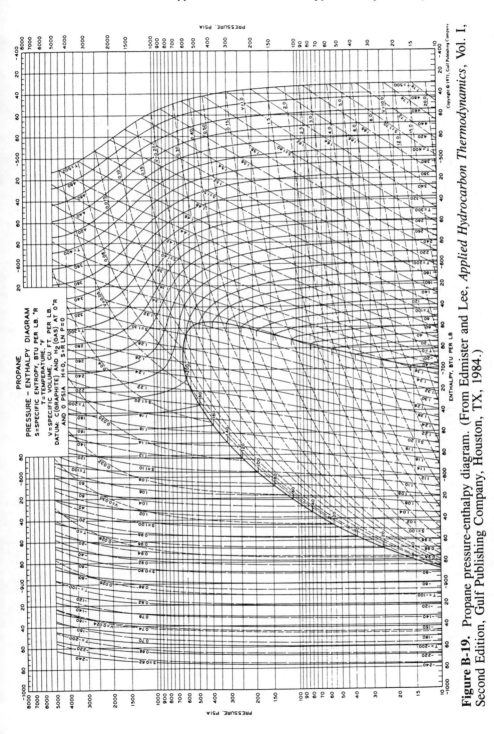

Figure B-19. Propane pressure-enthalpy diagram. (From Edmister and Lee, *Applied Hydrocarbon Thermodynamics*, Vol. I, Second Edition, Gulf Publishing Company, Houston, TX, 1984.)

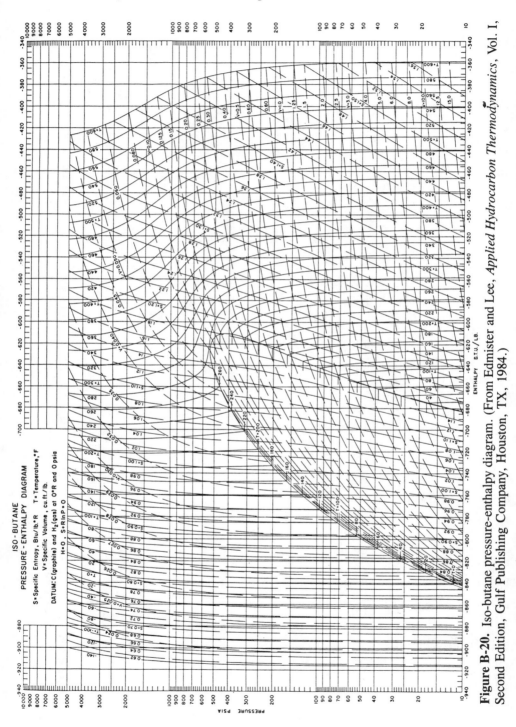

Figure B-20. Iso-butane pressure-enthalpy diagram. (From Edmister and Lee, *Applied Hydrocarbon Thermodynamics*, Vol. I, Second Edition, Gulf Publishing Company, Houston, TX, 1984.)

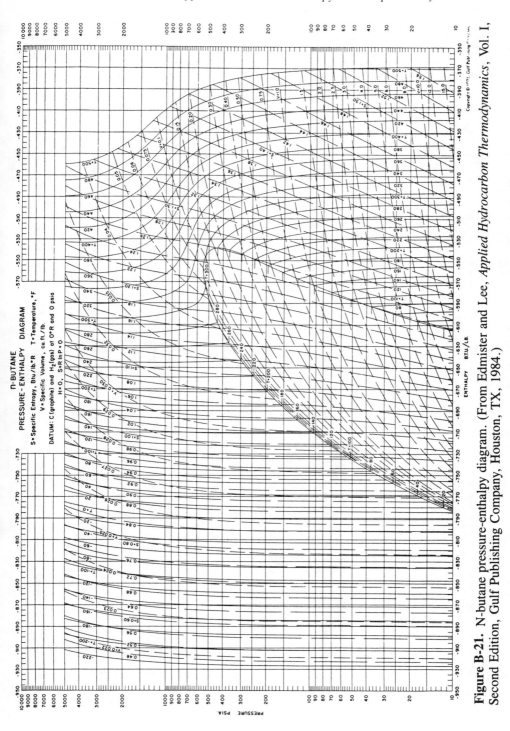

Figure B-21. N-butane pressure-enthalpy diagram. (From Edmister and Lee, *Applied Hydrocarbon Thermodynamics*, Vol. I, Second Edition, Gulf Publishing Company, Houston, TX, 1984.)

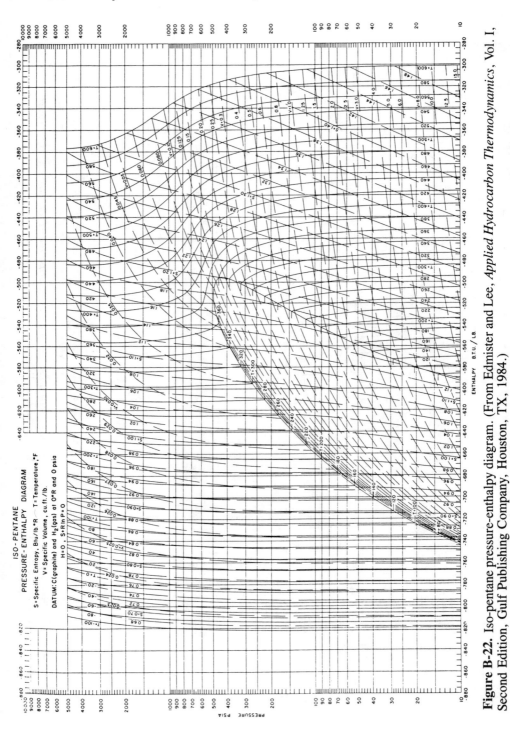

Figure B-22. Iso-pentane pressure-enthalpy diagram. (From Edmister and Lee, *Applied Hydrocarbon Thermodynamics*, Vol. I, Second Edition, Gulf Publishing Company, Houston, TX, 1984.)

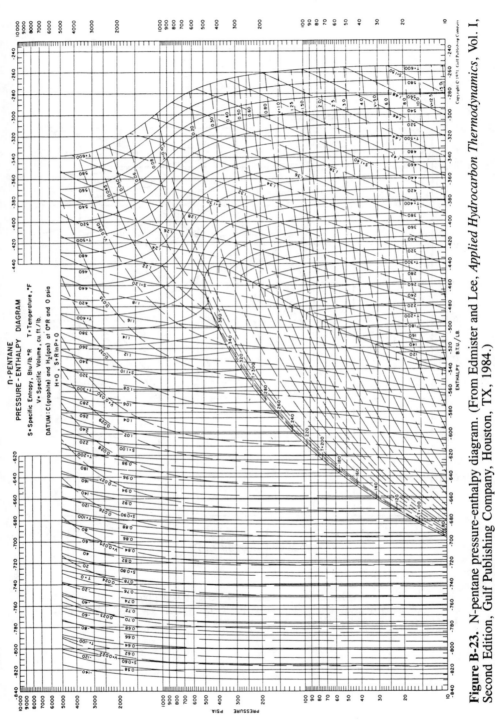

Figure B-23. N-pentane pressure-enthalpy diagram. (From Edmister and Lee, *Applied Hydrocarbon Thermodynamics*, Vol. I, Second Edition, Gulf Publishing Company, Houston, TX, 1984.)

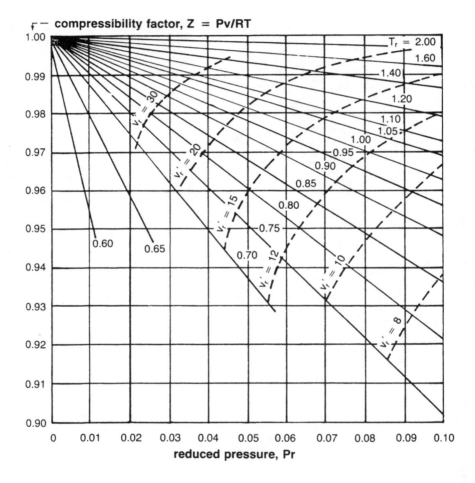

Figure B-24. Generalized compressibility chart. (Excerpted by special permission from Chemical Engineering, July 1959, copyright © 1954, by McGraw-Hill, Inc., New York, NY.)

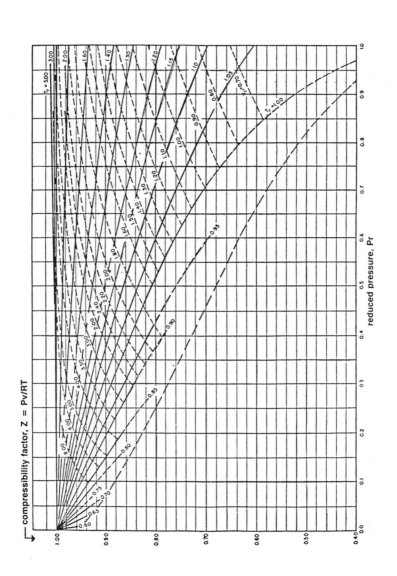

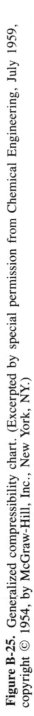

Figure B-25. Generalized compressibility chart. (Excerpted by special permission from Chemical Engineering, July 1959, copyright © 1954, by McGraw-Hill, Inc., New York, NY.)

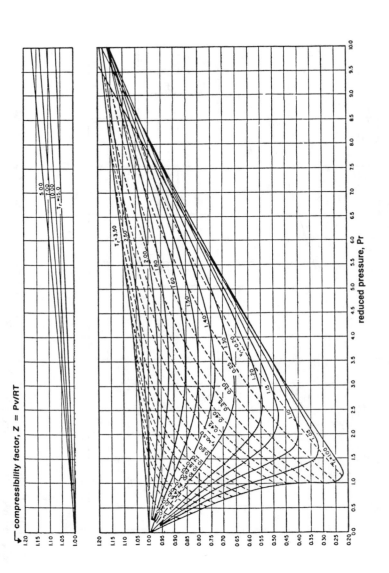

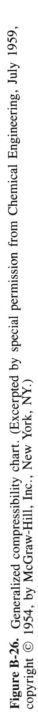

Figure B-26. Generalized compressibility chart. (Excerpted by special permission from Chemical Engineering, July 1959, copyright © 1954, by McGraw-Hill, Inc., New York, NY.)

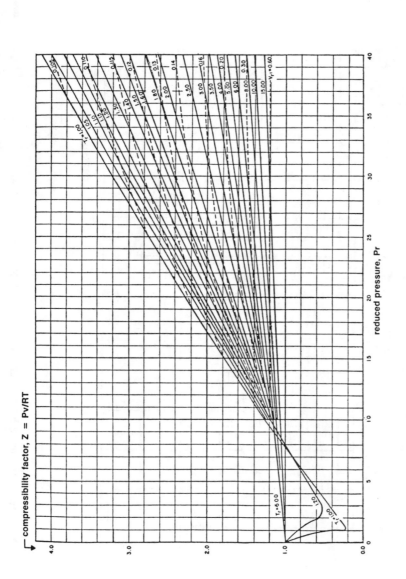

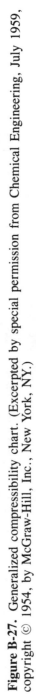

Figure B-27. Generalized compressibility chart. (Excerpted by special permission from Chemical Engineering, July 1959, copyright © 1954, by McGraw-Hill, Inc., New York, NY.)

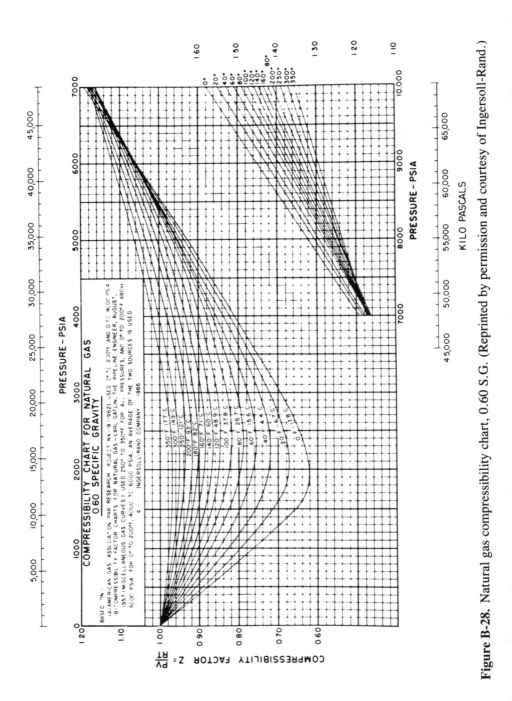

Figure B-28. Natural gas compressibility chart, 0.60 S.G. (Reprinted by permission and courtesy of Ingersoll-Rand.)

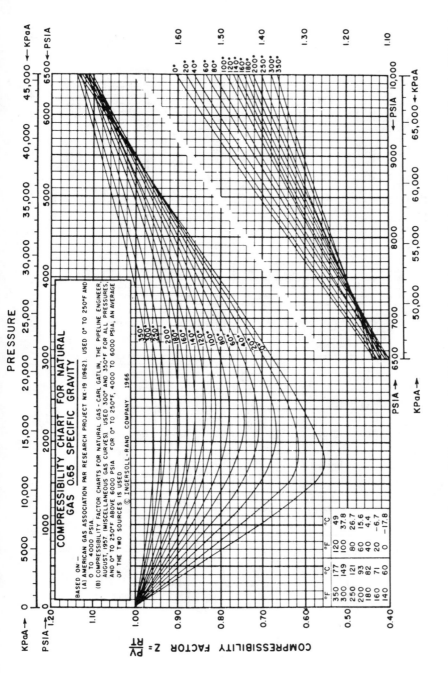

Figure B-29. Natural gas compressibility chart, 0.65 S.G. (Reprinted by permission and courtesy of Ingersoll-Rand.)

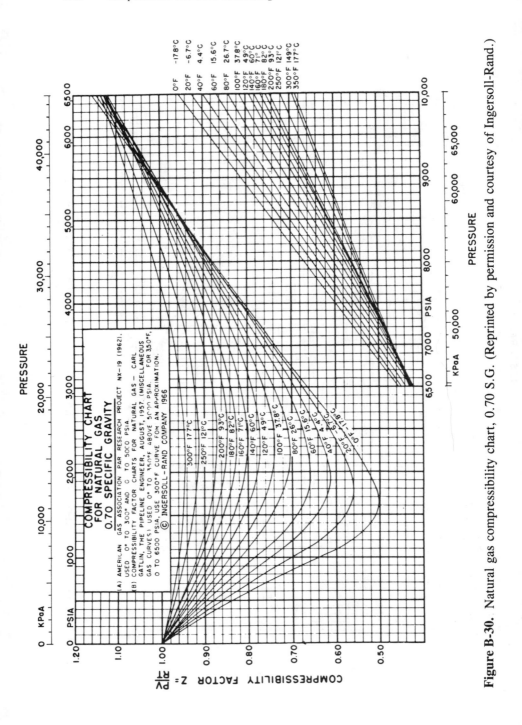

Figure B-30. Natural gas compressibility chart, 0.70 S.G. (Reprinted by permission and courtesy of Ingersoll-Rand.)

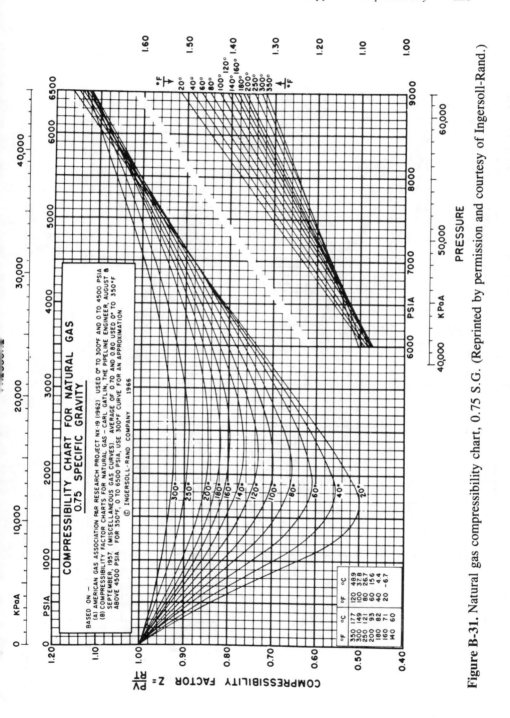

Figure B-31. Natural gas compressibility chart, 0.75 S.G. (Reprinted by permission and courtesy of Ingersoll-Rand.)

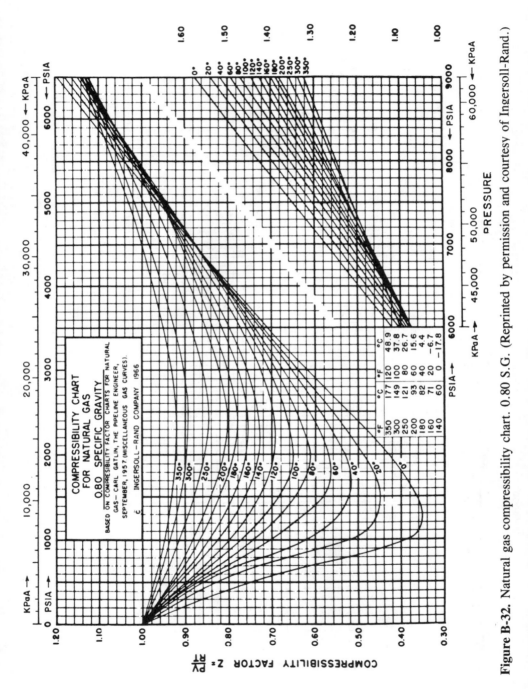

Figure B-32. Natural gas compressibility chart. 0.80 S.G. (Reprinted by permission and courtesy of Ingersoll-Rand.)

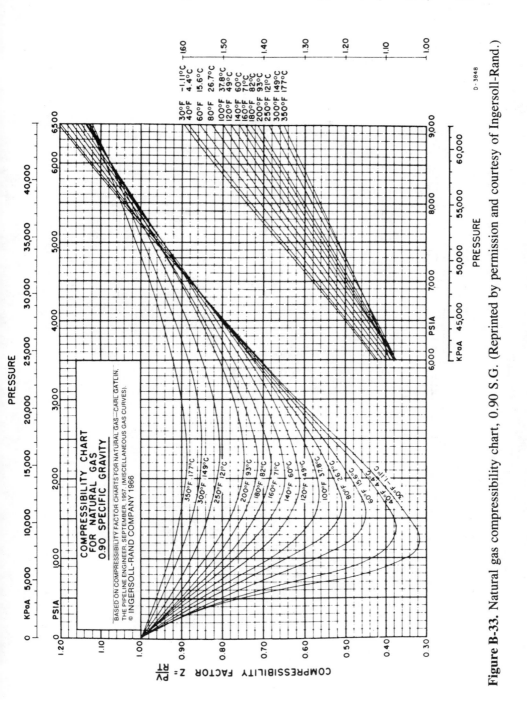

Figure B-33. Natural gas compressibility chart, 0.90 S.G. (Reprinted by permission and courtesy of Ingersoll-Rand.)

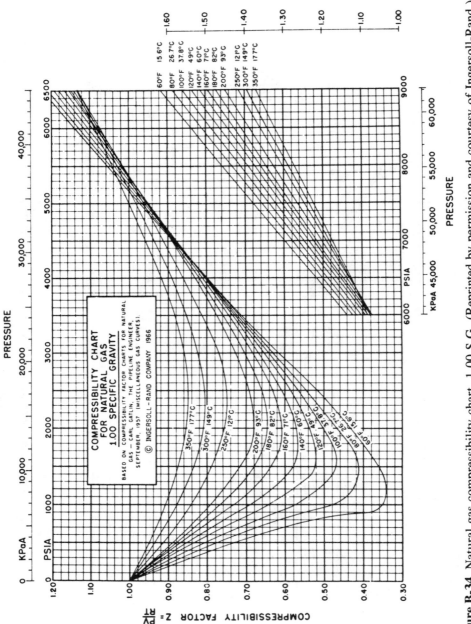

Figure B-34. Natural gas compressibility chart, 1.00 S.G. (Reprinted by permission and courtesy of Ingersoll-Rand.)

Appendix C

Physical Constants of Hydrocarbons

TABLE C-1
Physical Constants of Hydrocarbons

No.	Compound	Formula	Molecular weight	Boiling point °F, 14.696 psia	Vapor pressure, 100°F, psia	Freezing point, °F, 14.696 psia	Critical constants Pressure, psia	Temperature, °F	Volume, cu ft/lb
1	Methane	CH₄	16.043	−258.69	5000:	−296.46ᵈ	667.8	−116.63	0.0991
2	Ethane	C₂H₆	30.070	−127.48	(800)	−297.89ᵈ	707.8	90.09	0.0788
3	Propane	C₃H₈	44.097	−43.67	190.	−305.84ᵈ	616.3	206.01	0.0737
4	n−Butane	C₄H₁₀	58.124	31.10	51.6	−217.05	550.7	305.65	0.0702
5	Isobutane	C₄H₁₀	58.124	10.90	72.2	−255.29	529.1	274.98	0.0724
6	n−Pentane	C₅H₁₂	72.151	96.92	15.570	−201.51	488.6	385.7	0.0675
7	Isopentane	C₅H₁₂	72.151	82.12	20.44	−255.83	490.4	369.10	0.0679
8	Neopentane	C₅H₁₂	72.151	49.10	35.9	2.17	464.0	321.13	0.0674
9	n−Hexane	C₆H₁₄	86.178	155.72	4.956	−139.58	436.9	453.7	0.0688
10	2−Methylpentane	C₆H₁₄	86.178	140.47	6.767	−244.63	436.6	435.83	0.0681
11	3−Methylpentane	C₆H₁₄	86.178	145.89	6.098	—	453.1	448.3	0.0681
12	Neohexane	C₆H₁₄	86.178	121.52	9.856	−147.72	446.8	420.13	0.0667
13	2,3−Dimethylbutane	C₆H₁₄	86.178	136.36	7.404	−199.38	453.5	440.29	0.0665
14	n−Heptane	C₇H₁₆	100.205	209.17	1.620	−131.05	396.8	512.8	0.0691
15	2−Methylhexane	C₇H₁₆	100.205	194.09	2.271	−180.89	396.5	495.00	0.0673
16	3−Methylhexane	C₇H₁₆	100.205	197.32	2.130	—	408.1	503.78	0.0646
17	3−Ethylpentane	C₇H₁₆	100.205	200.25	2.012	−181.48	419.3	513.48	0.0665
18	2,2−Dimethylpentane	C₇H₁₆	100.205	174.54	3.492	−190.86	402.2	477.23	0.0665
19	2,4−Dimethylpentane	C₇H₁₆	100.205	176.89	3.292	−182.63	396.9	475.95	0.0668
20	3,3−Dimethylpentane	C₇H₁₆	100.205	186.91	2.773	−210.01	427.2	505.85	0.0662
21	Triptane	C₇H₁₆	100.205	177.58	3.374	−12.82	428.4	496.44	0.0636
22	n−Octane	C₈H₁₈	114.232	258.22	.0.537	−70.18	360.6	564.22	0.0690
23	Diisobutyl	C₈H₁₈	114.232	228.39	1.101	−132.07	360.6	530.44	0.0676
24	Isooctane	C₈H₁₈	114.232	210.63	1.708	−161.27	372.4	519.46	0.0656
25	n−Nonane	C₉H₂₀	128.259	303.47	0.179	−64.28	332.	610.68	0.0684
26	n−Decane	C₁₀H₂₂	142.286	345.48	0.0597	−21.36	304.	652.1	0.0679
27	Cyclopentane	C₅H₁₀	70.135	120.65	9.914	−136.91	653.8	461.5	0.059
28	Methylcyclopentane	C₆H₁₂	84.162	161.25	4.503	−224.44	548.9	499.35	0.0607
29	Cyclohexane	C₆H₁₂	84.162	177.29	3.264	43.77	591	536.7	0.0586
30	Methylcyclohexane	C₇H₁₄	98.182	213.68	1.609	−195.87	503.5	570.27	0.0600
31	Ethylene	C₂H₄	28.054	−154.62	—	−272.45ᵈ	729.8	48.58	0.0737
32	Propene	C₃H₆	42.081	−53.90	226.4	−301.45ᵈ	669.	196.9	0.0689
33	1−Butene	C₄H₈	56.108	20.75	63.05	−301.63ᵈ	583.	295.6	0.0685
34	Cis−2−Butene	C₄H₈	56.108	38.69	45.54	−218.06	610.	324.37	0.0668
35	Trans−2−Butene	C₄H₈	56.108	33.58	49.80	−157.96	595.	311.86	0.0680
36	Isobutene	C₄H₈	56.108	19.59	63.40	−220.61	580.	292.55	0.0682
37	1−Pentene	C₅H₁₀	70.135	85.93	19.115	−265.39	590.	376.93	0.0697
38	1,2−Butadiene	C₄H₆	54.092	51.53	(20.)	−213.16	(653.)	(339.)	(0.0649)
39	1,3−Butadiene	C₄H₆	54.092	24.06	(60.)	−164.02	628.	306.	0.0654
40	Isoprene	C₅H₈	68.119	93.30	16.672	−230.74	(558.4)	(412.)	(0.0650)
41	Acetylene	C₂H₂	26.038	−119ᵉ	—	−114.ᵈ	890.4	95.31	0.0695
42	Benzene	C₆H₆	78.114	176.17	3.224	41.96	710.4	552.22	0.0531
43	Toluene	C₇H₈	92.141	231.13	1.032	−138.94	595.9	605.55	0.0549
44	Ethylbenzene	C₈H₁₀	106.168	277.16	0.371	−138.91	523.5	651.24	0.0564
45	o−Xylene	C₈H₁₀	106.168	291.97	0.264	−13.30	541.4	675.0	0.0557
46	m−Xylene	C₈H₁₀	106.168	282.41	0.326	−54.12	513.6	651.02	0.0567
47	p−Xylene	C₈H₁₀	106.168	281.05	0.342	55.86	509.2	649.6	0.0572
48	Styrene	C₈H₈	104.152	293.29	(0.24)	−23.10	580.	706.0	0.0541
49	Isopropylbenzene	C₉H₁₂	120.195	306.34	0.188	−140.82	465.4	676.4	0.0570
50	Methyl Alcohol	CH₄O	32.042	148.1(2)	4.63(22)	−143.82(22)	1174.2(21)	462.97(21)	0.0589(21)
51	Ethyl Alcohol	C₂H₆O	46.069	172.92(22)	2.3(7)	−173.4(22)	925.3(21)	469.58(21)	0.0580(21)
52	Carbon Monoxide	CO	28.010	−313.6(2)	—	−340.6(2)	507.(17)	−220.(17)	0.0532(17)
53	Carbon Dioxide	CO₂	44.010	−109.3(2)	—	—	1071.(17)	87.9(23)	0.0342(23)
54	Hydrogen Sulfide	H₂S	34.076	−76.6(24)	394.0(6)	−117.2(7)	1306.(17)	212.7(17)	0.0459(24)
55	Sulfur Dioxide	SO₂	64.059	14.0(7)	88.(7)	−103.9(7)	1145.(24)	315.5(17)	0.0306(24)
56	Ammonia	NH₃	17.031	−28.2(24)	212.(7)	−107.9(2)	1636.(17)	270.3(24)	0.0681(17)
57	Air	N₂O₂	28.964	−317.6(2)	—	—	547.(2)	−221.3(2)	0.0517(3)
58	Hydrogen	H₂	2.016	−423.0(24)	—	−434.8(24)	188.1(17)	−399.8(17)	0.5167(24)
59	Oxygen	O₂	31.999	−297.4(2)	—	−361.8(24)	736.9(24)	−181.1(17)	0.0382(24)
60	Nitrogen	N₂	28.013	−320.4(2)	—	−346.0(24)	493.0(24)	−232.4(24)	0.0514(17)
61	Chlorine	Cl₂	70.906	−29.3.24)	158.(7)	−149.8(24)	1118.4(24)	291.(17)	0.0281(17)
62	Water	H₂O	18.015	212.0	0.9492(12)	32.0	3208.(17)	705.6(17)	0.0500(17)
63	Helium	He	4.003	—	—	—	—	—	—
64	Hydrogen Chloride	HCl	36.461	−121.(16)	925.(7)	−173.6(16)	1198.(17)	124.5(17)	0.0208(17)

TABLE C-1 (continued)
Physical Constants of Hydrocarbons

Specific gravity 60°F./60°F.a,b	lb/gal·a (Wt in vacuum)	lb/gal·a,c (Wt in air)	Gal/lb Mole*	Temperature Coefficient of density,a	Pitzer acentric factor (18)	Compressibility factor of real gas, Z 14.696 psia, 60°F.	Specific gravity Air = 1*	cu ft gas/lb*	cu ft gas/gal liquid*	Cp, Btu/lb/°F. Ideal gas	Cp Liquid	No.
0.3'	2.5'	2.5'	6.4'	——	0.0104	0.9981	0.5539	23.65	59.'	0.5266	——	1
0.3564^h	2.971^h	2.962^h	10.12^h	—	0.0986	0.9916	1.0382	12.62	37.5^h	0.4097	0.9256	2
0.5077^h	4.233^h	4.223^h	10.42^h	0.00152^h	0.1524	0.9820	1.5225	8.606	36.43^h	0.3881	0.5920	3
0.5844^h	4.872^h	4.865^h	11.93^h	0.00117^h	0.2010	0.9667	2.0068	6.529	31.81^h	0.3867	0.5636	4
0.5631^h	4.695^h	4.686^h	12.38^h	0.00119^h	0.1848	0.9696	2.0068	6.529	30.65^h	0.3872	0.5695	5
0.6310	5.261	5.251	13.71	0.00087	0.2539	0.9549	2.4911	5.260	27.67	0.3883	0.5441	6
0.6247	5.208	5.199	13.85	0.00090	0.2223	0.9544	2.4911	5.260	27.39	0.3827	0.5353	7
0.5967^h	4.975^h	4.965^h	14.50^h	0.00104^h	0.1969	0.9510	2.4911	5.260	26.17^h	(0.3866)	0.554	8
0.6640	5.536	5.526	15.57	0.00075	0.3007	——	2.9753	4.404	24.38	0.3864	0.5332	9
0.6579	5.485	5.475	15.71	0.00078	0.2825	——	2.9753	4.404	24.15	0.3872	0.5264	10
0.6689	5.577	5.568	15.45	0.00075	0.2741	——	2.9753	4.404	24.56	0.3815	0.507	11
0.6540	5.453	5.443	15.81	0.00078	0.2369	——	2.9753	4.404	24.01	0.3809	0.5165	12
0.6664	5.556	5.546	15.51	0.00075	0.2495	——	2.9753	4.404	24.47	0.378	0.5127	13
0.6882	5.738	5.728	17.46	0.00069	0.3498	——	3.4596	3.787	21.73	0.3875	0.5283	14
0.6830	5.694	5.685	17.60	0.00068	0.3336	——	3.4596	3.787	21.57	(0.390)	0.5223	15
0.6917	5.767	5.757	17.38	0.00069	0.3257	——	3.4596	3.787	21.84	(0.390)	0.511	16
0.7028	5.859	5.850	17.10	0.00070	0.3095	——	3.4596	3.787	22.19	(0.390)	0.5145	17
0.6782	5.654	5.645	17.72	0.00072	0.2998	——	3.4596	3.787	21.41	(0.395)	0.5171	18
0.6773	5.647	5.637	17.75	0.00072	0.3048	——	3.4596	3.787	21.39	0.3906	0.5247	19
0.6976	5.816	5.807	17.23	0.00065	0.2840	——	3.4596	3.787	22.03	(0.395)	0.502	20
0.6946	5.791	5.782	17.30	0.00069	0.2568	——	3.4596	3.787	21.93	0.3812	0.4995	21
0.7068	5.893	5.883	19.39	0.00062	0.4018	——	3.9439	3.322	19.58	(0 3876)	0.5239^v	22
0.6979	5.819	5.810	19.63	0.00065	0.3596	——	3.9439	3.322	19.33	(0.373)	0.5114	23
0.6962	5.804	5.795	19.68	0.00065	0.3041	——	3.9439	3.322	19.28	0.3758	0.4892	24
0.7217	6.017	6.008	21.32	0.00063	0.4455	——	4.4282	2.959	17.80	0.3840	0.5228	25
0.7342	6.121	6.112	23.24	0.00055	0.4885	——	4.9125	2.667	16.33	0.3835	0.5208	26
0.7504	6.256	6.247	11.21	0.00070	0.1955	0.9657	2.4215	5.411	33.85	0.2712	0.4216	27
0.7536	6.283	6.274	13.40	0.00071	0.2306	——	2.9057	4.509	28.33	0.3010	0.4407	28
0.7834	6.531	6.522	12.89	0.00068	0.2133	——	2.9057	4.509	29.45	0.2900	0.4332	29
0.7740	6.453	6.444	15.22	0.00063	0.2567	——	3.3900	3.865	24.94	0.3170	0.4397	30
					0.0868	0.9938	0.9686	13.53	——	0.3622	——	31
0.5220^h	4.352^h	4.343^h	9.67^h	0.00189^h	0.1405	0.9844	1.4529	9.018	39.25^h	0.3541	0.585	32
0.6013^h	5.013^h	5.004^h	11.19^h	0.00116^h	0.1906	0.9704	1.9372	6.764	33.91^h	0.3548	0.535	33
0.6271^h	5.228^h	5.219^h	10.73^h	0.00098^h	0.1953	0.9661	1.9372	6.764	35.36^h	0.3269	0.5271	34
0.6100^h	5.086^h	5.076^h	11.03^h	0.00107^h	0.2220	0.9662	1.9372	6.764	34.40^h	0.3654	0.5351	35
0.6004^h	5.006^h	4.996^h	11.21^h	0.00120^h	0.1951	0.9689	1.9372	6.764	33.86^h	0.3701	0.549	36
0.6457	5.383	5.374	13.03	0.00089	0.2925	0.9550	2.4215	5.411	29.13	0.3635	0.5196	37
0.658^h	5.486^h	5.470^h	9.86^h	0.00098^h	0.2485	(0.969)	1.8676	7.016	38.49^h	0.3458	0.5408	38
0.6272^h	5.229^h	5.220^h	10.34^h	0.00113^h	0.1955	(0.965)	1.8676	7.016	36.69^h	0.3412	0.5079	39
0.6861	5.720	5.711	11.91	0.00086	0.2323	(0.962)	2.3519	5.571	31.87	0.357	0.5192	40
0.615^k	——				0.1803	0.9925	0.8990	14.57	——	0.3966	——	41
0.8844	7.373	7.365	10.59	0.00066	0.2125	0.929(15)	2.6969	4.858	35.82	0.2429	0.4098	42
0.8718	7.268	7.260	12.68	0.00060	0.2596	0.903(21)	3.1812	4.119	29.94	0.2598	0.4012	43
0.8718	7.268	7.259	14.61	0.00054	0.3169	——	3.6655	3.574	25.98	0.2795	0.4114	44
0.8848	7.377	7.367	14.39	0.00055	0.3023	——	3.6655	3.574	26.37	0.2914	0.4418	45
0.8687	7.243	7.234	14.66	0.00054	0.3278	——	3.6655	3.574	25.89	0.2782	0.4045	46
0.8657	7.218	7.209	14.71	0.00054	0.3138	——	3.6655	3.574	25.80	0.2769	0.4083	47
0.9110	7.595	7.586	13.71	0.00057	——	——	3.5959	3.644	27.67	0.2711	0.4122	48
0.8663	7.223	7.214	16.64	0.00054	0.2862	——	4.1498	3.157	22.80	0.2917	(0.414)	49
0.796(3)	6.64	6.63	4.83	——	——	——	1.1063	11.84	78.6	0.3231^v(24)	0.594(7)	50
0.794(3)	6.62	6.61	6.96	——	——	——	1.5906	8.237	54.5	0.3323^v(24)	0.562(7)	51
0.801^m(8)	6.68^m	6.67^m	4.19^m	——	0.041	0.9995(15)	0.9671	13.55	——	0.2484(13)	——	52
0.827^h(6)	6.89^h	6.88^h	6.38^h	——	0.225	0.9943(15)	1.5195	8.623	59.5^h	0.1991(13)	——	53
0.79^h(6)	6.59^h	6.58^h	5.17^h	——	0.100	0.9903(15)	1.1765	11.14	73.3^h	0.238(4)	——	54
1.397^h(14)	11.65^h	11.64^h	5.50^h	——	0.246	——	2.2117	5.924	69.0^h	0.145(7)	0.325^h(7)	55
0.6173(11)	5.15	5.14	3.31	——	0.255	——	0.5880	22.28	114.7	0.5002(10)	1.114^h(7)	56
0.856^m(8)	7.14^m	7.13^m	4.06^m	——	——	0.9996(15)	1.0000	13.10	——	0.2400(9)	——	57
0.07^m(3)	——				0.000	1.0006(15)	0.0696	188.2	——	3.408(13)	——	58
1.140(25)	9.50^m	9.49^m	3.37^m	——	0.0213	——	1.1048	11.86	——	0.2188(13)	——	59
0.810(26)	6.75^m	6.74^m	4.15^m	——	0.040	0.9997(15)	0.9672	13.55	——	0.2482(13)	——	60
1.414(14)	11.79	11.78	6.01	——	——	——	2.4481	5.352	63.1	0.119(7)	——	61
1.000	8.337	8.328	2.16	——	0.348	——	0.6220	21.06	175.6	0.4446(13)	1.0009(7)	62
——												63
0.8558(14)	7.135	7.126	5.11	0.00335*	——	——	1.2588	10.41	74.3	0.190(7)	——	64

TABLE C-1 (continued)
Physical Constants of Hydrocarbons

No.	Compound	Net — Btu/cu ft Ideal gas, 14.696 psia (20)*	Gross — Btu/cu ft Ideal gas, 14.696 psia (20)*	Gross — Btu/lb liquid (wt in vacuum)	Gross — Btu/gal liquid*	Heat of vaporization, 14.696 psia at boiling point, Btu/lb	Refractive index, n_D 68°F.	Air required for combustion ideal gas* cu ft/cu ft	Flammability Lower	Flammability Higher	ASTM Motor method D-357	ASTM Research method D-908
1	Methane	909.1	1009.7	—	—	219.22	—	9.54	5.0	15.0	—	—
2	Ethane	1617.8	1768.8	—	—	210.41	—	16.70	2.9	13.0	+.05[f]	+1.6[l,f]
3	Propane	2316.1	2517.5	21513	91065	183.05	—	23.86	2.1	9.5	97.1	+1.8[l,f]
4	n-Butane	3010.4	3262.1	21139	102989	165.65	1.3326[h]	31.02	1.8	8.4	89.6[l]	93.8[l]
5	Isobutane	3001.1	3252.7	21091	99022	157.53	—	31.02	1.8	8.4	97.6	+.10[l,f]
6	n-Pentane	3707.5	4009.6	20928	110102	153.59	1.35748	38.18	1.4	8.3	62.6[l]	61.7[l]
7	Isopentane	3698.3	4000.3	20889	108790	147.13	1.35373	38.18	1.4	(8.3)	90.3	92.3
8	Neopentane	3682.6	3984.6	20824	103599	135.58	1.342[h]	38.18	1.4	(8.3)	80.2	85.5
9	n-Hexane	4403.7	4756.2	20784	115060	143.95	1.37486	45.34	1.2	7.7	26.0	24.8
10	2-Methylpentane	4395.8	4748.1	20757	113852	138.67	1.37145	45.34	1.2	(7.7)	73.5	73.4
11	3-Methylpentane	4398.7	4751.0	20768	115823	140.09	1.37652	45.34	(1.2)	(7.7)	74.3	74.5
12	Neohexane	4382.6	4735.0	20710	112932	131.24	1.36876	45.34	1.2	(7.7)	93.4	91.8
13	2,3-Dimethylbutane	4391.7	4744.0	20742	115243	136.08	1.37495	45.34	(1.2)	(7.7)	94.3	+0.3[f]
14	n-Heptane	5100.2	5502.8	20681	118668	136.01	1.38764	52.50	1.0	7.0	0.0	0.0
15	2-Methylhexane	5092.1	5494.8	20658	117627	131.59	1.38485	52.50	(1.0)	(7.0)	46.4	42.4
16	3-Methylhexane	5095.2	5497.8	20668	119192	132.11	1.38864	52.50	(1.0)	(7.0)	55.8	52.0
17	3-Ethylpentane	5098.2	5500.9	20679	121158	132.83	1.39339	52.50	(1.0)	(7.0)	69.3	65.0
18	2,2-Dimethylpentane	5079.4	5482.1	20620	116585	125.13	1.38215	52.50	(1.0)	(7.0)	95.6	92.8
19	2,4-Dimethylpentane	5084.3	5487.0	20636	116531	126.58	1.38145	52.50	(1.0)	(7.0)	83.8	83.1
20	3,3-Dimethylpentane	5085.0	5487.6	20638	120031	127.21	1.39092	52.50	(1.0)	(7.0)	86.6	80.8
21	Triptane	5081.0	5483.6	20627	119451	124.21	1.38944	52.50	(1.0)	(7.0)	+0.1[f]	+1.8[f]
22	n-Octane	5796.7	6249.7	20604	121419	129.53	1.39743	59.65	0.96	—	—	—
23	Diisobutyl	5781.3	6234.3	20564	119662	122.8	1.39246	59.65	(0.98)	—	55.7	55.2
24	Isooctane	5779.8	6232.8	20570	119388	116.71	1.39145	59.65	1.0	—	100.	100.
25	n-Nonane	6493.3	6996.5	20544	123613	123.76	1.40542	66.81	0.87[s]	2.9	—	—
26	n-Decane	7188.6	7742.1	20494	125444	118.68	1.41189	73.97	0.78[s]	2.6	—	—
27	Cyclopentane	3512.0	3763.7	20188	126296	167.34	1.40645	35.79	(1.4)	—	84.9[l]	+0.1[f]
28	Methylcyclopentane	4198.4	4500.4	20130	126477	147.83	1.40970	42.95	(1.2)	8.35	80.0	91.3
29	Cyclohexane	4178.8	4480.8	20035	130849	153.0	1.42623	42.95	1.3	7.8	77.2	83.0
30	Methylcyclohexane	4862.8	5215.2	20001	129066	136.3	1.42312	50.11	1.2	—	71.1	74.8
31	Ethylene	1499.0	1599.7	—	—	207.57	—	14.32	2.7	34.0	75.6	+0.3[f]
32	Propene	2182.7	2333.7	—	—	188.18	—	21.48	2.0	10.0	84.9	+0.2[f]
33	1-Butene	2879.4	3080.7	20678	103659	167.94	—	28.63	1.6	9.3	80.8[l]	97.4
34	Cis-2-Butene	2871.7	3073.1	20611	107754	178.91	—	28.63	(1.6)	—	83.5	100.
35	Trans-2-Butene	2866.8	3068.2	20584	104690	174.39	—	28.63	(1.6)	—	—	—
36	Isobutene	2860.4	3061.8	20548	102863	169.48	—	28.63	(1.6)	—	—	—
37	1-Pentene	3575.2	3826.9	20548	110610	154.46	1.37148	35.79	1.4	8.7	77.1	90.9
38	1,2-Butadiene	2789.0	2940.0	20447	112172	(181.)	—	26.25	(2.0)	(12.)	—	—
39	1,3-Butadiene	2730.0	2881.0	20047	104826	(174.)	—	26.25	2.0	11.5	—	—
40	Isoprene	3410.8	3612.1	19964	114194	(153.)	1.42194	33.41	(1.5)	—	81.0	99.1
41	Acetylene	1422.4	1472.8	—	—	—	—	11.93	2.5	80.	—	—
42	Benzene	3590.7	3741.7	17992	132655	169.31	1.50112	35.79	1.39	7.99	+2.8[f]	—
43	Toluene	4273.3	4474.7	18252	132656	154.84	1.49693	42.95	1.29	7.19	+0.3[f]	+5.8[f]
44	Ethylbenzene	4970.0	5221.7	18494	134414	144.0	1.49588	50.11	0.999	6.79	97.9	+0.8[f]
45	o-Xylene	4958.3	5210.0	18445	136069	149.1	1.50545	50.11	1.19	6.49	100.	—
46	m-Xylene	4956.8	5208.5	18441	133568	147.2	1.49722	50.11	1.19	6.49	+2.8[f]	+4.0[f]
47	p-Xylene	4956.9	5208.5	18445	133136	144.52	1.49582	50.11	1.19	6.69	+1.2[f]	+3.4[f]
48	Styrene	4828.7	5030.0	18150	137849	(151.)	1.54682	47.72	1.1	6.1	+0.2[f]	>+3.[f]
49	Isopropylbenzene	5661.4	5963.4	18665	134817	134.3	1.49145	57.27	0.889	6.59	99.3	+2.1[f]
50	Methyl Alcohol	—	—	9760	64771	473.(2)	1.3288(8)	7.16	6.72(5)	36.50	—	—
51	Ethyl Alcohol	—	—	12780	84600	367.(2)	1.3614(8)	14.32	3.28(5)	18.95	—	—
52	Carbon Monoxide	—	321.(13)	—	—	92.7(14)	—	2.39	12.50(5)	74.20	—	—
53	Carbon Dioxide	—	—	—	—	238.2[n](14)	—	—	—	—	—	—
54	Hydrogen Sulfide	588.(16)	637.(16)	—	—	235.6(7)	—	7.16	4.30(5)	45.50	—	—
55	Sulfur Dioxide	—	—	—	—	166.7(14)	—	—	—	—	—	—
56	Ammonia	359.(16)	434.(16)	—	—	587.2(14)	—	3.58	15.50(5)	27.00	—	—
57	Air	—	—	—	—	92.(3)	—	—	—	—	—	—
58	Hydrogen	274.(13)	324.(13)	—	—	193.9(14)	—	2.39	4.00(5)	74.20	—	—
59	Oxygen	—	—	—	—	91.6(14)	—	—	—	—	—	—
60	Nitrogen	—	—	—	—	87.8(14)	—	—	—	—	—	—
61	Chlorine	—	—	—	—	123.8(14)	—	—	—	—	—	—
62	Water	—	—	—	—	970.3(12)	1.3330(8)	—	—	—	—	—
63	Helium	—	—	—	—	—	—	—	—	—	—	—
64	Hydrogen Chloride	—	—	—	—	185.5(14)	—	—	—	—	—	—

Appendix D

Labyrinth and Carbon Ring Seal Leakage Calculations

Straight Labyrinth Seal

$$W = 5.76K \frac{A}{(1-\alpha)^{1/2}} \frac{P_1}{(RT_1)^{1/2}} \beta$$

where

  W = weight flow, lb/sec
  A = annular area, in^2

P_1 = upstream labyrinth pressure, psia

R = specific gas constant = 1,545/MW

T_1 = upstream labyrinth temperature, °R

$$\beta = \text{pressure factor} = \left[\frac{1 - r^2}{N + \ln 1/r} \right]^{1/2}$$

(see Figures D-1 and D-2)

r = labyrinth pressure ratio = P_2/P_1

P_2 = downstream labyrinth pressure, psia

N = number of restrictions

$$\alpha = \text{residual energy factor} = \frac{8.52}{\left(\dfrac{s-1}{c} \right) + 7.23}$$

(see Figure D-4)

s = axial distance between restrictions, in

1 = tip width of restriction, in

c = radial clearance, in

K = flow coefficient (see Figure D-3)

$$R_e = \text{Reynolds number} = \frac{24W}{\pi D g \, \mu}$$

Note: Values of g μ may be substituted directly from Figure D-6. Units are correct.

D = labyrinth diameter, in

μ = absolute viscosity, lb-sec/ft^2

g = gravitational constant, ft/sec^2

Staggered Labyrinth Seal

$$W = 5.76 A K \frac{P_1}{(RT_1)^{1/2}} \beta'$$

This formula differs from that of the straight seal in that the residual energy term $\sqrt{1 - \alpha}$ is omitted and β' is a function of the equivalent number of restrictions N'.

$N' = (N/2) \, y$

where

N' = equivalent number of restrictions
N = actual number of restrictions
y = a function of s/c (See Figure D-5)

It should be noted that for values of s/c greater than 15, the reduction in leakage attained by using a staggered labyrinth is very small. The reduction in flow varies from

10% at s/c = 15 to 8% at s/c = 50.

Carbon Ring Seal

$$W = \frac{1070 \, \alpha \, DhP_0}{(RT_0n)^{1/2}}$$

where

W = leakage, lb/min.
D = shaft diameter, in
h = radial clearance, in
P_0 = high side pressure, psia
T_0 = temperature, °R
R = specific gas constant
n = number of rings in series
α = parameter, function of μ & ρ_0 (see Figure D-7)

where

$\mu = P_n/P_0$ P_n = low side pressure, psia
$\rho_0 = .02(s/h)$ s = seal ring width, in

Equations and charts:
Courtesy of A-C Compressor Corporation.

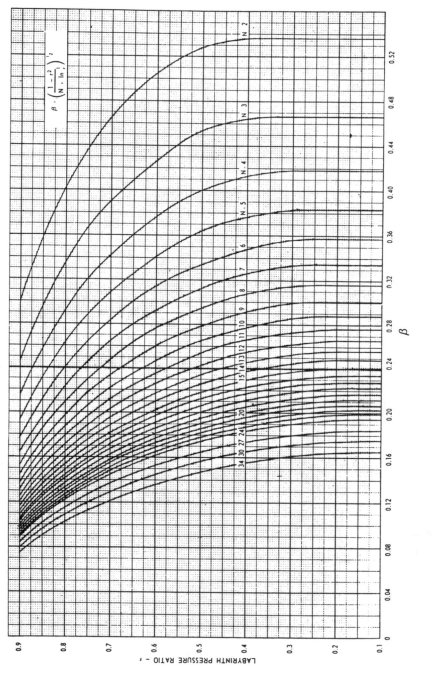

Figure D-1. Pressure factor vs. pressure ratio and number of restrictions for labyrinth seals. *(Reprinted by permission and courtesy of A-C Compressor Corporation)*

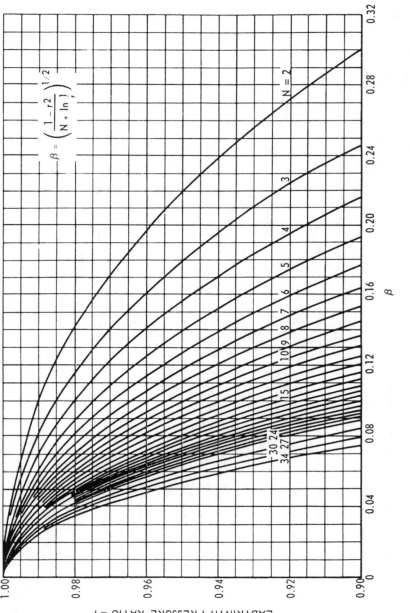

Figure D-2. General information for labyrinth seal. *(Reprinted by permission and courtesy of A-C Compressor Corporation)*

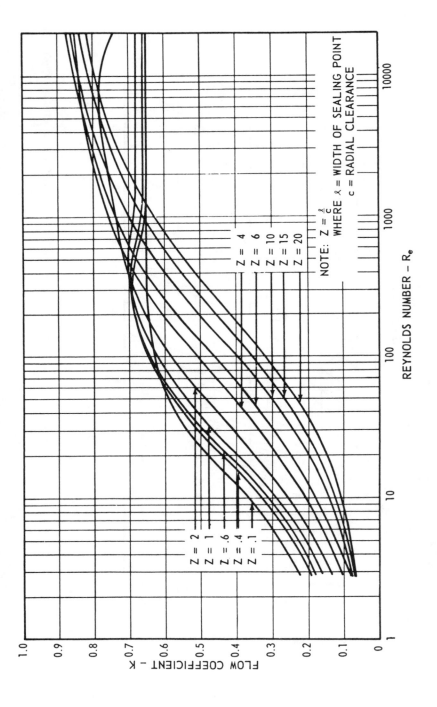

Figure D-3. Reynolds numbers vs. flow coefficient for annular orifices. *(Reprinted by permission and courtesy of A-C Compressor Corporation)*

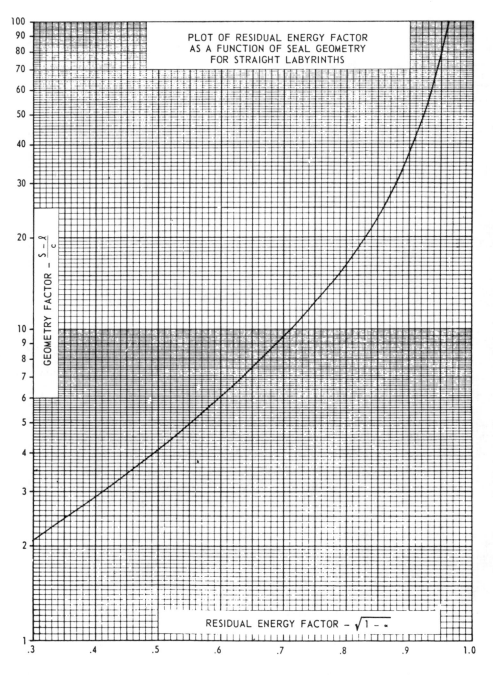

Figure D-4. Residual energy factors for straight labyrinths. (*Reprinted by permission and courtesy of A-C Compressor Corporation*)

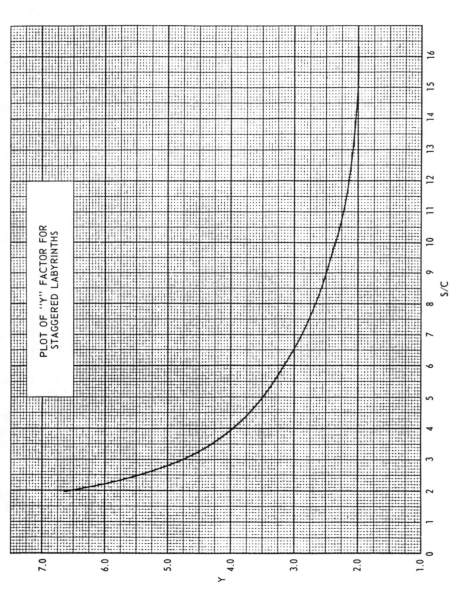

Figure D-5. "Y" factor for staggered labyrinths. *(Reprinted by permission and courtesy of A-C Compressor Corporation)*

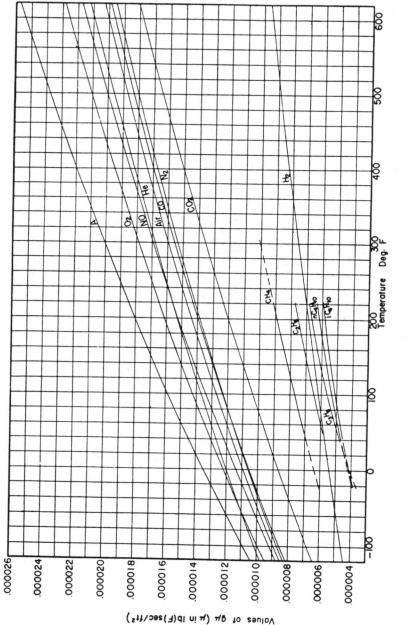

Figure D-6. Viscosities of selected commercial gases at one atmosphere. *(Reprinted by permission and courtesy of A-C Compressor Corporation)*

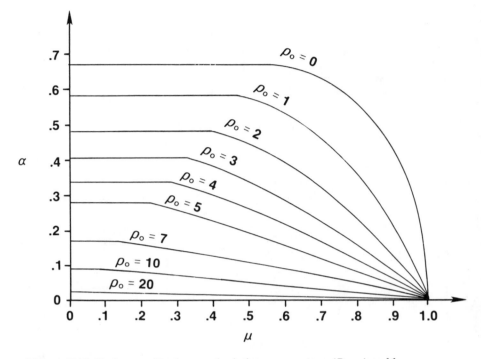

Figure D-7. Carbon seal leakage calculation parameters. (*Reprinted by permission and courtesy of A-C Compressor Corporation*)

Index

A

Academie des Sciences, 224
Accelerometers, 349
Accumulators, 317
Acoustic Velocity, 26, 39
Adiabatic, 30
Adiabatic efficiency
  dry, helical, 101
  sliding vane, 126
Adiabatic exponent for air-
  vapor mixtures, 494
Adiabatic head, 32
Adiabatic head
  coefficient, 156
Adiabatic head equation, 33
Adiabatic process, 31
Adiabatic shaft work, 36
AGMA, 118, 331
Air relative humidity, 20
Air specific humidity, 20
Airfoil
  CFD, 232
  controlled diffusion, 232
  nomenclature, 227
Airfoils, NACA
  65 series, 227
American Gear
  Manufacturers
  Association, 330
Angle engine compressor, 53
Angular acceleration, 392
Antisurge control, 363, 364

API, standard atmospheric
  conditions, 21
API-500, 257, 275
API-540, 257
API-612, 290
API-613, 328, 451, 479
API-614, 125, 304, 408, 449
API-616, 293
API-617, 193, 372, 373, 386,
  406, 411
API-618, 62, 66, 67, 68, 80,
  84, 85, 86, 405, 414, 457
API-619, 114, 413
API-670, 344
API-671, 335, 397, 451
API-677, 328
Area classification, 275
Arithmetic average
  roughness height, 331
ASME long radius flow
  nozzle, 431
ASME Power Test Code, 417
ASME Pressure Vessel
  Code, 315, 316
ASME, standard atmospheric
  conditions, 21
Aspect ratio, 227
Atlas Copco, 96
Automatic control systems,
  356, 357
Avogadro's Law, 16
Axial compressor, 13, 14
  application range, 14, 226
  balance piston, 252

  bearings, 251
  blade nomenclature, 226
  blading, 250, 363, 478, 482
  capacity control, 253
  casing connections, 248
  casings, 245
  compression cycle, 228
  curve shape, 236
  discharge temperature, 241
  fouling, 246
  history, 224, 225
  hub-to-tip ratio, 244
  maintenance, 248
  performance curve, 3
  pressure coefficient,
    238, 239
  reaction, 230, 231
  reliability, 477
  rotor, 248
  seals, 252
  shaft, 250
  shaft horsepower, 241
  sizing, 238, 241, 243
  soleplate, 451
  stagger, 232, 233, 234
  stators, 247
  surge, 234, 236
  surge control, 364
  velocity diagram, 231, 234
  volume, last stage, 240
Axial flow expansion
  turbines, 296
Axial shaft motion
  measurement, 354

Axial/centrifugal series
    arrangement, 237

B

Backlash, timing gears, 118
Back-pressure turbine, 282
Back-to-back
    arrangement, 142
Balance machines, 370, 372,
    374, 375
Balance piston, 208, 209
Balance piston, axial, 252
Balancing
    crankshaft, 74
    definition of terms, 369
    dynamic, 368
    field, 368, 377
    reciprocating cylinder, 368
    static, 369
Barrel type centrifugal, 135
Baseplates, 451
Beam type single stage, 137
Bearing coefficients, 385
Bearing specification
    information, 448
Bearing temperature sensors,
    344, 345
Bearings
    dam, 222
    journal, 197
    liner, 199
    magnetic, 204–8
    radial, 197
    rolling element, 116
    thrust, 116, 200
Bearings, gear box, 328
Benedict-Webb-Rubin,
    equation of state,
    26, 425
Bently Research
    Corporation, 389
Blade attachment, axial, 250
Blade maintenance, 253
Blade, axial, 250
Blades, stator, 250
Booster pumps, 312
Brushless excitation, 257
BWR equation, 26

C

Camber, 227
Campbell diagram, 251, 284
Capacity control

centrifugal, 220
    helical, 119
Carnot, 29
Carryover, 471, 479, 483
Casing stress, 406
Centrifugal compressor, 11,
    12, 14
    application range, 14, 133
    balance piston, 209
    balance piston losses, 165
    barrel type, 135
    beam type single stage, 138
    bearing housings, 204
    bearings, radial, 197–200
    bearings, thrust, 200
    capacity control, 220
    casing connections, 194
    casings, 192, 193
    choke, 186
    classification, 134
    compound, 139
    compression cycle, 147
    curve shape, 184
    diaphragms, 193
    discharge temperature, 160
    double flow
        arrangement, 140
    drive methods, 146
    efficiency, 156, 157
    entrance flow effects, 186
    fan laws, 183
    flow coefficient, 168, 173
    head coefficient, 155, 156
    head equation, 160
    head per stage, 160
    history, 133
    horizontal-split, 134
    impellers, 11, 197
    integrally geared, 134, 137
    losses, 164
    maintenance, 221
    mechanical running
        test, 408
    multistage
        arrangements, 137
    overhung type, 134
    performance curve, 3
    reaction, 157, 158
    reliability, 476
    seals, interstage, 209
    seals, shaft end, 211
    shaft sleeves, 197
    shafts, 197

sidestreams, 143, 149
single-stage overhung
    type, 137
sizing, 159, 162, 165
slip, 153
soleplate, 451
stage, 134
surge, 185
surge control, 364
vector triangles, 151, 153
vertical-split, 134
volume, last stage, 162
Centrifugal/axial series
    arrangement, 237
Characteristic curve,
    axial, 234
Characteristic curves, 3
Charles Law, 15
Check valves, 472
Chord, 226
Clearance
    axial rotor, 253
    bearing, 222
    front shroud, 222
    interstage seal, 222
Clearance pockets, 67
Clearance volume, 55, 100
Closed-loop tests, 417, 420
Coastdown lube oil
    tanks, 320
Code stamp, 315, 316
Compound
    compressor, 139
Compressed Air and Gas
    Institute, 57
Compressibility, 17
Compressibility charts
    generalized, 17, 517–20
    natural gas, 521–27
Compressibility of
    mixtures, 22
Compression path
    exponent, 35
Compressor flow, 21
Compressor head, 30
Compressors
    application ranges, 3
    classifications, 5
    corrosive gases, 480
    fouling gases, 474
    gas sonic velocity, 481
    misapplication, 481
    polymerizing gases, 481

pulsations, 481
type comparison, 474
wearing parts, 474
Condensation in expansion
turbines, 299
Condensing turbine, 282
Continuous mode, 2
Continuous monitoring, 356
Control
antisurge, 363
automatic, 356, 357
manual, 356
pressure at constant
speed, 361
pressure at variable
speed, 357
volume, at constant
speed, 362
volume, at variable
speed, 359
weight flow, at constant
speed, 361
weight flow, with variable
stator vanes, 361
Control arrangements, 357
Control systems, 357
Controls, specification
information, 452
Conversion factors, 491–92
Cooler
transfer valve, 317
Coolers, lubrication
systems, 315
Cooling, 41, 78
Cooling pressure drop, 176
Cooling, reciprocating
cylinders, 66
Coordination meeting, 457
Coupling
alignment, 337
gear shaft, 331
hubs, 332, 334
lubrication, 337, 340
misalignment, 333
ratings, 334
spacers, 334
torsional tuning with, 330
Couplings
flexible, 333
flexible element, 338
gear, 337, 340, 341
limited end-float, 341

lubrication, pump, 309
reliability, 486
torsional damping,
396, 397
Couplings specification
information, 451
Critical pressure, 35
Critical temperature, 15
Crosshead type,
reciprocating, 51
Current-source inverter, 278,
280, 281
Cylinder liners,
reciprocating, 67

D

Dalton's Law, 18
Damped unbalance response,
385, 387, 388
Dampener
pulsation, suction, 78
Damping coefficients, 385
Deflection, differential
foundations, 471
Degassing drum, 322
Disc couplings, 338
Distance piece, 73, 446
Division 1 motor
enclosures, 260
Double-acting, 4
Double-flow
arrangement, 140
Double-helical gear, 330, 332
Doughty, 397
Driver specification
information, 450
Drivers
electric motor, 257–77
expansion turbine,
296–300, 299
gas engine, 292, 293
gas turbine, 293, 294, 296
lubrication, pump, 309
steam turbine, 278,
282–92
Dry gas seals, 117, 215–20
barrier seal, 219
buffer gas, 324
double opposed, 217
primary seal gas, 323
seal gas supply, 324
secondary seal gas, 324

separation seal, 219
single, 217
systems, 323
tandem, 217
triple, 217
Dynamic balancing, 370
Dynamic compressors, 11
Dynamometers, 425

E

Edmister and McGarry, 35
Efficiencies, volumetric
reciprocating,
discharge, 88
Efficiency
axial, effect of stagger,
234, 236
centrifugal stage, 158, 159
helical, 101
liquid piston, 130
reciprocating cylinder, 57
sliding vane, 126
steam turbine, 286
straight lobe, 121, 124
Efficiency multiplier
low pressure
reciprocating, 60
specific gravity
reciprocating, 60
Efficiency, volumetric
helical, 102
reciprocating, 56
Ejector, 10
Elastomer compression
couplings, 398
Electric Motors, 257
induction,
re-energized, 479
Enclosures, motor, 257
Energy equation, general,
29, 30
Energy of a system, 27
Energy transfer, axial, 230
Engine type,
reciprocating, 52
Enthalpy, 29, 30, 31
Enthalpy variation through
an axial, 229
Entropy, 29, 31, 35
Equations of state, 26, 27
Benedict-Webb-Rubin
(BWR), 26

Leland-Mueller rule, 26
Martin-Hou, 27
Redlich-Kwong, 27
Erosion, 484
Excitation frequencies, gear
box, 330
Exciter, 266
Expansion Turbine, 296–300
applications, 300
condensation, 299
operating limits, 297
power recovery, 297
refrigeration cycles, 298
reliability, 480
types, 296
Experience
influence on reliability, 482
Explosion-proof motors, 260
Explosion-relief device, 446
Extraction turbine, 282

F

Failure analysis
failure mode and effect,
467
fault tree, 467
Weibull, 467
Fan laws, 184
Field balancing, 377
Field test planning, 428, 431
Field testing, 428, 434
Filter transfer valves, 317
Filters
lubrication system, 316
motor cooling air, 276
Filtration, axial
applications, 246
First law of
thermodynamics, 27
Flexible element couplings,
338, 340, 341
Flow coefficient
centrifugal, 163
seal leakage, 537
valve, 316
Flow imbalance, double-flow
centrifugals, 144
Flow measurement, 345
Flow meters, 431
Flow nozzle, 431
Flow terminology
conventions, 21

Flow velocity, 26
Flushing lubrication
systems, 463
Flywheels, 77
Force-ventilated motors, 276
Fouling, 484
Fouling, axial, 246
Foundations, 470–71

G

Gas, 15
Gas chromatographs, 349
Gas composition, 432
Gas engines, 292
Gas expander, 147
Gas leak test, 459
Gas mixture, 18, 26, 27, 35
compressibility, 20
molecular weight, 19
use in testing, 425
Gas turbine, 53, 147, 292–96
economics, 295
reliability, 478
sizing and application, 296
types, 296
Gases
closed-loop tests, 417
viscosities, 540
Gauges, ring and plug, 335
Gay-Lussac, 15
Gear coupling teeth, 337
Gear meshing frequency, 330
Gear service
classification, 330
Gear teeth, 331
Gear unit
applications, 330
bearings and seals, 332
design and application, 330
housing, 333
lubrication, 333
reliability, 479
rotors and shafts, 331
Gear unit specification
information, 450
Gears
split, 118
timing, 118
Generalized compressibility
charts, 17, 517–20
Goodman diagram, 251
Guide vanes, 148

H

Hardening, gear teeth, 331
Hardness, gear coupling
teeth, 336
Head coefficient, 156
Head equation, adiabatic, 33
Head equation, polytropic, 34
Head, centrifugal, 156
Head, reciprocating, 58
Heat run test (dry), 413
Helical compressor, 5, 7
adiabatic efficiency, 101
application range, 7, 95
asymmetric profile, 96
bearings, 116
capacity control, 95
casings, 114
circular profile, 95
cooling, 111
discharge temperature
(dry), 117
discharge temperature
(flooded), 113
displacement, 93
dry, 7, 100
dry compression, 100
flooded, 7, 112
flooding fluid, 111
history, 95
liquid injection, 111, 113
lubrication (flooded), 108
movable slide stop, 121
noise, 95
oil flooded, 95, 109
oil separation
(flooded), 113
operating principle, 96
power, 109
pressure ratio, 95, 98
reliability, 475
rotors, 111, 115, 116
seal strip clearance
(dry), 116
seals, 116
slide valve, 95
soleplate, 451
suction throttling, 119
timing gears, 117
tip speed ratio, 100
turn valve, 120
variable volume ratio, 122

volume ratio, 99
volumetric efficiency, 100
Holroyd rotor cutting
machine, 95
Holset type coupling,
398, 399
Horizontally split
centrifugal, 134
Howden Company, 95
Humidity, 20
Hunting tooth design, 330
Hydraulic hub removal, 335
Hydrostatic testing, 405

I

Impeller
centrifugal compressor
types, 12
closed, 151
double-flow, 140
mixed flow, 14
open, 151
radial flow, 14
semi-open, 151
Impeller diameter
estimation,
centrifugal, 162
Impeller overspeed test, 406
Incidence angle, 237
Incipient surge, 186, 366
Induction motor, 146, 257
Induction-type turbine, 282
Inert gas-filled motors, 261
Inlet correction factors for
piping, 189
Input shaft power, 35
Inspection and testing, 444
Instrumentation
flow, 345
molecular weight, 349
pressure, 342
reciprocating oil system, 78
rod drop monitor, 348
speed, 347
temperature, 343, 344
torque, 346
vibration sensors, 349
Instrumentation history, 342
Instrumentation specification
information, 452
Insulation, motor, 257, 258

Integral engine
compressor, 53
Integrally geared centrifugal,
134, 138
Intercooling, 42, 43, 46
Intercooling, pressure
drop, 176
Interference fit
axial shaft to rotor, 250
Intermittent mode, 2, 4
Inverter, variable frequency
drive, 277
Isentropic, 33
Isentropic exponent, 34
ISO 9000, 488
Isothermal compression, 42,
43, 45

K

Kay's Rule, 20, 26
Keyed hubs, 335
Keyed shafts, 335
Kingsbury, 201

L

Labyrinth piston, 49
Labyrinth seal leakage, 532
Lantern ring, 74
Lapping block set, 335
Lateral critical speeds, 384
Leakage, seal, 532, 533, 534
Leland-Mueller rule, 26
Lift coefficient, 226
Liquid
effect on expansion
turbines, 296
effects on centrifugals,
187, 472, 484
Liquid piston compressor,
9, 10
application range, 125
bearings, 125
efficiency, 127
mechanical
construction, 129
operation, 130
seals, 130
Lissajous figure, 378
Loop testing, 417
Lube and seal system
specification
information, 449

Lubrication of couplings, 337
Lubrication pumps and
drivers, 309
Lubrication systems
check valves, 319
commissioning, 462, 463
coolers, 315
degassing drum, 322
filters, 316
gear box, 328
oil flooded helical, 109
overall system review, 323
overhead tanks, 319
piping, 322
pressure control
valves, 313
pressurized, reciprocating,
77, 304
reliability, 485
relief valves, 313
reservoir, 307, 308
ring-oiled bearing, 303
seal oil drainers, 321
sliding vane, 126
startup control, 313
straight lobe
applications, 121
transfer valves, 317
types, 302
with seal and control
oil, 304
Lubrication, reciprocating
cylinder, 49, 78
frame, 77
Lubricator, mechanical
type, 78
Lysholm, 95

M

Mach number, 26, 186, 427
effect on axials, 231
rotor tip, 100
Magnetic bearings, 204–8
air gap, 207
auxiliary bearing, 207
electromagnets, 205
laminated sleeve, 205
load capacity, 206
sensors, 206
Manual control systems, 356
Marine type gear
coupling, 336

Martin-Hou equation, 27
Mass flow rate, 16
Materials of construction
    accumulators, 317
    axial blades, 250
    axial casings, 247
    axial rotors, 248
    centrifugal casings,
        192, 193
    centrifugal impellers, 195
    centrifugal shafts, 197
    coupling elements, 338
    crankshafts, 74
    helical, 118
    liquid piston, 130
    lubrication pumps, 310
    lubrication system,
        coolers, 315
    piston rings, 68
    reciprocating cylinders, 70
    sliding vane, 126
    straight lobe, 121
    timing gears, 117
Materials, specification
    information, 447
Matthews, 430
Mechanical contact
    seals, 304
Mechanical running test,
    408, 413
Meridional flow vector, 152
Meter run requirements, 431
Method of piping, 190
Michell bearing, 201
Mini lube system,
    reciprocating, 78
Mixed-flow, 11, 14
Mixed-flow impeller, 14
Mixture compressibility, 20
Mode shapes, 386, 388
Moisture corrections, 21
Mole, 16
Mole fraction, 18
Molecular weight, 1, 19, 22,
    24, 37
    effects on centrifugal
        sizing, 159
    effects on dry helical, 100
    recorded ranges, 1
Molecular weight
    measurement, 345

Molecular weight of gas
    mixtures, 19
Molecular weight, effect on
    centrifugal sizing, 159
Mollier charts, 27
Monitoring system, 356
Motor, 146
    enclosure, 260
    equations, 267
    insulation, 257
    locked rotor torque, 270
    selection, 270
    service factor, 262
    starting characteristics, 270
    starting time, 273, 274
    synchronous vs
        induction, 265
    variable frequency drives,
        277, 280
    voltage, 258
Motors
    induction, 259
    lubrication pump, 309
    synchronous, 259
Mounting plate,
    specification
    information, 451, 452
Multistage, 4
Multistage centrifugal
    arrangements, 133
Myklestad, 385

N

NACA 65 Series airfoils, 227
National Advisory Committee
    for Aeronautics
    (NACA), 225
National Aeronautics and
    Space Administration
    (NASA), 225
National Electrical Code, 257
Naval Boiler and Turbine
    Laboratory, 337
NEMA, 115, 262, 335
NEMA SM-23, 194
Newton's law, 391
Nilson, Hans, 95
Nilson, Rune, 96
Noise
    dry, helical, 118
    flooded, helical, 109

Nomenclature
    axial, 226
    centrifugal, 192
Non-perfect gas, 34
Nozzle velocity, 39
Nozzles, axial compressor,
    fitting problems, 247

O

Oil film seals, 304
On stream, 144
Open-loop testing, 417
Operation
    off-design, 484
Overhead tank, lube and seal
    system, 319
Overhung centrifugal, 134

P

Partial pressure, 18
Particle size, 484
Perfect gas equation, 15, 32
Performance curve, 3
    axial, 232
Performance maps, gas
    turbine, 294
Performance testing, 420
Periodic monitoring, 354, 355
Physical constants of
    hydrocarbons, 495–96
Pipe strains, 473
Pipeline and Compressor
    Research Council, 85
Piping, 473
Piping arrangements,
    189, 190
Piping of axials, 248
Piston engine, 53
Piston rings, 68, 69
Piston speed, 57, 58
Pitch, 226
Pitch line velocity, 330
Polytropic compression
    exponent, 160
Polytropic efficiency, 34
Polytropic exponent, 34
Polytropic head
    coefficient, 156
Polytropic head equation, 34
Polytropic shaft work, 36

Positive displacement, 2, 3, 5
  performance curve, 3
Power, 35
Power factor, motor, 257
Power, field testing, 434
Preload, 387
Pressure
  differential straight
    lobe, 123
  seal oil, 304
Pressure coefficient, 230
Pressure control at constant
  speed, 361
Pressure control at variable
  speed, 357
Pressure control valve, 313
Pressure effect on valve
  loss, 62
Pressure enthalpy diagram
  ammonia, 497
  carbon dioxide, 498
  ethane, 510
  ethylene, 509
  HFC-134a, 499
  hydrogen, 504, 505
  iso-butane, 513
  iso-pentane, 515
  methane, 508
  n-butane, 514
  nitrogen, 506
  n-pentane, 516
  oxygen, 507
  propane, 512
  propylene, 511
  R-12, 500, 501
  R-22, 502, 503
Pressure factor
  labyrinth seal, 535
Pressure oriented instrument
  piping, 343
Pressure ratio, 7, 98, 133
Pressure ratio per stage, 60
Pressure testing, 405
Pressure,instrumentation, 342
Pressurized cooling fluid
  system, 78
Procurement steps, 438
  award contract, 457
  bid evaluation, 455
  bid or quotation, 455
  coordination meeting, 457
  engineering reviews, 458

inquiry specification, 443
  inspections, 459
  installation and startup, 462
  pre-award meeting, 456
  preliminary sizing, 440
  purchase specification, 457
  shipment, 461
  site arrival, 462
  tests, 459
  vendor information, 454
Progressive balancing, 374
Prohl, 385
Propagating stall, 237
Proximity sensors, 350
Proximity transducers, 350
Pseudocriticals, 20
Psychrometric chart, 20,
  495–96
PTC-9, 417, 425
PTC-10, 417, 426
PTC-19.1, 431
PTC-19.7, 425
Pulsation control, 84
Pulsation snubbers, 85
Pulse-width-modulated
  unit, 278
Pump couplings, 312
Pump sizing, 312
Pumps
  lubrication, 309

Q

Quality
  role in reliability, 487

R

Radial bearing temperature
  sensor, 345
Radial bearings
  cylindrical, 249
  dam type, 199
  straight cylindrical, 198
  three-lobe, 332
  tilting pad, 116, 200, 252,
    332, 388
Radial flow, 11
Radial flow expansion
  turbines, 296
Radial flow impeller, 11
Radial shaft vibration
  measurement, 353
Radial vibration probes, 204

Ratio of specific heats, 18
Reaction, 13, 230
  axial, 230
  axial, 224
  centrifugal, 157
Real gas compression
  exponent, 34, 160
Reboiler, 41
Reciprocating compressor, 4
  application range, 5,
    52–53
  arrangements, 4, 52
  bar over test, 414
  bearings, 74
  capacity control, 81
  clearance pockets, 68
  clearance volume, 55
  compression cycle, 56
  condensation effects, 66
  connecting rod, 75
  cooling, 78
  crankshaft, 74
  crosshead, 74
  crosshead type, 51
  cylinder displacement, 55
  cylinders, 67
  cylindrical efficiency, 59
  double acting piston, 49
  drive methods, 53
  efficiency volumetric, 56
  engine type, 53
  ideal indicator diagram, 54
  labyrinth piston, 49
  lubrication, 69
  mini lube system, 49
  nonlubricated cylinder, 49
  piston rods, 69, 70
  pistons, 67, 69
  power, 59
  pulsation control, 83–84
  reliability, 474
  rod packing, 73
  sizing, 67
  standing waves, 482
  static gas test, 414
  tail rod, 69
  temperature, discharge, 59
  trunk piston, 50
  valve loss, 62
  valves, 69–73
  valves, concentric ring, 70
  valves, poppet, 71

valves, ported plate, 70
valves, rectangular
element, 70
volumetric efficiency, 56
Reciprocating engine
driver, 292
Reciprocating shaking
forces, 368, 378
Redlich-Kwong equation, 27
Reduced pressure, 17
Reduced temperature, 17
Refrigeration, expansion
turbine cycles, 298
Relative humidity, 20, 21
Relief valves, 308, 313
Reservoir, lube system, 303
Residual unbalance, 371
Resilient couplings, 396, 397
Resistance temperature
detector, 343
Reverse rotation, 472
Reversible process, 29
Reynolds number, 426
Reynolds numbers
flow coefficient, seals, 537
Rider rings, 68, 348, 445
Rigid rotor modes, 388
Ring-oiled bearing, 303
Robust design, 468–70
Rod drop monitor, 348
Rod packing, 73
Rod reversals, 69
Rotary compressor
arrangements, 94
features common,
93, 467
Rotary compressors, 5
Rotary shaking force, 382
Rotating stall, 237
Roto dynamics
aerodynamic cross
coupling, 477
squeeze film dampers, 477
Rotor balancing, 370
Rotor construction, axial,
248, 250
Rotor dynamics, 383–84
bearing damping, 477
Rotor slip, 100
Rotor stability, 477, 482
testing, 413
Rotor tip Mach number, 7

Royal Aircraft
Establishment, 225
RTD, 343
Run time, 467, 469, 488

S

Saybolt Seconds Universal,
333
Schibbye, 96
Schultz, 426
Seal clearance, 222
Seal leakage, 213
Seal leakage calculations
carbon ring, 533
flow coefficient, 537
staggered labyrinth, 533
straight labyrinth, 532
Seal oil drainers, 321
Seal oil overhead tank, 319
Seal specification
information, 449
Seals
back diffusion, 326
carbon ring, 117, 211
dry gas, 117, 215–20
gear box, 328
labyrinth, 117, 209
liquid film, 117, 213
liquid film, 213
mechanical contact,
117, 214
Second law of
thermodynamics, 29
Seismic sensors, 350
Seismic transducers, 349
Self equalizing bearing, 202
Service factor
gear, 330
motor, 262
Shaft vibration measurement
methods, 350
Shaking forces
reciprocating, 368, 378
rotary, 382
Shop performance test, 416
Shrink fitted hubs, 335
Shrouds, stator blades, 248
Sidestream arrangements, 143
Sidestream compressors,
423, 424
Sidestreams, 13
Simon and Bulskamper, 426

Single-acting, 4
Single-stage, 4
Sizing, axial, 238, 241,
243, 245
Sizing, centrifugal, 159, 162
Sizing, preliminary, 440
Sliding vane compressor, 9
adiabatic efficiency, 126
application range, 128
bearings, 129
compression cycle, 126
cooling jacket, 128
displaced volume, 126
drivers, 129
lubrication, 128
mechanical
construction, 129
mechanical losses, 128
power, 128
pressure ratio, 126
seals, 129
sizing, 126
vane wear, 128
volumetric efficiency, 127
Slip
centrifugal, 153
motor, 267
Sohre, 468
Soleplates, 452
Solidity, 226
Sonic velocity, 39
Sour oil pots, 214
Southern Gas Association, 85
Spacers, coupling, 334
Spare rotor test, 414
Specific diameter, 156
Specific gas constant, 16
Specific gravity, 19, 21
effect on valve loss, 62
Specific heat at constant
pressure, 19
Specific heat ratio, 18, 22,
26, 32, 34
Specific humidity, 20
Specific speed, 156
Specific volume, 27
Specification content and
objectives, 441
Specification writing
basic design information,
443, 444
bearing information, 448

control and
  instrumentation
  information, 452
coupling information, 451
driver information, 450
gear unit information, 450
general information, 444
guarantee and
  warranty, 454
inspection and testing
  information, 453
lube and seal
  information, 449
materials information, 447
mounting plate
  information, 451
outline, 443
shaft end seal
  information, 448
Speed measurement, 347
Speed, equivalent test, 420
Speed, field testing, 434
Speed-torque curve, 272
Spiral-lobe, 5, 8
SRM, 95
SRM compressor, 7
Stage pitch, 149
Stage, axial, 226
Stage, centrifugal, 134
Stagger angle, 227
Staggered
  labyrinth, seal "Y"
  factor, 539
Staggered labyrinth seal
  leakage, 533
Stagnation, 30
Staight lobe compressor
  tip speed, 124
Stall, 186
Stall, rotating, 237
Standard atmospheric
  conditions, 21
Standby pump, 310
Starling BWR, 26, 425
Static balance, 369
Static cooling system, 80
Static gas test, 414
Stator blade linkage, 254
Stator blades, 250
Stator vane control, axial, 253
Stator vanes, 253
Steam cylinder, 53

Steam turbine, 53, 146,
  282–92, 479
back pressure, 282
blade deposits, 479
condensing, 282
efficiency, 288
extraction, 282
induction-type, 282
partial admission, 288
rating, 290
reliability, 478
selection variable, 275, 285
speed, 278
stage losses, 286
steam temperatures, 284
steam velocity, 288
trip and throttle valve, 479
Step unloading system, 80
Stiffness coefficients, 385
Stodola slip, 153, 155
Stonewall, 186
Straight labyrinth seal
  leakage, 532
  residual energy factors, 538
Straight lobe compressor, 8
application range, 8
applications, 124
bearings, 125
compression cycle, 121
differential pressure, 123
discharge
  temperature, 124
drivers, 125
horsepower, 124
lubrication, 125
mechanical
  construction, 125
mechanical losses, 124
output volume, 123
seals, 125
sizing, 124
slip, 122
slip speed, 123
volumetric efficiency, 122
Straightener row, 228
Straightener vanes, 148
Straightening vanes, 191
Straight-through multistage
  centrifugal, 137
Strainers, suction, 66, 308
Stress, axial blade, 251
Suction drum, 187, 471

Surge
axial, 234
centrifugal, 185
control, 363
effect on foundations, 382
Svenska Rotor Maskiner
  AB, 95
Synchronous motor, 257
Synchronous motors,
  146, 257

T

Tail rods, 69
Tandem-cylinder
  arrangement, 52
Tapered hubs, 335
TEFC, 259
Temperature
axial discharge, 241
centrifugal discharge, 160
reciprocating discharge
  limit, 73
Temperature control, 41
Temperature
  instrumentation, 342
Temperature measurement in
  motors, 262
Test
abnormalities, 426
classes, 417
codes, 420
correlation, 425
gas purity, 423
instrument location, 432
instrumentation, 425
loop, 421
sidestream, 424
uncertainty, 431
witnessing, 408
Testing objectives, 403
Tests
bar over,
  reciprocating, 413
dry helical heat run, 413
field performance, 429
gas leak test, 459
hydrostatic, 405, 459
lubrication system, 415
mechanical running,
  408, 413
rotor stability, 413

running, 460
shop performance, 416
spare rotor, 414
static gas, 414
Texas A & M University,
  389
Theoretical work, 30
Thermocouple, 343
Thermodynamics
  first law of, 27
  second law of, 29
Thermosyphon system, 80
Thermowells, 432
Thrust balance, 143
Thrust bearing
  flat land, 200
  tapered land, 116, 200
  tilting pad, 201, 252
  tilting pad, Michell, 201
  tilting pad, self equalizing,
    202
Thrust bearing temperature
  sensor, 345
Thrust friction losses, 409
Thyristors, 266
Timing gears, 95
Torque meter, 346
Torque meters, 425
Torsiograph, 331
Torsional criticals, 336
Torsional damping and
  resilient coupling, 396
Torsional excitations, 118,
  391
Torsional resonant response,
  390
Torsional shop test, 390
Torus type couplings, 398,
  399
Total enthalpy, 30
Tournaire, 224
Transducer, 342
Transfer valves, 317, 416
Trunk piston, reciprocating,
  50
Turboexpanders, 296, 298

U

Unbalance
  dynamic, 368
  residual, 371, 372, 375,
    377
  residual, 374
  static, 369
Unbalance, response
  programs, 372
Uninterruptable power
  supply (UPS), 311
Universal gas constant, 16
University of Virginia, 389
Unloader
  plug type, 81
  plunger type, 81
  port type, 81
  valve, 81
Uranium hexafluoride, 1

V

Vacuum breaker, 291
Vacuum pump, 9
Vacuum service, 10
  ejectors in, 11
  liquid piston, 9
  liquid piston, 130
  sliding vane, 10
Valve design, 62
Valve loss, reciprocating, 62
Valve unloader, 81
Valve velocity, 62
Valves
  reciprocating, cylinder,
    69–73
Vane angle, 152
Vapor, 15, 17, 20
Vector triangle, centrifugal,
  151, 153
Velocity diagram, axial, 228
Velocity head, 36, 38
Velocity sensors, 352
Velocity transducers, 349
Velocity variation through
  an axial, 229

Vendor data, 444, 454
Vertical-split centrifugal, 134
Vibration
  lateral, 384
  sources and effects, 369
  steam turbine, blade, 284
  torsional, 400
Vibration amplitude, 411
Vibration sensors, 353
Vibrational stress, axial
  blading, 250
Viscosities
  selected gases, 540
Voltage, motor, 258
Voltage-source inverter, 278
Volume bottle, 85
Volume bottle sizing, 87, 88
Volume control at constant
  speed, 359
Volume control at variable
  speed, 359
Volume ratio, 98, 420
Volumetric efficiency, 56
  helical, 100
  reciprocating, 57
  sliding vane, 127
  straight lobe, 122

W

Wear bands, 68, 348, 445
Weather-protected, Type I
  (WP-I), 259
Weather-protected, Type II
  (WP-II), 259
Weight flow control at
  constant speed, 362
Weight flow control with
  variable stator vanes,
    361
Wiesner, 153, 426
Wong, 426
Work, 59, 109
Work input coefficient,
  actual, 154
Work input coefficient,
  ideal, 155